CRC SERIES IN AGING

Editors-in-Chief

Richard C. Adelman, Ph.D.
Director
Aging Research Program
Temple University School of Medicine
Philadelphia, Pennsylvania

George S. Roth, Ph.D.
Research Biologist
Gerontology Research Center
National Institute on Aging
Baltimore City Hospitals
Baltimore, Maryland

HANDBOOK OF BIOCHEMISTRY IN AGING

Editor
James Florini
Department of Biology
Syracuse University

HANDBOOK OF PHYSIOLOGY IN AGING

Editor
Edward J. Masoro
Department of Physiology
University of Texas Health Sciences Center

HANDBOOK OF IMMUNOLOGY IN AGING

Editors
Takashi Makinodan and Marguerite M. B. Kay
Geriatric Research Education and Clinical Center
Wadsworth Veterans Administration Hospital

IMMUNOLOGY OF AGING

Editors
William H. Adler and Albert A. Nordin
Gerontology Research Center
National Institute on Aging
Baltimore City Hospitals

Additional topics to be covered in this series include Cell Biology of Aging, Microbiology of Aging, Pharmacology of Aging, Evolution and Genetics, Animal Models for Aging Research, Hormonal Regulatory Mechanisms, Detection of Altered Proteins, Insect Models, Lower Invertebrate Models, Morphological Techniques for Gerontology, Testing the Theories of Aging, and Nutritional Approaches to Aging Research.

Senescence in Plants

Editor:

Kenneth V. Thimann, Ph.D.

Professor of Biology
The Thimann Laboratories
University of California
Santa Cruz, California

CRC Series in Aging

Editors-in-Chief

Richard C. Adelman, Ph.D.

Director
Aging Research Program
Temple University School of Medicine
Philadelphia, Pennsylvania

George S. Roth, Ph.D.

Research Biochemist
Gerontology Research Center
National Institute on Aging
Baltimore City Hospitals
Baltimore, Maryland

CRC Press, Inc.
Boca Raton, Florida

Direct all inquiries to CRC Press, 2000 N.W. 24th Street, Boca Raton, Florida, 33431.

International Standard Book Number 0-8493-5803-5

Library of Congress Card Number
Printed in the United States

PREFACE

Many authors dealing with animals still refer to the aging of the organism, whereas the majority of writers on plants use the term senescence. This difference in usage is curious because it was the zoologist P. B. Medawar who provided a convenient distinction between the two terms, defining aging as all those processes of accruing maturity which occur in time, without reference to death as a consequence, and senescence as describing those deteriorative changes which lead sooner or later to the natural death of an organism or some part of it.[1] It is of interest to reflect upon why botanists and zoologists have tended to continue in the use of the two terms. It might be a wholly trivial matter arising from the fact that those working with animals have not been much bothered with problems of definition. It may be, however, that there is a real difference in the success with which these definitions can be applied in animals and plants. For example, in the case of plants we can map out the distribution of senescence in the differentiation of cells, tissues and organs at each successive stage in the life cycle,[7] but in animals senescence processes are less clearly seen.

There are a few points of similarity between some aspects of senescence in animals and plants: we may note the selective senescence of organs, as in the loss of the tail in the metamorphosis of a tadpole or the withering of the cotyledons in a developing seedling; again there are clearly defined patterns of senescence of specific groups of cells in the differentiation of the limbs of animals[4] and of the conducting tissues in vascular plants. When we consider senescence in the whole organism the problem changes. There are just a few animals (the Pacific salmon is one) which behave like monocarpic plants, so that after the eggs are laid the whole population succumbs to an astonishing synchronous death which appears to be under hormonal control, but this is exceptional in animals.

In most higher animals the deterioration of the whole organism is a gradual process and it is not easy to pinpoint the crucial events. In man, for example, one may find accounts of declining bodily function with advancing age, affecting almost every conceivable system from the wrinkling of the skin to declining renal function, cardiac malfunctioning and decreased physical and mental activity, but the incidence of these various symptoms is very uneven in different individuals so that doubts arise as to which ones are really causal to the ultimate death of the person. I suspect that there are deeper reasons for the difficulties in defining and analyzing senescence in animals which have to do with what I will call the tightness of their organization. If we take for example an ancient and decrepit oak, remove a twig and cause it to root, we can raise a new tree with a vigor and growth rate equal to that of a seedling. The senescence of the parent was a facet of the whole tree, related to such factors as the decay of its trunk and the constraints imposed by such factors as the amount of dependent non-green tissue which may come to exceed the photosynthetic capacity of the canopy. The growth of individual shoots is only loosely subject to the rest of the tree in respect of factors such as supplies of water and nutrients; loss of function in another part of the tree might depress the growth rate of the branches by virtue of the effect on supplies, but does not cause irreversible damage to the living cells in the branches. In animals the situation is different; there is a much greater variety of cells and organs and there are many more hormones which, through the agencies of the nervous and circulatory systems, serve to integrate the functions of the different parts. It is a consequence of this closer integration in animals that minor changes taking place in any of the several organs can set up perturbations in the other organs, so that in this way effects can accumulate and be amplified and so contribute to the progressive decline of the whole organism.

If challenged to pinpoint the most fundamental differences underlying senescence in animals and plants, I would say that on present evidence they appear to be concerned with differences in the relative stabilities of the genomes in the two groups and with the organisation of meristems in plants. The sharp increase in the incidence of congenital defects in children born to older women and the greatly increased frequencies of neoplasms which are consistent features of advancing age, are caused by damage to the genome and defects in the regulation of gene expression. One of the most exciting findings in recent years which points to the role of the genome in mammalian senescence is the remarkable correlation which has been demonstrated between lifespan and the capacity for DNA repair.[3] Plants do not appear to be encumbered with these problems of genetic instability; although their DNA repair mechanisms are less well understood than those of animals, they appear to be efficient. Thus several instances are known of plant clones which have retained their characteristics and vigor through centuries of vegetative propagation, and when plant cells are brought into culture many cell types can be maintained for long periods without loss of totipotency (the ability to regenerate new plants). It seems probable that in the case of vegetative propagation, the multicellular nature of the meristems of most vascular plants permits selection to take place against cells which become defective so that they no longer contribute to the differentiation of new tissues and organs.

When we turn to consider the regulation of senescence in plants it is interesting to note how often (as will be seen by reference to the chapters in this book) one comes back to the involvement of plant hormones or growth substances. The pioneering work of Chibnall[1] showed that roots produce substances which can delay the senescence of leaves; from the mass of work which followed this discovery, some patterns are beginning to emerge. Auxins, cytokinins and gibberellins are frequently found to retard senescence while abscisic acid and ethylene tend to promote senescence. There is doubtless more to be found out about the interactions between the different growth substances and inhibitors and how these influence events, but the great stumbling block to progress in this aspect of the study of senescence is the lack of knowledge of how these substances act.

The majority of studies on the regulation of senescence are carried out with a few species of economic importance; mostly short-lived plants which have been selected over the centuries for high growth rates under conditions of relative abundance of water and mineral nutrients. However, over much of the land surface of the earth, supplies of water and mineral nutrients are sub-optimal for much of the time, and in many cases these constraints have served as agents of natural selection which have profoundly modified the various manifestations of senescence. In areas where rainfall is unpredictable we encounter most frequently those species which have, in the language of the strategist, abandoned any attempt to track the fluctuations of water supply and adopted the "trivial" solution of carrying through their life cycle exclusively in the period following rain. The quantitative implications of the tying of senescence of the whole plant to the process of reproduction can be conveniently approached by the use of models.[8] It is worth noting in passing that although the ephemeral habit is a successful adaptation which has evolved in species from a number of families, it carries with it the requirement for concomitant longevity in the seeds; moreover it generally appears to have been achieved through in-breeding and a reduction in the size and complexity of the genome[6] which may herald a blind alley in the wider evolutionary sense.

Over great areas of the earth's surface, in peatlands, the podzolised soils of the temperate zones and laterites of the tropical regions, the supply of nutrients is suboptimal for the growth of plants. In many of these areas the prevailing aspect of the vegetation is evergreen, although deciduous species may occur spasmodically. The senescence of leaves in these evergreen communities is generally found to be delayed until the new set of leaves is established; the significance of this timing is probably a highly efficient internal transfer of inorganic nutrients from the old leaves to the new, thereby short-circuiting the need for the plant to recover the nutrients in the face of competition on the forest floor. Under these circumstances there are evidently mechanisms whereby the growth of the new set of leaves is geared to some very early stage in the senescence of older leaves. It also frequently happens that in evergreen forests the seedlings of the dominants must persist for many years in the shade afforded by their parents. It appears that these plants can replenish their carbohydrate supplies when circumstances occasionally permit, but for the rest of the time they conserve their reserves by adopting very low rates of respiration. It is instructive to punch discs from the leaves of such species and float them on water in a darkened incubator at 25°C; in many cases the discs can be retrieved, healthy and green, a month later; similar treatment of discs from leaves of cereals, tobacco or cotton would of course lead to their death within 3 to 4 days.

I have made this brief reference to the adaptive aspects of senescence in order to emphasize that the signals which elicit senescence processes are many and various, and we should not be persuaded into hasty generalizations concerning the ways in which they are mediated on the evidence obtained using leaf discs from a handful of cultivated species.

By now we take for granted the fact that senescence may occur in one part of a plant but not in another; for example in a flower the petals and stamens senesce while the ovary starts to grow. It is perhaps not so often appreciated that within the cell there is a similar range of phenomena in which the cell may vary the order and extent to which it loses its various constituents. The point can be illustrated by four examples:

1. In the case of the formation of heartwood from sapwood, the xylem parenchyma first embark on the biosynthesis of an assemblage of secondary metabolites and develop cellular outgrowths (tyloses) into the adjacent tracheary elements; they then export most of their mineral nutrients to the younger sapwood zone and finally die.
2. In the course of the differentiation of xylem elements, specific regions of the primary walls are hydrolyzed, the tonoplast then breaks down, followed by degradation of the cytoplasmic membranes, and the nucleus is one of the last constituents of the protoplast to go.
3. In the case of the formation of functional phloem, the tonoplast, endoplasmic reticulum, ribosomes, and nuclei are broken down while the plastids, mitochondria and plasma membrane are retained.
4. In the senescence of leaves it is commonly found that the chloroplasts are the first organelles to show clear evidence of degeneration.

In many herbaceous plants the degeneration of the chloroplasts is a process which takes place gradually as the leaf passes into the shade of other leaves which are formed above it. In many species this senescence of the chloroplasts remains reversible so that in the event of the removal of the upper layers, as by the grazing of a herbivore, the photosynthetic activity of the senescing member may be restored within a couple of days.

As attempts are made to elucidate the biochemistry of senescence, recognition of these subtleties in the patterns of subcellular changes becomes important. In some cases we find increased proteolytic activity associated with senescence, in other cases not. In some cells and tissues the selective breakdown of lipids seems to afford a key to many of the most significant physiological changes.[5] Indeed, it is now evident that so far as the natural senescence of plant cells is concerned, there are many different ways of getting dead. In a thermodynamic sense, life may be viewed as poised on a threshold of instability and senescence as the final descent into chaos; but for the plant physiologist it becomes increasingly clear that many of the manifestations of senescence are facets of development carried out in an orderly fashion and serving a wide variety of adaptive functions.

I am honored by the Editor's invitation to write this introduction to what is clearly becoming an increasingly popular topic. It is one of the many aspects of plant physiology to which he has made outstanding contributions. The importance of the subject is perhaps best measured in terms of the many scientific problems which it poses, but we should not lose sight also of its practical applications in forestry, the post-harvest storage of seeds, green vegetables, and cut flowers, and in agricultural and horticultural practices where manipulation of the lifespan of the plant or its organs is a desired end.

Harold W. Woolhouse
University of Leeds
Leeds, Yorkshire
England

REFERENCES

1. **Chibnall, A. C.,** *Protein Metabolism in the Plant,* Yale University Press, New Haven, 1939.
2. **Medawar, P. B.,** *An Unsolved Problem of Biology,* 1952, Reprinted as *The Uniqueness of the Individual,* Methuen, London, 1957, pp. 44—70.
3. **Hart, R. W. and Setlow, R. B.,** Correlation between deoxyribonucleic acid excision repair and life-span in a number of mammalian species, *Proc. Natl. Acad. Sci. U.S.A.,* 71, 2169—2173, 1974.
4. **Saunders, J. W. and Fallon, J. F.,** Cell death in morphogenesis, *Major Problems in Developmental Biology,* Locke, M., Ed., Academic Press, New York, 1966.
5. **Simon, E. W.,** Phospholipids and membrane permeability, *New Phytol.,* 73, 377—420, 1974.
6. **Stebbins, G. L., Jr.,** *Variation and Evolution in Plants,* Columbia University Press, New York, pp. 1—643, 1950.
7. **Woolhouse, H. W.,** Longevity and senescence in plants, *Sci. Prog. (London),* 61, 123—147, 1978.
8. **Woolhouse, H. W.,** Senescence processes in the life cycle of flowering plants, *BioScience,* 28, 25—31, 1978.

CONTRIBUTORS

Michael J. Dalling, Ph.D.
Senior Lecturer in Agronomy
School of Agriculture and Forestry
University of Melbourne
Parkville, Victoria, Australia

Karl Esser, D. Phil.
Professor für Allgemeine Botanik
Direktor des Botanischen Gartens
Ruhr-Universität Bochum
Bochum, Germany

Graeme J. T. Frith, Ph.D.
Director
Horticultural Research Institute
Knoxfield, Victoria, Australia

Eliezer E. Goldschmidt, Ph.D.
Associate Professor
Department of Horticulture
The Hebrew University of Jerusalem
Rehovot, Israel

Abraham H. Halevy, Ph.D.
Head, Department of Ornamental Horticulture
Professor of Horticulture and Plant Physiology
The Hebrew University of Jerusalem
Rehovot, Israel

A. Carl Leopold, Ph.D.
William C. Crocker Scientist
Boyce Thompson Institute
Cornell University
Ithaca, New York, USA

Shimon Mayak, Ph.D.
Senior Lecturer
Department of Ornamental Horticulture
The Hebrew University of Jerusalem
Rehovot, Israel

Larry D. Noodén, Ph.D.
Professor of Botany
Department of Botany
University of Michigan
Ann Arbor, Michigan, USA

Daphne J. Osborne, Ph.D., D. Sc.
Head, Developmental Botany
Weed Research Organization
Oxford, England

Michael J. C. Rhodes, Ph.D.
Head, Post-harvest Biochemistry Section
Food Research Institute
Norwich, England

Kenneth V. Thimann, Ph.D.
Professor of Biology Emeritus
The Thimann Laboratories
University of California
Santa Cruz, California, USA

Paul Tudzynski, D. Rer. Nat.
Lehrsthul für Allgemeine Botanik
Ruhr-Universität Bochum
Bochum, Germany

T. A. Villiers, Ph.D.
Chairman and Professor
Department of Biology
University of Salford
Salford, England

TABLE OF CONTENTS

Chapter 1

AGING AND SENESCENCE IN PLANT DEVELOPMENT

A. C. Leopold

TABLE OF CONTENTS

I. INTRODUCTION: TURNOVER IN BIOLOGY

Life may be compared with a candle flame; a unit of matter with characteristics that are persistent in time, although the component parts are being continuously exhausted and renewed. The simile can be extended to say that the life of an individual is that of a population of cells, tissues, organelles, and substrates — which are being continuously exhausted and renewed. Again, the life of the biosphere is that of a population of organisms that is being continously exhausted and renewed. A central function of aging and senescence in plants is the provision of the means by which turnover is assured.

One might begin by asking why the turnover of organisms should be such a pervasive characteristic of the biosphere. An essential part of the answer is that evolution has its very foundation in the dynamics of organism turnover. Furthermore, the rate of evolutionary change can be represented as the product of the extent of variability of a population of organisms times the turnover rate; thus taxa that have a high degree of variability between individuals and also a short life span should have the greatest capacity for evolutionary change.

Once one accepts the essentiality of turnover for evolutionary change or adaptation, then one might ask how turnover is actually achieved. Populations of wild animals commonly experience organism removal through either predation or stress (including stress by disease or parasites), and less commonly animal populations are thinned out by the erosion of aging. In plants, it is markedly less common for organisms to be removed by predation (or the removal of individuals by other organisms) — although it does occur in heavily grazed regions, for example. But more generally, plant organisms succumb to stress (environmental or disease), to an internally regulated senescence, or to the gradual deteriorative effects of aging.

Some extra attention might be appropriate here to the issue of internally regulated or programmed senescence. It is repeatedly apparent that among animal physiologists there is little credence given to the concept of genetically programmed senescence, whereas among plant physiologists the subject is highly visible and the object of serious attention by numerous researchers, as can be seen in many chapters of this book. Many organisms die in what must be an internally programmed senescence; among plants there are common examples such as the simultaneous death of entire populations of soybeans, corn, and the small grains; among animals, the death of the entire breeding population of salmon is a well-known example, and the near-simultaneous death of entire populations of the Ephemerid mayflies and of the male or drone honeybee at the termination of copulation with the queen bee will serve as examples in the insect kingdom. While examples of apparently programmed senescence are markedly more abundant in the plant kingdom, it is curious to see the concept almost ignored by the animal scientists; in a comprehensive book on aging, Strehler[13] mentions programmed senescence only within the context of the death of individual cells, such as in the course of embryo ontogeny; Sacher[12] asserts that "there is no evidence for senescence genes", which may be true for mammals, but certainly not for all vertebrates (witness the salmon pattern of synchronous death at the time of spawning); Timiras[14] discounts the possibility of any programmed senescence. In spite of the lack of conviction by the animal scientists, it seems that there is no alternative explanation of the synchronous death of entire populations of some species of animals and plants than that there is a genetically programmed induction of death. If we accept the existence of programmed senescence in biology, then a general statement of the causes of turnover of organisms can include the consequences of predation, of stress, of programmed senescence, or of deteriorative aging.

The time range for turnover of different biological entities varies from minutes to millennia; expressing the turnover time in the language of biochemistry as the half-life (that is, the time over which half of a given population would be expected to be lost), one can draw a diagram of the half-life of various biological entities as a function of the degree of complexity, as shown in Figure 1. For example, the half-life of metabolic substrates might be estimated at about 10 to 100 min; enzymes persist for about 5 to 500 hr; organelles might have half-lives of from 10 to 100 days, judging from the persistence of mitochondria and chloroplasts; cells themselves might remain alive for 10 to 1000 days in most plants; most organisms may range from 1 to 500 years; plant species may show wide differences in half-lives ranging from 1,000 to 10,000 centuries; and finally, floras or associations of species might persist over a range of 10 to 100 million years. Collectively, these estimates suggest a further generalization; that is that biological units with higher degrees of complexity generally have longer half-lives. This observation is indicated in Figure 2 as the inverse of entropy in the ordinate of the graph. Of course this generalization of the greater half-life expectancy for units with lower entropy is only an approximation.

The concept that entropy plays a role in aging and longevity has been discussed by Sacher[12] with respect to the role of the metabolic rate of an organism as an influence on longevity; he suggests that the metabolic rate (as a factor tending to increase entropy) might relate to the speed of physiological decline in aging. Failla[3] proposed that in animal systems longevity is inversely related to some function of 'vitality', which might be interpreted as relating to the metabolic rate. Strehler[13] suggests the 'vitality' is the maximal rate at which repair can be achieved in an organism following a physiological challenge, and that aging represents a linear decline in this function with time.

In discussing aging and senescence, then, it is reasonable to begin with the fact that the biosphere is a dynamic system in which the components are more-or-less continuously undergoing turnover — not only at the organism level, but also at the suborganismal level and at the superorganismal level. Evolution is a construction based on this turnover, and the biological organizations of higher degrees of complexity tend to have slower turnover rates. And finally, aging and senescence are basic elements in the overall process of turnover.

II. AGING AND SENESCENCE

Because the terms aging and senescence are such general terms used in both scientific and social contexts, there has arisen considerable confusion within scientific circles about the precise usages of the terms (See Preface by Woolhouse). It would seem reasonable for the purposes of the present book to conform to the meanings proposed by Medawar[9] and restated by Leopold.[5] *Senescence* may be defined as the deteriorative processes that are natural causes of death. *Aging,* by contrast, refers to processes of accruing maturity with the passage of time. By these definitions, then aging includes a much wider span of physiological changes, some of which may lead to the weakening of the organism while others may be quite neutral with respect to the capability of the biological organism to survive. Senescence, on the other hand, refers to those changes that provide for the endogenous regulation of death. As examples, the physiological changes in a plant — which cause its conversion from a seedling to a juvenile plant, from a juvenile plant to a mature plant, or the *gradual* decline in vigor with increasing age, would be examples of aging. On the other hand, physiological changes involved in the autumnal coloring (and subsequent death) of leaves, or in the dieback of tulips and other spring bulbs in early summer, or in the yellowing and death of annual and biennial species after the completion of fruiting or after a brief period of drought would be examples of senescence. Referring to Figure 2 (curve B), the gradual loss of survivorship in the later phases of life (as in the case of man, and trees and other perennials) may be a natural expression of the slowly accumulating changes with aging — changes that are not in themselves naturally lethal, but which lead to lower viability. The abrupt termination of life at a given stage in the life cycle (curve C), as in the case of the annual grains and weeds, or following a brief environmental stress, would be examples of senescence.

A. Longevity

Longevity is a natural correlate of aging and senescence. The ranges of longevity over the plant kingdom are enormous, ranging from a few weeks (in the case of desert annuals) to perhaps 4000 years (in the case of bristlecone pines). Less readily measured are the life spans for clonal species such as the grass clones in the prairies of the U.S.; it has been estimated that these clones might have persisted since the time of the last glacial period[12a] — perhaps 15,000 years. A similar longevity has been suggested for the clones of aspen trees in the north central states, which may have persisted since glacial times. The persistence of individual plants through cloning has been utilized in the grape industry, where individual vines have been propagated by graftage for centuries.

When tissues or organs are removed from the parent plant they may show extensive persistence or longevity. Leaves can be removed from the parent plant and rooted, and in this condition (without a growing point) they can not only escape the time of senescence (which the parent plant might have had), but they can persist for a long time, and even regenerate new buds and recommence a new life cycle. When lines of cells are placed in tissue culture, they again may show division and growth over a period of many years, the best example being that of excised tomato roots that have been maintained in tissue cultures for decades.[18]

For many species, longevity is linked to flowering and fruiting. This is most evident in the case of annuals, which experience a uniform death of the entire population at the time of fruit maturation, (e.g., soybeans, corn, small grains, and many garden annuals). The uniform death of entire stands of wheat plants in the month of July is surely not attributable to the end of the frost-free growing season. The onset of senescence in the small grain species is often a cause of loss of yield, especially when the crop is late in maturing its fruiting heads. As yet, the agronomist has no effective means of delaying the onset of senescence other than through the use of genetic lines — which have longer periods before senescence. Other cases of species that experience death upon the completion of fruiting are the biennial plants such as carrots, beets, and onions, which grow vegetatively for one year, and then flower, fruit, and die in the second growing season. Other species may have longer longevity, such as the century plant and some ornamental bamboos — which may remain in the vegetative state for decades, after which they become reproductive and then die. Perennial trees and forbs present a more diffuse pattern of longevity, proceeding through periods of reproductive growth, which may continue for many decades before the gradual decimation of the population by death.

In general, it might be said that the shortest-lived plant species tend to have longevities limited by programmed senescence, whereas the perennials tend to have much less precise limits of longevity and tend to die through a gradual attrition associated with aging.

While the term longevity is ordinarily considered to relate to organisms, there is also a relevance to the life span of organs, tissues, and cells. As an ordinary component of the development of perennials such as trees, there is a programmed death of leaves —

after one growing season in the case of deciduous trees, and after 2 to 4 years in the case of evergreens. For examples of limited longevity in tissues, one need only consider the effective life spans of tissues such as the xylem and phloem. Limited longevity of individual cells (or clusters of cells) may be observed as natural events, even in meristems; for example, the formation of compound leaves is ordinarily accomplished by the death of clusters of cells at the periphery of the earliest leaf meristems; a similar cellular senescence is involved in the formation of the digits in the foot of a bird, which arise through the death of small clusters of cells in the earliest embryonic foot. The various examples of limited longevity of organs, tissues, and cells provide further evidence of the capacity of an organism to impose a programmed senescence and death on its various component parts.

B. Aging Phenomena

Among the most conspicuous gradual changes with accruing maturity is the decline in growth rate. In trees this is evident as a decline in height increments, as illustrated in Figure 3. Wareing[15] has pointed out that trees often show other characteristic changes with aging, in addition to the decline in height and girth increments; these include a gradual loss of apical dominance and a gradual loss of geotropic orientation of the stems. Each of these aging characteristics can be alleviated by graftage or by the regeneration of the plant through cuttings. Thus it is reasonable to presume that the aging characteristics mentioned above may be consequences of increasing limitations on physiological effectiveness associated with the structural mass, and not consequences of loss of totipotency of individual cells.

Another type of aging is seen in the deteriorative changes in clonal material. Many crops that are regularly propagated by cloning (such as potato, pineapple, and sugarcane) show a gradual deterioration of vigor and productiveness of individual clones. Such declines in performance are not alleviated by further cloning or by graftage, and it would seem probable that this type of aging includes the gradual accumulation of either genetic errors and/or of virus infections.

Probably a separate type of aging is shown by the deteriorative changes in the meristems through the life cycle of a plant. Such changes have been detected through the graftage of apices between young and old plants.[8] Not only do aged apices carry the characteristics of slow growth to a young plant to which they are grafted, but they can actually have deleterious effects upon the growth rate of the stock plant bearing them. Apical meristem aging may occur at the time of the ripening of fruits, but the graftage work shows that this type of aging may also be imposed by temperature stress; it is persistent even after the meristem is grafted to a non-aged plant, indicating that this type of aging is related to some deteriorative processes in the meristem itself.

Most plant tissues show a slow (but steady) decline in vigor, even in conditions of tissue culture. An example of tissue culture aging is given in Figure 4. In the case of numerous types of callus tissues in culture, there is a gradual loss of their capability for regeneration of organs, such as roots and shoots.[10] Not all culture lines show the growth decline, but it is frequently observed. The gradual loss of regenerative capabilities in tissue culture has placed heavy limitations on the extent to which tissue cultures can be utilized in genetic engineering.

Seeds represent an important agricultural example of aging. The gradual decline in germinability and in the subsequent vigor of the resultant seedling are widely recognized in agricultural practice. The seeds of most agronomic crops deteriorate at a rate sufficiently rapid to make them a poor planting material after the first year. The deteriorative changes with aging are illustrated for soybean seeds in Figure 5.[7] Such changes involve a wide spectrum of physiological and anatomical characteristics, and are discussed in Chapters 2 and 3.

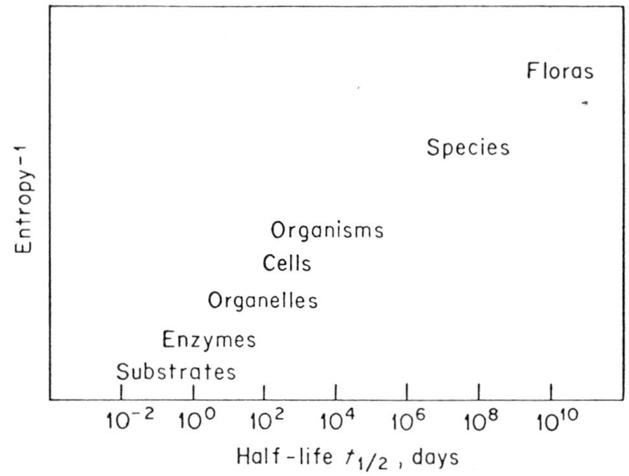

FIGURE 1. A diagrammatic representation of the ranges of half-lives for various biological entities (From Leopold, A. C. and Kriedemann, P. E., *Plant Growth and Development,* 2nd ed., 1975. With permission.)

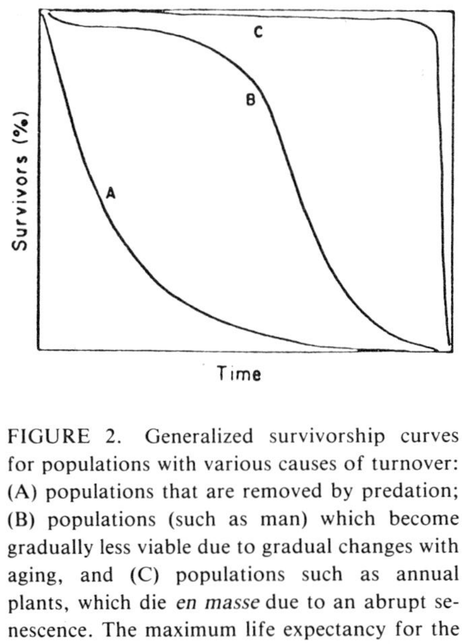

FIGURE 2. Generalized survivorship curves for populations with various causes of turnover: (A) populations that are removed by predation; (B) populations (such as man) which become gradually less viable due to gradual changes with aging, and (C) populations such as annual plants, which die *en masse* due to an abrupt senescence. The maximum life expectancy for the species is plotted as the length of the abscissa. (From Leopold, A. C., *Science,* 134, 1727—1732, 1961. With permission.)

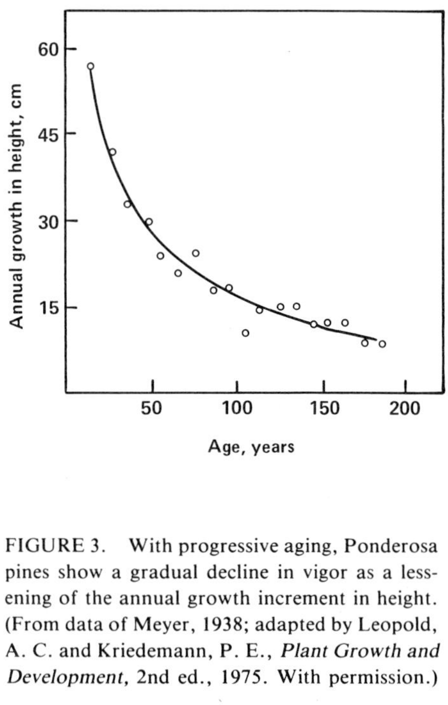

FIGURE 3. With progressive aging, Ponderosa pines show a gradual decline in vigor as a lessening of the annual growth increment in height. (From data of Meyer, 1938; adapted by Leopold, A. C. and Kriedemann, P. E., *Plant Growth and Development,* 2nd ed., 1975. With permission.)

Mention should be made here of the capability for rejuvenation of some aged plant materials. Several types of rejuvenation treatments are commonly utilized by horticulturists, mostly involving pruning. Thus, systematic removal of the buds from an aged apple tree can restore the capability for the rooting of cuttings.[16] The pruning of branches is a widely practiced form of rejuvenation of growth rate in tree and vine crops. The removal of growing apices of a bean seedling can prevent the senescence of the older leaves,[5] and can bring about the rejuvenation of leaves that have already

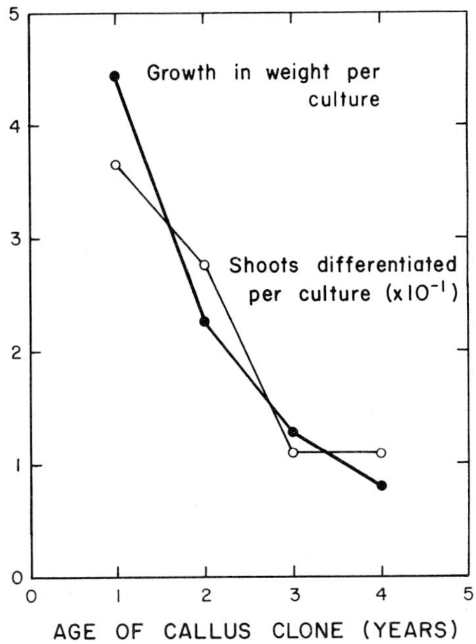

FIGURE 4. Aging of tobacco callus in tissue culture is evidenced as a decline in the capability for growth and a decline in the capability for differentiation of shoots. (From Murashige, T. and Nakano, R., *Am. J. Bot.*, 52, 819—827, 1965. With permission.)

become partly senescent.[4] The pruning of many types of vines and woody plants can cause a partial reversion to a more juvenile form (see review by Leopold and Kriedemann[7]).

C. Senescence Phenomena

The dynamics of senescence becomes especially striking in the survivorship curves for senescing (compared to non-senescing) species. If one plots the maximal life expectancy for a species on the abscissa, and the percent survivorship on the ordinate as in Figure 2, one will find that many species of plants, and especially of animals subject to predation, will follow a survivorship curve in which there is approximately a constant proportion of the population being removed per unit time (Figure 2, Curve A). This population shows neither aging nor senescence. Another type of survivorship is shown in Curve B (Figure 2); in this case, for which man will serve as a good example, most members of the population are shown to survive up to about half of the maximal age for the species, and after that time there will be an increasing tendency towards death; this population shows a loss of survivorship by aging. A third type of survivorship curve is shown in Curve C (Figure 2), in which most members of the population survive for the entire expected life span, after which there is a sudden and complete loss of the entire population; this population shows a loss of survivorship by senescence.

The senescence of organisms occurs most often in conjunction with the completion of flowering and fruiting — the so-called monocarpic condition. The phenomenon is described in detail in Chapter 10.

Senescence in plants shows a wide range of patterns; the total death of the entire

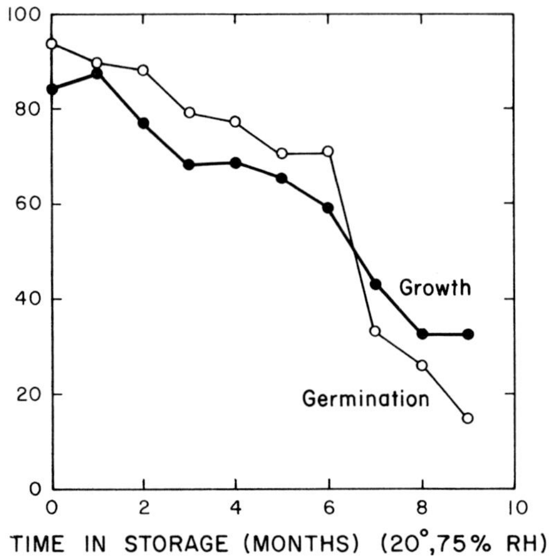

FIGURE 5. Aging of soybean seeds in storage is evidenced in the decline in germinability and in the decline in growth rate after germination. (From Byrd, H. W. and Delouche, J. C., *Proc. Assoc. Off. Seed Anal.*, 61, 41—57, 1971. With permission.)

plant at the end of the reproductive phase is the most extreme case; somewhat less drastic is the death of all above-ground parts *(Top senescence* in Figure 6). This type of dieback is characteristic of many forbs and other perennials, including the spring bulbs. Less drastic is the senescence of the entire leaf array at the end of the growing season, leaving the stems and roots alive *(Deciduous senescence,* Figure 6). Still less drastic is the progressive senescence of leaves along the stem, the dieback of the oldest leaves in the normal development of annual plants *(Progressive senescence,* Figure 6). These patterns illustrate that senescence can be a quantitative phenomenon, the extent of which is apparently under regulatory control within the developmental program of the plant. It is reasonable to presume that these various types of senescence patterns may well have similar underlying controls.

The senescence of leaves normally follows set patterns such as those illustrated in Figure 6. But in addition to the normal periods of leaf senescence and the normal progressive senescence of the oldest leaves, brief periods of stress to the plant can result in marked increases in the extent of leaf senescence. The familiar yellowing and drop of leaves (starting with the oldest) under nutrient deficiencies is a common example of such a stress response. Similar responses of enhanced leaf senescence can occur following stress by temperature extremes, drought, or light deficiencies. The imposition of such stresses to hasten leaf senescence is commonly utilized in physiological studies of senescence, as will be described in Chapter 5. The time course of normal leaf senescence is rather rapid; as illustrated in Figure 7, after a leaf reaches full size, it does not undergo evident changes until just before the termination of its effective life, at which time there is an abrupt and rapid deterioration of its chlorophyll content, and a change in color from green to yellow. It might be noted that the chlorophyll decline makes an easy and convenient measure of the progress of leaf senescence, and leaves are widely used in experiments on senescence for this reason. In connection with the changes in leaf color, it might be mentioned that the autumnal coloration of leaves associated with deciduous senescence is a conspicuous characteristic of senescence, which has no apparent advantage to the plant.

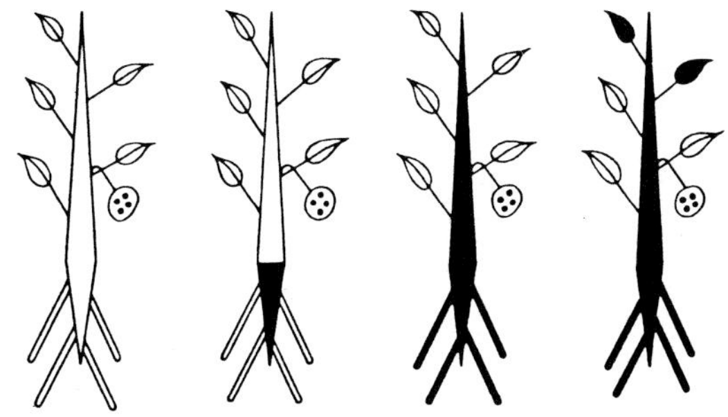

FIGURE 6. Representation of various patterns of senescence in plants; from left to right: the senescence of the entire plant, the senescence of only the aboveground parts, the senescence of deciduous leaves, and the progressive senescence of leaves up the stem. Dying parts are unshaded. (From Leopold, A. C., *Science,* 134, 1727—1732, 1961. With permission.)

The senescence of flowers ordinarily occurs abruptly after the completion of pollination. The physiology of this event will be discussed in Chapter 7.

The senescence of fruits is a characteristic event occurring after fruit maturation, and involves numerous physiological events with similarities to those occurring in leaves and flowers, as will be discussed in Chapters 8 and 9. The senescence of fruits often involves the development of animal attractants, which serve to encourage the collection and transportation of the seeds by animals. Thus the bright colors and sweet tastes* of senescing fruits may be considered to be adaptations, which facilitate the distribution of seeds.

In most spermatophytes, the senescence of any given organ is usually associated with the development of a separation or abscission layer, by which the senescing organ is shed or pruned from the plant. Many of the hormonal changes that are associated with organ senescence find close parallels with the hormonal changes associated with the development of abscission (see review by Nooden and Leopold[11]).

The senescence of leaves and stems in response to environmental stress is, of course, an obvious positive adaptation with survival value. After an experience of water stress, for example, a plant may exercise the senescence of the oldest leaves and sometimes of branches as well, serving to reduce the water evaporative surfaces of the plant.

The influences of environmental factors on the rate of senescence development in leaves and other organs are conspicuous characteristics, often used in experimental and practical work on senescence. The speeding effects of darkness on leaf senescence are widely used in studies of senescence; the retarding effects of low temperature and of elevated carbon dioxide concentrations in the atmosphere are widely used in the preservation of flowers and fruits. The regulation of leaf senescence by pathogenic microorganisms is often distinctive and forms an integral component of the host/parasite interaction. The capability of environmental factors for regulating the progress of senescence is markedly more evident in plants than in animals.

A more abstract positive function of senescence may be the impact on natural selection and evolutionary adaptation. As pointed out in Section 1 on biological turnover,

* The hydrolysis of polysaccharides to produce reducing sugars is one of the processes common to senescing leaves and to maturing and senescing fruits. Ed.

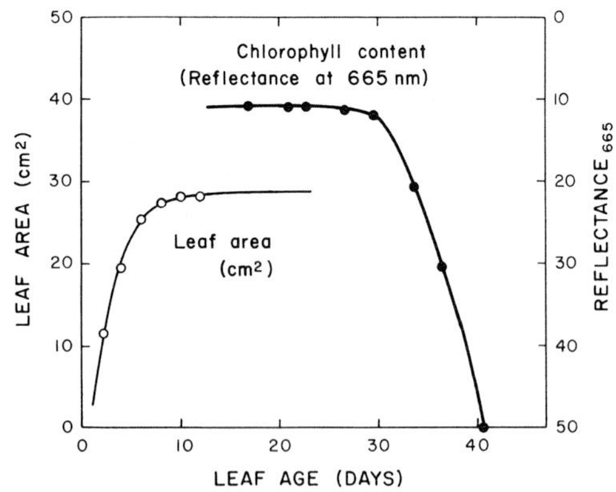

FIGURE 7. Senescence of bean leaves occurs after the growth has been completed, and is evidenced by a rapid decline in chlorophyll content. (Modified from Leopold, A. C., *"Regulateurs Naturels de la Croissance Vegetale,"* 1964, 705—718. With permission.)

evolutionary change will be a function of the variability of a population multiplied by its turnover rate; species with a high degree of variability and a rapid turnover rate (such as many monocarpic weeds) will have the potential for maximal rate of adaptation to new environmental situations. It is noteworthy that annual weeds are some of the most aggressive in adapting to new environmental situations, and they show the monocarpic senescence characteristic very generally.

In short, it is evident that the capability for senescence offers numerous types of advantages to the plant. The widespread occurrence of senescence in organisms, in organs, in tissues, and in cells, and the rapid onset of death in these various situations indicates that through evolutionary time plant species have developed elaborate capabilities for regulating senescence through genetic programs.

III. SENESCENCE AND SURVIVAL

Among zoologists it is generally considered that the death of individuals has negative or only diffuse positive values; older components of the population are sloughed off, the population size is reduced, and in the case of man at least, individuals with the most learning are removed. The situation on the plant side, however, offers a very different picture, in which the senescence of organisms and of organs provides distinct advantages in terms of ecological adaptation, natural selection, and the efficiency of internal physiological functioning.

In many species, seasonal adaptation is achieved largely through senescence. Familiar examples are the spring bulbs, which complete their flowering and fruiting cycle and then die back through top senescence at the time when the intense competition for light develops with the outgrowth of foliage in other species in the spring. Similar adaptations to spring and early summer are seen in the small grains. Species such as soybeans, maize, and many annuals of the temperate climates are also adapted to terminate their life cycle before the onset of frost; this termination involves overall senescence in the case of monocarpic species, top senescence in the case of perennial forbes, and deciduous senescence in the case of woody perennials. Another type of seasonal adaptation is the senescence of some species before the onset of an annual drought; species of rice growing in the monsoon regions of Southeast Asia are adapted to undergo senescence before the wet season gives way to an annual drought season.

The senescence of leaves and of branches represents an adaptation, which provides for self-pruning and improved competition for light in many instances. Thus leaves in positions allowing poor light interception may be pruned off, and lateral branches of trees may be pruned off by natural abscission, allowing for better light competition by the plant.[2]

IV. CONCLUSION

Returning to the simile of life as a candle flame, persistent in characteristics, but constantly being exhausted and renewed, it can be said that the flame characteristic applies to all levels of the biosphere — from cells and tissues, through the organism level, and to the highest orders of organization into floras. Senescence may be considered to be a device by which the turnover characteristic can be actually exploited to provide advantages to the plant. Aging can be considered to be a more passive manner of the development of turnover.

The capability for programmed senescence has been utilized frequently by plants. Examples of the positive utilization of senescence by plants have been suggested, including adaptability to environmental and seasonal events, and at the cellular level, where senescence has been utilized in ontogenetic development.

It has been suggested by Strehler[13] that longer life spans can be equated with evolutionary advantages; the widespread utilization of senescence mechanisms by plants suggests that quite the reverse may be true in certain instances, and that evolutionary and adaptive values can be achieved through the foreshortening of life by senescence.

In contrast to the senescence systems, aging appears to provide a passive set of changes that may gradually increase the vulnerability to death, especially noticeable in organisms that are not removed from the living biosphere by predation or by active senescence. It may be, then, that aging provides a final guarantee of the ubiquity of turnover in biology.

REFERENCES

1. **Byrd, H. W. and Delouche, J. C.,** Deterioration of soybean seed in storage, *Proc. Assoc. Off. Seed Anal.,* 61, 41—57, 1971.
2. **Chaney, W. R. and Leopold, A. C.,** Enhancement of twig abscission in white oak by ethephon, *Can. J. For. Res.,* 2, 492—495, 1972.
3. **Failla, G.,** The aging process and somatic mutations, in *"The Biology of Aging,"* Strehler, B. L., Ed., American Institute Biological Science, Washington, 1960, 170—175.
4. **Krul, W. R.,** Nucleic acid and protein metabolism of senescing and regenerating soybean cotyledons, *Plant Physiol.,* 54, 36—40, 1974.
5. **Leopold, A. C.,** Senescence in plant development, *Science,* 134, 1727—1732, 1961.
6. **Leopold, A. C.,** Kinins and the regulation of leaf aging, in *"Regulateurs Naturels de la Croissance Vegetale,"* CNRS, Paris, 1964, 705—718.
7. **Leopold, A. C. and Kriedemann, P. E.,** *Plant Growth and Development,* 2nd ed., McGraw-Hill, New York, 1975.
8. **Lockhart, J. A. and Gottschall, V.,** Fruit-induced and apical senescence in *Pisum sativum, Plant Physiol.,* 36, 389—398, 1961.
9. **Medawar, P. B.,** *The Uniqueness of the Individual,* Basic Books Inc., New York, 1957, 191 pp.
10. **Murashige, T. and Nakano, R.,** Morphogenetic behavior of tobacco tissue cultures and implication of plant senescence, *Am. J. Bot.,* 52, 819—827, 1965.

11. **Nooden, L. D. and Leopold, A. C.**, Phytohormones and the endogenous regulation of senescence and abscission, in *"Plant Hormones and Related Compounds,"* Letham, D. S., Ed., Elsevier-North Holland, Amsterdam, 1978, 329—369.
12. **Sacher, G. A.**, Longevity and aging in vertebrate evolution, *BioScience,* 28, 497—501, 1978.
12a. **Stebbins, G. L.**, *Variation and Evolution in Plants,* Columbia University Press, New York, 1957, 643 pp.
13. **Strehler, B. L.**, *Time, Cells, and Aging,* Academic Press, New York, 1977, 456 pp.
14. **Timiras, P. S.**, Biological perspectives on aging, *Am. Sci.*, 66, 605—613, 1978.
15. **Wareing, P. F.**, Growth and its coordination in trees, in *"Physiology of Tree Crops"* Luckwill, L. C. and Cutting, C. V., Eds., Academic Press, London, 1970, 1—21.
16. **Wellensiek, S. J.**, Rejuvenation of woody plants by formation of sphaeroblasts, Proc. Kon. Ned. Akad. Wet. Ser. C, 55, 567—573, 1952.
17. **White, P. R.**, Potentially unlimited growth of excised plant callus in an artificial nutrient, *Am. J. Bot.*, 26, 59—64, 1939.
18. **White, P. R.**, *A Handbook of Plant Tissue Culture,* Jaques Cattell Press, Lancaster, Pa., 1943, 1—277.

Chapter 2

SENESCENCE IN SEEDS

Daphne J. Osborne

TABLE OF CONTENTS

I. INTRODUCTION

Seeds are truly remarkable assemblies of living cells, each seed containing a potential new individual of its species. They are adapted to survive a wide range of unfavorable conditions for growth, whilst holding the records for long-term cellular longevity. Morphologically, mature seeds are formed in such a way that we cannot tell with certainty if they are alive or dead until we try to germinate them. Special physiological and biochemical features have evolved that favor extended life-spans and afford protection to the cells under conditions that are not conducive to growth. This chapter discusses various aspects of cellular senescence in seeds and relates them to the maintenance of embryo viability and the performance at germination.

II. THE LONGEVITY STRATAGEM

Following pollination of the flower and fertilization of the ovule an embryo grows within the subtending tissues of the fruit to develop to a certain critical stage of differentiation of roots and shoots. Then, upon a signal that is not yet understood, an organized dehydration occurs to an extent that would spell serious impairment of cellular function or death in most other plant tissues.[1,2] In seeds, however, this dehydration leads to a cessation of cell growth so that the embryo is arrested, fully viable, in its developmental tracks.[3] A period of quiescence ensues, continuing until water becomes available and the appropriate conditions for renewed metabolic activity, growth, and development are re-established.

A. Adaptations for Survival

The extent to which the embryo is differentiated when development is arrested and the extent to which dehydration is tolerated vary greatly between different species. The small and simple seeds of the Orchidaceae, for example, are essentially undifferentiated — except for a small group of meristematic cells at one end. They possess little in the way of food reserves and have evolved obligate and symbiotic associations with specific fungi from which they obtain the nutrients that tide them over their early stages of embryo development. The chances of seed survival are therefore low, and the rates of senescence are high.

A well advanced embryo with ample food reserves does not necessarily ensure a seed's potential longevity.[4] Some very large seeds from the trees of tropical rain forests, or the seeds from certain fleshy fruits such as citrus, are fully differentiated when they are shed, and possess cotyledons with considerable storage reserves. The cells are, however, not resistant to the stress of air drying, so that, again, the period of longevity of the seed is relatively brief and can be measured in days or weeks. From an evolutionary point of view this is a reasonable adaptation for seeds that are shed into conditions where dehydration is unlikely to occur. For the forester the storage of such seed poses considerable problems.

Even shorter-lived are seeds of many plants or trees adapted to aquatic environments. Those of river banks (willows)[5] or estuarine habitats (mangroves) grow at once if they fall into the water, but those that fall on dry land can desiccate within a few hours to moisture contents below their critical levels for survival. Indeed, in some species, (e.g., the red mangrove, *Rhizophora mangle*) the embryo continues to differentiate and grow without pause, so that although a seed structure is formed, the embryo never develops the physiological adaptations that permit dehydration and extended cell longevity. Consequently, a seedling grows on the mother plant before the seed itself is shed — a condition equivalent to vivipary in animals. For plants, this situation is the exception rather than the rule, and few physiological or biochemical studies have been carried out with viviparous species.

Those plants that have perforce evolved successful stratagems for quiescence and survival to combat extended periods of extreme cold, heat, or dehydration produce seeds with the longest potential longevity. At the time of shedding most of them have well-developed embryos of low moisture content (±10%), and many possess thick seed coats impervious to water or gases. In addition, they may be in a state of dormancy, a condition in which a supply of water alone is not sufficient to initiate their germination. Most of the physiological and biochemical information available on survival and senescence of seeds concerns the major crop and weed species, many of which have developed some or all of the cellular adaptations necessary for extended seed longevity.

B. Adaptations to Dehydration

There are relatively few cells in a plant that possess the physical and chemical characteristics and ultrastructural organization that permit them to lose water yet retain the functional integrity of their constituent organelles and macromolecules. The tissues that develop these features are common only to seeds,[3] the leaves of certain mosses,[6] and the shoots of the so-called resurrection plants.[7] They can all withstand desiccation to less than 10% moisture content and yet survive, with viability unimpaired, to regain full metabolic function when water is again available. But whereas resurrection plants and mosses can accommodate such dehydration from water contents of 80% or more in a matter of hours, seeds must desiccate more slowly over a period of days in order to maintain a living embryo.

It seems, therefore, that some process of metabolic change must first occur in the embryo in order to prepare the cells for the period of dehydration that is to follow. Although the stimulus that initiates the change remains unknown, it is clear that as soon as the moisture content of a seed starts to decrease, cell division in the embryo stops and there is a progressive attenuation of nucleic acid and protein synthesis. As measured by the incorporation of radioactive precursors, a halt in the replication of DNA occurs as an early event in dehydration,[8] to be followed by the cessation of RNA and protein synthesis.[9] Ultrastructural changes include the reduction in the number of cristae in the mitochondria, a contraction of the visible endoplasmic reticulum, a compression of dictyosomes to stacks of flattened vesicles, the disappearance of vacuoles, and a condensation of the nuclear chromatin.[10] During the last stage of dehydration, it is quite usual for lipid bodies to become distributed in a closely packed and ordered array adjacent to the plasmamembrane, suggesting that some charged attraction develops between the single osmiophilic layer of the droplet and the enclosing unit membrane of the cell.[10] As a result of these changes in cytoplasmic organization, cells are rendered some of the potentially longest-lived of any biological tissue, so that the embryo of a seed can remain viable (or living) and capable of germination for very extended periods of time.

C. Longevity of Seeds

Records are always of interest (Table 1), but the critical scientific evidence for the oldest known viable seed is difficult to obtain. Stories of the successful germination of the Emmer wheat from the ancient Egyptian silos at Fayum, or from the tomb of Tutankhamen at Thebes, are certainly untrue. Carbon dating has confirmed their age (6000 and 4000 years, respectively), but from the tests of their viability carried out by Professor John Percival of the University of Reading, and Sir Wallis Budge of the British Museum when the grain was first brought to Britain,[11] it was quite clear that all the material was already dead. More recently, electron microscopy of other grains from those stocks has shown complete disruption of cellular organization,[12] to an extent that would be quite unlikely to have occurred during the time interval from their discovery in Egypt to their attempted germination in England. Similarly, biochemical tests demonstrated a complete absence of any high molecular weight nucleic acid components;[13] again, an unlikely situation for seeds that had recently been viable. Reports of germination of the even older seeds of *Lupinus arcticus,*[14] fround frozen and buried in the Pleistocene silt in rodent burrows in the Arctic Yukon are equally doubtful, for although artifacts in these burrows were carbon dated at 14,000 years, the ages of the seeds, themselves, were never determined.

A claim of 1700 years survival in the soil,[15] again based on the dating of associated archaeological remains and not of the seed, has been made for *Chenopodium album* and *Spergula arvenis*, and samples of viable seeds of the Indian lotus *Nelumbo speciosum* from the bed of a Manchurian lake,[16] which were once dated as 1000 years,[17] were retested and then dated at 100 years.[18]

The authentic oldest living seeds are probably those of *Canna compacta* discovered as part of a rattle necklace in the pre-Inca tombs of Santa Rosa de Tastil in Argentina.[19] It is presumed that the Canna lily seeds had been inserted into the young growing fruits of a Juglans species, so that at maturity when the walnut fruit had healed, each Canna seed was entirely enclosed within a hardened shell. Material from the shells was carbon dated at 620 ± 60 years,[19] and the Canna seed must therefore have been as old or older. When germinated in 1968, the Canna seed grew very slowly and with difficulty, the roots exhibiting abnormal geotropic behavior;[20] all symptoms typical of the germination of an aged seed. This famous plant and its progeny now survive in the grounds of the Botany Department in the University of La Plata, Argentina.

Other authentic aged seeds that remained viable are those from herbarium sheets in the British Museum. In 1940, during an air raid in the Second World War, the museum was damaged by fire and the sheets inadvertently doused with water. Seeds of *Albizzia julibrissin*[18] collected in China in 1793 still germinated. A seed of *Nelumbium* sp. from the Hans Sloane collection of 1705 was germinated in 1942.[21] In 1856, packets of seeds from over 600 species were sent from Kew Gardens to Australia to stock the proposed Botanic Garden in Melbourne. They remained dry and unopened until 1906 when Professor A. J. Ewart tested them and found that many of them germinated.[5] Included among these were seeds of *Goodia latifolia* and *Hevea heterophylla*, which were then 105 years old. Ewart then drew attention to the fact that hard coated seeds with more or less impermeable seed coats (mainly Leguminoseae, Malvaceae, or Myrtaceae) appeared to be the longest lived.

Longevity and immortality have concerned man from the earliest times, and the Greeks were no exception. Theophrastus (285 B.C.) in his discourse into plants[22] observed that, "Of seeds, some have more vitality than others as to keeping" and, "In Cappadocia at a place called Petra they say that seed" (presumably wheat) "remains even after forty years fertile and fit for sowing". These comments accord with our present knowledge of 30 to 40 years for wheat, but whereas many of the life spans recorded in the past were based primarily on man's memory, modern information is based upon precise experimentation.

One of the first to design a long-term experiment to test the duration of seed viability was Dr. W. J. Beal. In 1879[23] he placed 20 samples of fresh seeds of 21 different plant species in moist sand in unstoppered bottles, and buried them in an inverted position 20 cm below the soil in the grounds of Michigan State University, U.S.A. Every 5, and later 10 years a bottle was withdrawn and the viability of the seeds tested. *Oenothera biennis, Rumex crispus* and *Verbascum blatteria* survived for 70 years,[24] and those of *Verbascum* still germinated in the most recent assay after 90 years.[25] From these

TABLE 1

Authentic Longevity of Seeds Maintained Dry

	Years	Ref.
Cassia bicapsularis	115	109
Albizzia julibrissia	147	18
Cassia multijuga	158	109
Mimosa glomerata	221	110
Nelumbo sp.	c.237	21
Canna compacta	c.600	19

Authentic Longevity of Seeds Maintained Imbibed

	Years	Ref.
Fraxinus excelsior	6	4(p. 324)
Avena fatua	9	34
Verbascum blattaria	90	25
Nelumbo sp.	c.100	16, 18

Authentic Longevity of Seeds Submerged in Water[111]

4.25 years	7 years
Asclepias Syriaca	*Juncus bufonius*
Plantago Rugelii	*Juncus tenuis*
Solidago rugosa	*Sium cicutaefolium*
Verbena urticaefolia	*Syntherisma sanguinalis*

experiments and those of a similar kind set up by Duvel[26] at Rosslyn, Virginia in 1905 with weed and crop seeds, it is clear that most of the crop seeds failed to survive even a year. The crop and weed seeds with the longest survival not only tended to be hard-coated (as Ewart had found), but also possessed a deep and long-term dormancy.

D. The Regulation of Life-Spans

It is clear that seed survival is both genetically and environmentally controlled. The type of seed is determined genetically, and, depending upon the history of the seed during its development and the temperature, moisture content, and the availability of oxygen in the gaseous environment during storage, seeds can survive for longer or shorter periods of time. Senescence of seeds is demonstrably a non-stochastic event, and Roberts[4] has derived a simple mathematical equation expressing some of the factors determining viability:

$$\log P_{50} = Kv - C_1 m - C_2 t$$

where P_{50} is the time taken for 50% of the seed to lose viability, m = moisture content, t = temperature (0°C) and K, v, C_1, and C_2 are constants. Using this formula, quite accurate predictions can be made for the expected viability and life span of a variety of crop and weed species.

That the rate of loss of viability (or rate of senescence) can be greatly modified by the degree of hydration has already been described, and one of the best experiments to illustrate this was carried out by Mackay and Flood in 1968.[27] One of the species selected was rye *(Secale cereale)*. Newly harvested grain of 99% viability from a 1964 harvest was artificially dried, and samples with moisture contents of 15, 12, 11, 10, and 5% were prepared and enclosed in sealed polythene bags, and stored in the cool below 20°. Five years later the grain was sampled for germination tests and the stocks were found to have 0, 15, 45, 48, and 64% viability, respectively.

Many seeds store well at less than 5% moisture content, but a lower limit of hydration is reached at which damage is incurred. It has been proposed that the shells of crystalline water that are bound to the protein or nucleic acid macromolecules in the cytoplasm protect their structural configuration.[28,29] Certain seeds (wheat) can be dehydrated to 0.5% moisture content before critical damage occurs. With further dehydration, loss of viability is incurred, presumably due to withdrawal of the critical bound water with the ensuing loss of integrity of biopolymers.[30]

Most dry seeds remain viable for long periods if stored at −5°C or at −20°C. Even lower temperatures are readily tolerated, for Lipman[31] exposed a variety of different seeds to temperatures close to absolute zero for more than 42 hr without harm. In my own laboratory, isolated embryos of rye have been stored for more than 10 weeks in liquid nitrogen at −196°C without impairment of viability. Storage of dry seed at normal temperatures under vacuum can also aid seed longevity, but is clearly not as important as temperature and moisture content. The rate at which cellular dysfunction and loss of viability occurs, therefore, is not primarily an oxygen-dependent process.

Living systems exist in a high degree of order or a low state of entropy, so where metabolic activity is minimal (as in a dry seed) decrease in order and a rise in entropy must occur with time. A background level of potentially lethal stochastic occurrences such as hits by cosmic radiation and peroxidative events in membrane lipids resulting from free radical formation must provide a slow, but ever-increasing level of cellular lesions and a rise in entropy.[32] In the very long term, a seed must inevitably accumulate a lethal level of such lesions. Long before that happens, however, degradative biochemical changes take place that pre-empt stochastic events as the causative lesions leading to seed senescence.

III. SENESCENCE DRY OR SENESCENCE IMBIBED?

There is debate as to whether a seed stored dry survives longer and undergoes senescent change more slowly than a seed that is fully imbibed. The conditions are so different that a direct comparison is difficult. Furthermore, only seeds that are dormant can remain imbibed without germinating. For most crop plants dormancy is considered to be an unacceptable attribute, and as far as possible has been bred out of the genetic complement. The imbibed seed must therefore either germinate, or degenerate and die. The question is of greatest interest when we come to consider the biochemical processes involved in seed senescence. Comparisons have been made for a few dormant seed species, and of particular relevance are those for thermodormant lettuce seed. When maintained fully imbibed at 30°C (a temperature imposing dormancy) they remained fully viable for 105 days, whereas dry seed kept at the same temperature were dead within 42 days.[33] For the wild oat (*Avena fatua*), the weight of evidence suggests that, compared with an imbibed seed, the dry seed may have as long or a slightly greater longevity: in arable soils a maximum survival of 9 years has been recorded,[34] though an average of about 4 to 5 years is more common. In the dry state at room temperature, Thurston[35] reported 80 to 90% germination at 4 years, although some seeds are known to have survived for 25 years.[36] Important, however, is the fact that although seed of *A. fatua* survives anaerobic submergence up to 6 days at 24°C without apparent harm,[37] periods of longer than 36 days lead to complete loss of viability.[38] The maintenance of viability in the imbibed seed is therefore an oxygen-dependent process.

Clearly, seeds can become senescent either while they remain dry or when they are imbibed. If dry they can be stored anaerobically in liquid nitrogen; if imbibed they require oxygen. There seem, therefore, to be two means by which a seed may reduce the rate of senescence change; either by remaining dry and metabolically arrested, so taking a long time to accumulate the lethal level of hydrolytic enzyme-mediated events that lead to loss of cellular integrity, or by remaining imbibed and metabolically active, so that the rate of biochemical turnover is sufficiently high to maintain a condition approaching homeostasis — during which senescent change tends not to occur. Most plants have evolved one or other stratagem with great success, and each will be discussed in detail.

A. Senescence Changes in Dry Seeds

1. Ultrastructural Changes

From external appearance or by dissection, it is impossible to tell by eye whether a dry seed is viable or not. At the ultrastructural level, however, it is evident that there are some quite consistent differences between the live and dead seeds,[39] which are presented in detail in Chapter 3. Here we may note only that electron micrographs of cells of non-viable (or dead) embryos of rye[40] show that the outer membrane of many mitochondria is extended to form a "blister" and is dissociated from a broken inner membrane. The plasma membrane is frequently contracted from the cell wall, and the presence of electron-dense material in the area between the two indicates that the semi-permeability and molecular integrity of the plasma membrane have been disrupted. The nuclear membrane, too, may show irregularities and breaks not present in cells of a viable embryo. Although some of the membrane changes seen in a dry non-viable embryo may be exacerbated by aqueous fixation procedures, they clearly reflect a difference existing between the membranes present in viable and non-viable cells.

2. Germination Performance

Information on the state of senescence of a dry seed can be more properly assessed biochemically, or by the performance at germination when water is supplied. Provided there are no restrictions to embryo growth, (e.g., dormancy) and the conditions are suitable, newly harvested seeds germinate with speed, and respiratory activity,[41] ATP generation,[42] protein synthesis,[43] and all classes of RNA synthesis[44] can commence within minutes. In rye, for example, it has been demonstrated by autoradiography that [^{3}H] uridine is incorporated within 10 min into an RNAase-degradable product located within the nucleus.[45] Fractionation of radioactive nucleic acid components by polyacrylamide gel electrophoresis has shown that heterodisperse high molecular weight RNA and 4S and 5S RNA synthesis commences within 20 min of imbibition.[44] Precursor ribosomal RNA (31S) and 25S rRNA commence by 40 min, and the production of 18S rRNA by 60 min.[44] As the cells of the embryo expand and elongate, a critical stage in reactivation of embryo development is reached when the first DNA replication occurs.[46] The rate at which this sequence of events takes place at germination is a measure of the extent of the senescent changes that have accumulated in the seed during the earlier dry state.[46] Seeds that have been stored for long periods of time (even in favorable conditions) germinate more slowly than fresh seed, and initially show low rates of protein and RNA synthesis.[47] In addition, the critical time at which DNA replication is initiated becomes progressively delayed.[46] The old seed, though viable, expresses all the symptoms of what the agriculturalist calls loss of "vigor".[4] If the senescent changes are sufficiently extensive, the rates of synthesis of new cytoplasmic components remain at such a low level that the embryo fails to establish as a seedling. It seems that all cells of a seed are equally susceptible to senescent change. Autoradiographic evidence

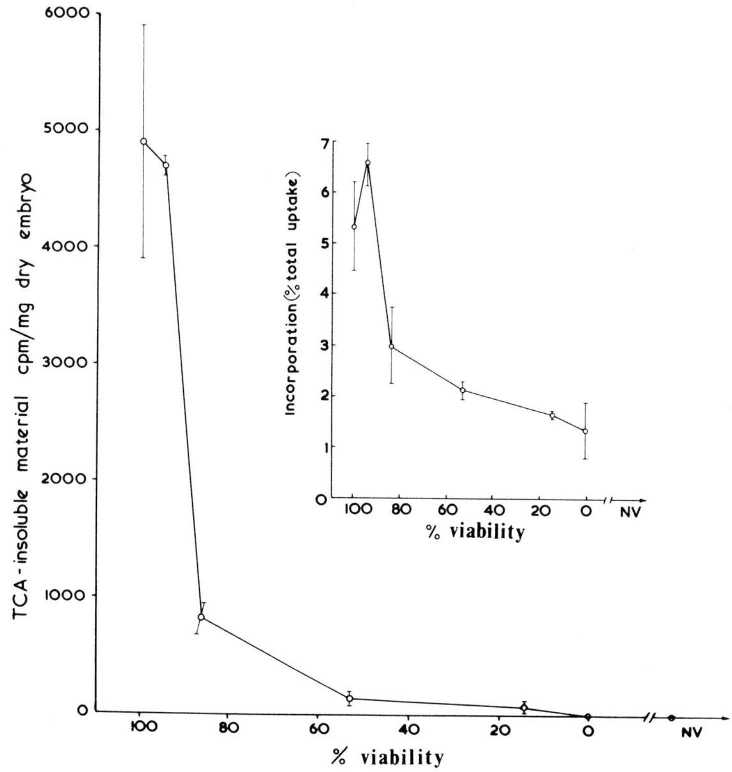

FIGURE 1. Decline in protein synthesis with senescence and loss of viability in rye embryos in the first 2 hr of germination (incorporation of amino acids into TCA-insoluble material). (From Sen, S. and Osborne, D. J., *Biochem J.*, 166, 33—38, 1977. With permission.)

from grain counting has shown a failure of incorporation of radioactive precursors with loss of viability in all tissues of rye embryos.[47,48]

In barley, aleurone cells exhibit impaired hydrolytic α-amylase activity with reduced germination performance of the embryo. Neither the production of α-amylase, nor germination, is improved by the addition of gibberellin,[49] indicating that a low level of hormone does not limit the performance. Further evidence that each part of the seed deteriorates on storage comes from embryo "transplant" experiments in wheat.[50] When embryos of low vigor were attached to aleurone parts of high vigor (or the reverse transplant made), aged material was not improved and vigorous material was impaired.

The extent to which an embryo can incorporate amino acids into protein in the very early hours of germination is a reliable measure of both viability and vigor (or degree of senescence), in many different kinds of seeds.[11,51,52,53,54] The curve (Figure 1) obtained for stocks of rye grain of different percentage viability shows how, although the excised embryo may still possess the capability for germination, the rate of amino acid incorporation by those that are viable in a 15% viable stock is very low, compared with that from a 95% viable stock. Although from conductivity experiments it is clear that the plasma membrane integrity is also modified, the incorporation values obtained are not attributable to different levels of amino acid uptake by the different samples.[47] Judging from such experiments, the greatest impairment of protein synthesis (or the greatest rate of senescence change) occurs very early in the life span of the dry embryo, during the loss of viability from 95 to 86%. In embryos isolated from seed that has just reached the condition at which it cannot germinate (0%), or from seed that has already remained non-viable for several years (NV), there is no synthesis of TCA-insoluble polypeptide material at all.[40,47] The major question of seed senescence is, therefore, "What are the degradative events that can occur in a dry seed that lead to an impaired synthesis of protein at germination?".

There is no evidence that a single cellular lesion determines cellular senescence. In dry seeds there is an occurrence of a variety of degradative events, all of which could be contributory factors in the loss of vigor and subsequent biochemical performance of an embryo or seed at germination. The major lesions that have been recognized and studied are set out in Table 2.

3. Membrane Changes

The progressive loss of integrity of the plasma membrane is readily demonstrated in aging seed by the extent of leakage of cytoplasmic components to the external medium when the dry seeds are put into water.[55,56] All seeds show some loss of solutes as measured by the rise in conductivity of the external solution, but in germinating embryos of high vigor, sugars, amino acids, and other metabolites are subsequently reabsorbed by active uptake. Depending upon the increasing degree of senescence of the embryo and progressive impairment of the membranes, the initial loss of cytoplasmic solutes is correspondingly greater, and the extent of subsequent uptake less. One of the events in a germinating embryo, therefore, is the apparent restoration of the semi-permeable property of the plasma membrane and repair of lesions in membrane integrity that occurred when the seed was dry. That membrane synthesis and repair does occur at imbibition is apparent from ultrastructural studies. In maize, for example, lysosomal engulfment of damaged organelles, (e.g., mitochondria) and their subsequent renewal follows closely upon imbibition.[57,58,59]

It has been proposed that loss of membrane integrity in the dry state could involve the formation of free radicals and the peroxidation of unsaturated fatty acid components. Indeed, the detrimental involvement of free radicals could be inferred from the reported decrease in unsaturated fatty acids during aging in embryonic axes of pea seeds; the levels of trienoic acids, (e.g., linolenic) decreasing more rapidly than those of dienoic acids, (e.g., linoleic).[60] A declining ability of a germinating seedling to repair the membrane lesions that occur in dry seeds would be a major factor leading to poor germination.

4. Mitochondrial Function

The decreased respiratory activity of aged seed at imbibition must have important consequences for limiting synthetic processes, and can be correlated with the poor integrity of mitochondrial membranes observed in electron micrographs of sections of "aged" or non-viable material.[40] A comparison of mitochondria extracted from 4-day seedlings of soya bean of one seed harvest with those from another stored from a harvest 3 years earlier, (but both of the same percentage viability) showed that the P/O ratio from the older stock was half that of the new seed.[61] Not only is total respiratory activity reduced, therefore, but the extent of respiratory uncoupling is increased. Electron micrographs of the pellets of these extracted mitochondria showed that those from the older seed source possessed swollen cristae and lacked their outer membranes. Such results are further evidence that age-related deleterious membrane changes do occur in mitochondria during seed senescence.

A general event in aging seeds is the progressive loss of activity of mitochondrial dehydrogenases with seed vigor,[62] culminating in the complete loss of activity at the stage at which the embryo can no longer synthesize protein or germinate. Failure of

TABLE 2

Location of Senescent Lesions that can Lead to Failure of Protein Synthesis at Germination

Membranes	Mitochondria	Ribosomes	Post-ribosomal supernatant components	Nuclei
Loss of cellular compartmentalization	Loss of organelle function: a. Respiratory enzymes b. Nucleotide-generating systems	Loss of integrity of ribosomal RNA	Failure of enzyme function a. Aminoacyl-tRNA synthetases b. Transfer enzymes I and II, (binding enzymes) I, (translocase enzymes) II	Loss of integrity of DNA template. Failure of RNA transcription Failure of synthesis of messenger RNA Impairment of function of long-lived messenger RNA

dehydrogenase function is the basis of an *in situ* test for viability in the early hours of imbibition;[63] whereas living cells will reduce a solution of the colorless 2,3,5-triphenyltetrazolium chloride to the stable, red-colored non-diffusible formazan within the tissues, the dead cells remain colorless.

In high vigor embryos, the activity of high energy nucleoside triphosphate generating systems and the rise in ATP levels during the first 20 min of germination can be dramatic.[42] It is not surprising, therefore, that at imbibition any loss of mitochondrial function should be associated with low levels of ATP production[64] and reduced rates of ATP and GTP-dependent events in protein synthesis.[52]

5. Lesions in Components of the Protein Synthesizing System

Although ATP generation is reduced with loss of viability,[64] and protein synthesis does not occur in an embryo that cannot germinate,[40,52,53] failure of mitochondrial function and ATP formation cannot be the major lesion in seed senescence, for polyribonucleotide synthesis can still take place.[65,66,67] When embryos of rye from 0% viability stocks, (i.e., those that have just lost the capacity to germinate and synthesize protein) are supplied with [^{3}H] uridine and the embryos subsequently fixed and sectioned for autoradiography, a radioactive RNase-sensitive material can be found located within the nucleus.[46] When extracted and fractionated by electrophoresis on polyacrylamide gels, the products were identified as low molecular weight (4-5S) RNA.[67,68] The failure of protein synthesis and lack of synthesis of stable high molecular weight RNA in 0% viable material suggests that some biochemical lesions occur in the components of the RNA and protein synthesizing system of the cell during the dry state. One possibility is that either stored or newly synthesized messenger RNA can no longer be translated or transcribed. The lack of any new high molecular weight RNA greater than 4-5S in the 0% viable material has a special significance, for it implies that neither new messenger RNA nor new ribosomes are synthesized, and poses the question of whether (in a senescent seed) this is due to failures in the mechanism of transcription from the genome, an extensive and rapid degradation of RNA by nucleases, or an impairment of the nuclear DNA template.

6. Integrity of Ribosomes and Cytoplasmic Factors

The weight of evidence suggests that a fully functional protein synthesizing system is present in the dry seed after the embryo is dehydrated at maturation.[69,70] On the basis of experiments with cell-free systems prepared from dry embryos of wheat,[71,72] it was proposed that the early protein synthesis at germination could be coded for by long-lived messenger RNAs and be translated on old ribosomes that are stable to storage in the dry seed. More recent evidence suggests an alternative; namely, that the early protein synthesis could be coded for by newly synthesized messenger RNA molecules transcribed at imbibition.[47,73] Because the assembly of newly synthesized ribosomes and their transfer to the cytoplasm apparently takes longer than the activation of protein synthesis, it is clear that old ribosomes can be the sites of early protein synthesis.[47]

That lesions do occur both in ribosomes and in components of the post-ribosomal supernatant fraction containing tRNA, aminoacyl-tRNA synthetases, and transferase enzymes was first demonstrated from studies with cell-free systems[52,53,54] isolated from dry rye embryos of different percentage viability. Comparisons were made of polyphenylalanine synthesis by homologous and heterologous cell-free systems containing ribosomes and supernatant fractions from 95% viable or non-viable embryo preparations and primed with the synthetic messenger poly-U. When the embryo supernatant from non-viable embryos was mixed with ribosomes extracted from 95% viable embryos, the incorporation of [^{14}C] phenylalanine was only 26% of that obtained with a homologous preparation of supernatant and ribosomes from 95% viable material. Addition of more 95% supernatant to the heterologous system restored activity to approximately 79% of the 95% homologous system.[74] Clearly a major lesion in the protein synthesizing system that increases with loss of viability is associated with components present in the supernatant fraction. Further lesions are associated with the ribosomes themselves, for the activity of non-viable ribosomes for sustaining [^{14}C] polyphenylalanine synthesis in the presence of 95% supernatant cannot be raised to more than 60 to 79% of the 95% viable homologous ribosomal and supernatant system.[53,74] Using similar techniques, investigations with other seeds (pea[75] and wheat[76]) have confirmed the presence of lesions in the ribosomes and post-ribosomal supernatant fractions from seeds of reduced vigor.

a. Integrity of Ribosomes

The progressive loss of integrity of ribosomal RNA in dry seed has been demonstrated during senescence in dry rye embryos. Extraction of total RNA or of rRNA from ribosomes from a range of embryos of different percentage viability, followed by fractionation of the high molecular weight RNA components by electrophoresis of polyacrylamide gels, has shown that 18S rRNA molecules suffer increasing degradation with loss of viability, yielding fragments appearing to be of defined molecular weight.[74] A progressively increased number of breaks with decreasing viability were revealed in 25S rRNA by disrupting the hydrogen bonding in helical regions of the RNA, (e.g., by heating a solution of the RNA, followed by rapid cooling) before fractionation by electrophoresis on polyacrylamide gels. Although most of the ribosomes from 98% viable embryos contain undegraded RNA, a proportion of the embryos having lower viabilities contain ribosomes with fragmented 25S and 18S rRNA, the proportion of impaired ribosomes being higher in the lower viability stocks.[45,53] There is, however, no evidence of a loss of components present in the protein-RNA complex of the ribosome, for all ribosomes sediment at the same rate in sucrose density gradients.[74] Since the rRNA obtained by direct extraction from the dry embryo and the rRNA obtained by extraction from ribosomes first isolated from the dry embryos both exhibit a similar pattern of fragmentation following electrophoresis on polyacrylamide gels, it seems likely that cleavage of the rRNA molecules at specific sites is attributable to the continuous and progressive action of nucleases present in the dry seed. Extensive purification by aqueous procedures can lead to further degradation of ribosomes of rye,[74] and if non-viable peas are imbibed for 8 or 16 hr before extraction of the RNA, a greater fragmentation of their 25 and 18S rRNA is very evident.[75] Both results suggest the presence and continuing action of ribonucleases both in vitro and in vivo. Since no obvious differences have been discovered between the activities of ribonucleases in the preparations from the embryos of different viabilities (at least until they become non-viable),the extent of degradation of ribosomal RNA in the dry state must depend upon the rate at which the nucleases can degrade RNA. This could be determined by the conditions of humidity and temperature during storage.

Ribonucleases, unlike dehydrogenases,[62] are highly stable during the extended storage of dry seed. In authentic 103-year-old wheat (which must have been non-viable for at least 60 years) active ribonucleases could still be extracted.[45] In the archaeological wheat from Egypt, however, no nuclease activity could be detected.[45] During and after loss of viability in a seed it seems that a continuous (but low) rate of nuclease-mediated degradation of ribosomal RNA can continue in the dead cells. Eventually this leads to complete degradation of all RNA to small molecular fragments or nucleotides. Assays carried out in the author's laboratory on a number of these ancient seeds have failed to detect the presence of any high molecular weight RNA or DNA.[13]

b. Transfer RNA, Aminoacyl-tRNA Synthetases and Transfer Enzymes

The separation and assessment of the activity of post-ribosomal supernatant factors essential for active peptide bond formation has been carried out in cell-free systems with preparations from seeds of different percentage viability. Some components retain full activity at loss of viability, whilst others are impaired. For example, in rye, phenylalanyl-tRNA and phenylalanyl-tRNA synthetase appear to be equally active in sustaining polyphenylalanine synthesis in preparations from embryos of any of the viabilities tested in homologous or heterologous systems coded for by Poly U. Failure of polypeptide synthesis in cell-free systems prepared from rye of low viability must therefore occur at some stage coincident with, or subsequent to, the binding of aminoacyl-tRNA to the ribosome.[52] Species may differ in the relative degree of integrity of these components of the protein synthesizing system, for some decrease in the activity of phenylalanine tRNA synthetase has been reported for aged pea seeds.[66]

Studies of the transfer enzymes in rye, pea, and wheat all indicate that the translocase or elongation factors (Transferase enzyme II) concerned with transfer of the amino acid from the aminoacyl to the peptidyl site on the ribosome show only slight impairment with loss of viability. However, the GTP-dependent binding of aminoacylated phenylalanyl-tRNA to ribosomes, determined by Transferase I, decreased in proportion to the loss of activity in the homologous system for polyphenylalanine synthesis.[52] The fall in Transferase I activity also parallels the loss of protein synthesizing activity in embryos[52] or seeds[66,76] in vivo and closely reflects the extent of cellular senescence in the seed.

7. Messenger RNA, Old or New

a. Long-lived Messenger RNA

The concept that messenger RNA synthesized during the development of the seed should remain stored within the cells during the period of dehydration and quiescence, and that the same messenger RNA should at once become available to direct protein synthesis when water is again provided at germination, has considerable attraction and theoretically permits the commencement of protein synthesis in the absence of new mRNA synthesis.[69,70]

Using incorporation of labeled amino acid precursors into acid-insoluble products as a measure of protein synthesis in cell-free systems, it has been shown that fractions containing mRNA template activity can be isolated from a variety of different dry seeds, including wheat,[77] rice,[78] and pea.[79] However, amino acid incorporation into unidentified acid-insoluble products in a cell-free system, or into products separable into discrete fractions by electrophoresis in polyacrylamide gels, does not provide unequivocal evidence for a similar messenger effectiveness for translation into defined proteins in vivo. Nor does it indicate whether the proteins, if formed, are important to the reactivation of synthetic activity at embryo germination.

The presence of polyadenylated sequences of up to 200 to 250 nucleotides long at the 3′OH terminus of many mRNA molecules from animals and plants has permitted their extraction, separation, and subsequent examination for template activity. Since polyA-rich RNA binds readily at high ionic strength to oligo-dT cellulose, it is readily separated from other non-polyadenylated RNA molecules. From the many preparations of polyA-rich RNA that have been obtained from a variety of dry seeds[80,81] (where it seems to be predominantly located in the nuclei),[82,83] there is no doubt that it has high template activity in cell-free systems and codes for numerous and different discrete proteins.[84,85] A comparison of the polyA-rich RNA fraction from 90% viable embryos with that from embryos that have just become non-viable (0%), showed that both sustained protein synthesis in a wheatgerm cell-free system. The template activity per microgram of polyA-rich RNA from the 0% preparation was approximately 69% of that from the viable (90%) preparation.[67] There are, however, some differences between the two fractions. Analysis of their base composition indicates a higher proportion of adenosine in the 0% material, suggesting that the non-adenylated 5′ terminus may be preferentially cleaved.[67] Fractionation of the two samples of polyA-rich RNA on sucrose density gradients, followed by hybridization of the separated fractions to [^{3}H]Poly U, has demonstrated that the 0% material is of lower molecular weight.[68] Both results are suggestive evidence that although the polyA-rich RNA molecules from 0% embryos can function as templates in cell free systems, they may well be foreshortened or impaired, and therefore ineffective as templates for translation in vivo.

However, though experiments with rapeseed oil suggest that some poly A-rich RNA will code for seed protein in cell-free systems, there is no evidence that seed protein is synthesized at germination, even in a fully viable seed.[84] This does not exclude the possibility that a non-polyadenylated mRNA fraction could code for certain proteins synthesized on imbibition in vivo. However, 0% embryos do not synthesize proteins in vivo at all, so it must be concluded that factors other than a lack of functional long lived mRNA are likely to be limiting to germination in a senescent seed. For example, few polysomes are present in dry seeds,[86] and in low vigor embryos polysome assembly on the old impaired ribosomes could be the main reason why the mRNA fails to support translation. This might be particularly important if the long lived mRNA molecules also had suffered some degree of degradation or conformational change.

b. Newly Synthesized Messenger RNA

With loss of viability and the general decline in capacity for protein synthesis in the early hours of germination, there is a parallel, but lesser depression in the rates of synthesis of all classes of RNA.[47] In rye this is accompanied by a reduced rate of processing of precursor ribosomal RNA (31S) to 25S and 18S components,[45,47] and at 0% viability by a synthesis of RNA molecules no greater in molecular weight than 4 to 5S.[67,68] The synthesis and degradation of this RNA and its failure to be utilized in protein synthesis by 0% viable material[47] could be due to a variety of causes. One possibility is that this is a "nonsense" RNA transcribed from an impaired template. Given that a transcriptional mechanism *is* operative with functional polymerases, it is of particular interest to consider the integrity and biochemical status of DNA during seed senescence.

The total level of DNA present in the nucleus of cells of rye embryos remains constant during senescence, and for at least a year after loss of viability.[53] The molecular integrity of the DNA does not, however, remain unchanged during this period, for the amount of high molecular weight material that can be recovered by spooling techniques decreases with loss of viability (10.2 mg of DNA per gram in 90% viable dry embryos compared with 3.1 mg of DNA per gram of non-viable embryos).[53] Breaks in DNA could contribute to the much reduced template activity, and possibly also to faulty transcription. Breaks might also be the cause of the absence of high molecular weight RNA products at transcription: the experiments with 0% viable rye embryos, which do not synthesize protein, but do synthesize low molecular weight (4 — 5S) polyribonucleotides located in the nucleus, do imply that the DNA template may be highly fragmented, though it cannot be excluded that the polynucleotides represent fragments of rapidly degraded high molecular weight transcripts[47,67] (see the following section).

If newly synthesized RNA is required for early protein synthesis, then any failure in the DNA template or loss in fidelity of transcription must lead to eventual cell death. Certainly, the products transcribed by 0% viable rye embryos[47,67] or pea embryonic axis[83] do not code for protein synthesis in vivo and are associated with cell death.

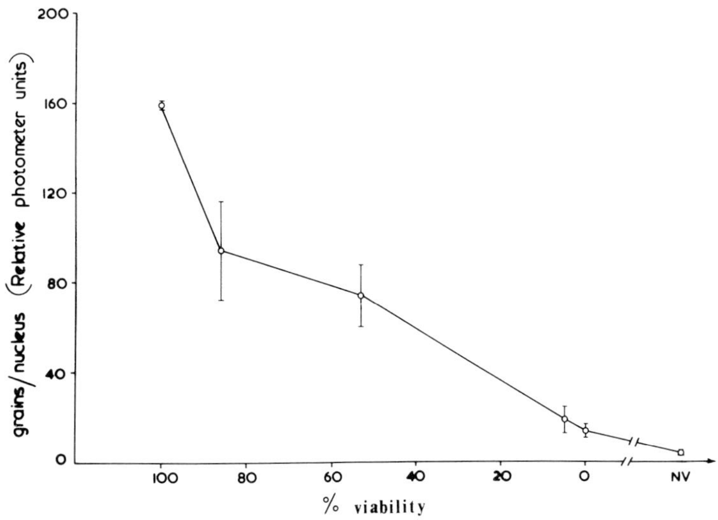

FIGURE 2. Decline in total RNA synthesis with senescence and loss of viability in rye embryos in the first 3 hr of germination (autoradiography of [^{3}H] uridine incorporation into nuclei). (From Sen, S. and Osborne, D. J., *Biochem J.*, 166, 33—38, 1977. With permission.)

8. Integrity of DNA

De Vries[87] in 1901 first questioned that changes in genetic material might be a cause of poor germination, and account for the high proportion of abnormal plants produced from old seed. Not until the 1930s were experiments planned to determine whether the extent of mutation was related to seed age. The correlation then became very clear. There was a greater frequency of cytologically aberrant mitosis in root tips on germination with increasing seed age,[4,88,89] and a higher number of chromosomally aberrant pollen grains were produced by plants from aged or "accelerated" aged seeds.[90] Chromosomal aberrations probably give a low estimate of the extent of genetic damage, as they are visible only when the cells divide at mitosis or during meiosis in pollen formation. It is interesting, therefore, that seeds stored under nitrogen show fewer chromosomal disturbances,[91] while increased oxygen concentration increased the numbers of aberrant chromosomes.[92,93] It was not clear, however, whether damage to DNA occurred *in situ* during senescence of the dry seed, or was an early event that followed imbibition.

Experiments on the isolation of DNA from dry viable or non-viable rye embryos have shown that the nucleosomal structure of DNA from embryos that have been dead for 7 years is in no way seriously impaired; the fragments produced by micrococcal nuclease digestion, including the nucleosome monomer core, being indistinguishable from those of 100% viable material, (i.e., 200 base pairs and 140 base pairs, respectively)[94] (Figure 4). Separation of DNA from nuclei isolated from the two samples of embryos and then lysed on alkaline sucrose density gradients, under conditions that precluded degradation by nucleases, showed that the DNA from non-viable embryos is of an overall lower mean molecular weight[95] (Figure 3).

Electrophoretic separation of the DNA from nuclei of non-viable and 95% viable embryos, which is lysed directly onto polyacrylamide gels, showed that the DNA of 95% nuclei migrates only a short distance into the gels, but DNA from the nuclei of non-viable material possesses an additional disperse DNA component that spreads throughout the gel, and is of a higher molecular weight than the nucleosome monomer

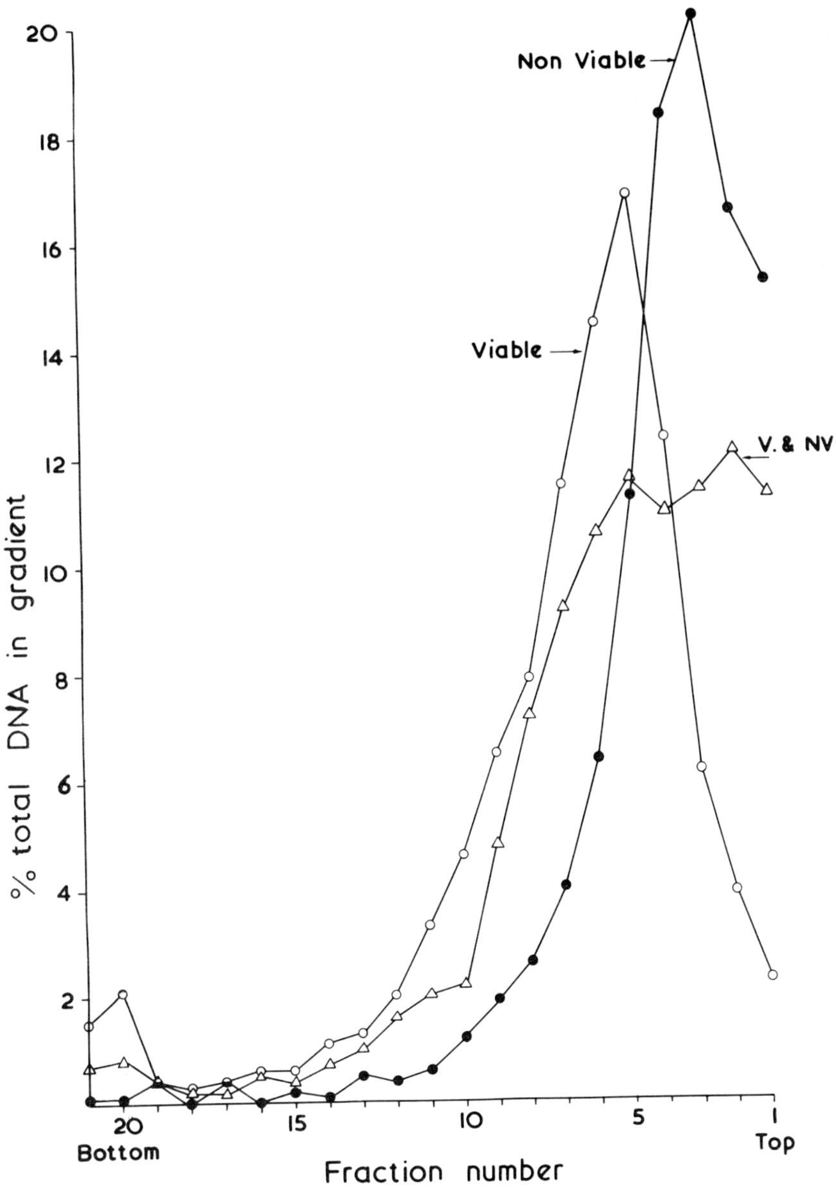

FIGURE 3. Alkaline sucrose density gradient analysis of DNA from viable (V) and non-viable (NV) dry rye embryos. (From Cheah, K. S. E. and Osborne, D. J., *Nature (London)*, 272, 593—599, 1978. With permission.)

(Figure 5). It was calculated that a proportion of the DNA is degraded from a length of at least 3000 base pairs to molecules as small as 200 by comparison with the nucleosome monomer. Although these fragments could represent either single or double stranded breaks, they were clearly not of any specific or repeating size class. This suggests that the degradation of the nuclear DNA represents a random cleavage of the molecule and not a precisely located series of breaks.

Determinations of the incidences of single stranded breaks in nuclear DNA *in situ* in fixed and sectioned preparations of 95% and 0% viable embryos can be made by

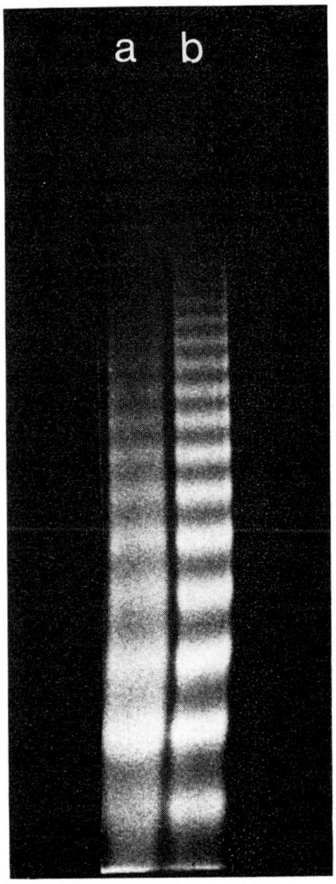

FIGURE 4. Fractionation by polyacrylamide gel electrophoresis of DNA fragments produced by micrococcal nuclease digestion of purified nuclei from a) non-viable and b) viable embryos of rye. (From Cheah, K. S. E. and Osborne, D. J., *Biochem J.*, 163, 141—144, 1977. With permission.)

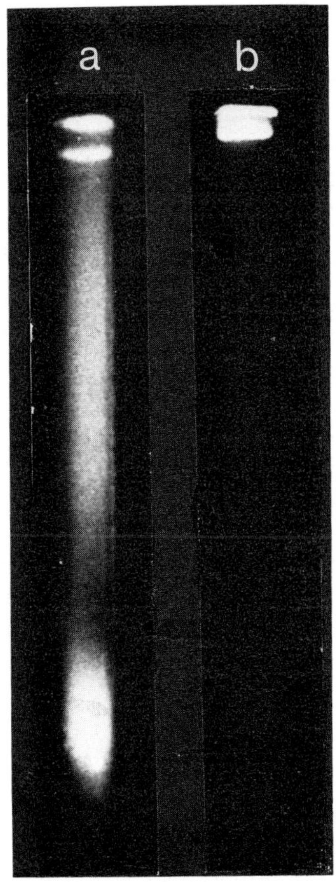

FIGURE 5. Separation on polyacrylamide gels by electrophoresis of high molecular weight DNA from purified nuclei of a) non-viable and b) viable embryos of rye. (From Cheah, K. S. E. and Osborne, D. J., *Nature (London)*, 272, 593—599, 1978. With permission.)

following the incorporation of [^{3}H]-dCTP to 3′OH termini of single-stranded end-breaks in the presence of added calf thymus deoxyribonucleotidyl transferase.[96]

Grain counts of sectioned material incubated and subsequently processed for autoradiography showed that a significantly higher level of grains were present over nuclei in the 0% viable material.[95] This may be taken as evidence of a significantly higher level of single stranded 3′OH termini in the 0% material, and therefore of larger numbers of breaks in the DNA.

Since total levels of DNA do not decline in the nuclei of embryos during senescence,[53] the most likely cause of loss of DNA integrity is the *in situ* activity of endodeoxyribonucleases, possibly with some additional degradation due to exonuclease activity. If such breaks resulting from endonuclease action occur before the embryos become 0% viable, then they could contribute to a progressive failure of the DNA template to code for complete RNA transcripts at germination.

Tests for DNAases in rye embryos showed their activity to be higher in the nonviable than the 95% viable dry material.[53] Since it is most unlikely that enzyme synthesis occurs in the dry embryo, this difference must result either from some form of activation of pre-existing latent enzyme, or from a preferential stability or extractability of nuclease from non-viable material. Experiments suggest that preparations of 95% viable embryos contain a heat-labile inhibitor of DNAase that may become inactivated with loss of viability.[95] If the endodeoxyribonucleases are associated with the nuclear chromatin, even in seeds with moisture contents of 10% or less, slow enzymic DNA fragmentation could occur.

There is limited evidence for DNA repair systems in plants. In root tips (of germinating onions), the number of chromosomal aberrations was found to decline following repeated cell divisions,[97] suggesting either a failure of divisions in cells with impaired nuclei, or a possible DNA repair process in the newly imbibed cells.[57,58,59] In viable rye embroys,[98] increasing levels of non-replicative incorporation ^{3}H-Me-thymidine into DNA is an early event (within 60 min) in germination in response to increasing doses of γ -irradiation to them in the dry state. This indicates a potential for repair of single-stranded DNA breaks. Repair of chemically-induced single-stranded breaks has been shown also in barley embryos.[98a,98b] The repair of irradiation-induced chromosomal disturbances may have been demonstrated in imbibed but in dormant lettuce seed.[98] Repair of lesions seems unlikely, however, during dry storage of seeds, and an accumulation of breaks in DNA could eventually reach a level too high to be repaired at germination. Furthermore, if the DNA repair system were composed of some of the more labile proteins of the cell, then in embryos of low viability, repair could be defective or completely lacking—with little chance of the appropriate enzymes being synthesized anew. It may be pertinent, therefore, that in most germinating seeds, the onset of DNA replication before cell division becomes progressively delayed, the lower the viability of the material.[46] This could be a reflection of deficiencies in repairing the DNA template prior to DNA replication and of the loss of DNA polymerase activity as reported for ageing embryos of wheat[98d] and in nuclei of 5-year-old barley seeds.[98e]

In pea, a rise in deoxyribonuclease activity is associated with the initiation of DNA replication[99] and may be concerned with the formation (by endonuclease cleavage) of initiation sites for synthesis of limited length transcripts of DNA. An inhibitor of this DNAase activity must be present in viable seeds and has been demonstrated in imbibed material up to the time of the replication event.[99] Further evidence that a DNAase is involved in early DNA replication comes from work with wheat;[100] here, all the early DNA synthesized prior to replication was of low molecular weight, some presumably being transcribed from low molecular weight DNA templates. It could be envisaged that the presence of a stable, endodeoxyribonuclease in the dry seed, together with a labile inhibitor of the enzyme, the presence of a stable endodeoxyribonuclease in the dry seed, together with a labile inhibitor of the enzyme, could result in viable seed with suppressed endonuclease until the time of DNA replication at germination. During seed senescence, the nuclease inhibitor could undergo slow inactivation so that eventually the endonuclease would be activated before germination. DNA could then suffer fragmentation and degradation in the dry seed. If the DNA repair enzymes are as labile as the inhibitor of DNAase then, the eventual irreparable loss of DNA integrity is a program for cell death.

B. Changes in Imbibed Seeds

The only seeds that do not grow when placed in water are those that are senescent or dormant. When a dead seed or a 0% viable seed (i.e., one incapable of protein synthesis) is placed in water, the relatively stable hydrolytic enzymes present in the cells continue to function and fragmentation of macromolecules and cytoplasmic disorganization proceeds progressively. The seed becomes essentially "deader than dead". However, if a dormant seed is placed in water all the normal metabolic processes are reactivated with the exception of those directed to cell growth and DNA replication. A dormant embryo can sustain this imbibed state of arrested development for months or years with the normal biochemical events, e.g., respiratory activity,[101] protein synthesis,[37,102,103] RNA synthesis,[104] and lipid metabolism[105] in continuous operation. An active polysome formation,[86] the synthesis of polyA-rich RNA,[81] and a constant turnover of membrane protein[37] and phospholipid[106] are characteristic of the dormant imbibed condition.

The question that can be asked is "Does senescence occur during the dormant imbibed state?" If the metabolic turnover is such that a true homeostatic condition obtains, then there can be no accumulation of lesions or loss of integrity of cytoplasmic organization, for constant turnover should maintain an error-free system. No specific repair processes need be invoked to account for the maintenance of full cellular function. With all the evidence for an active metabolic turnover of the normal maintenance processes that ensure cellular integrity, a progression of senescence events is difficult to envisage. Why then, do imbibed seeds fail to survive indefinitely?

One possibility is that storage reserves become depleted, although there is no evidence that this is so. Another possibility is that in the absence of other degenerative changes in the cells, the lethal accumulation of DNA lesions might take much longer to occur than in those of a dry seed. Indeed, experiments with dormant lettuce[98] have shown that seeds stored dry for various lengths of time show increasing numbers of chromosomal aberrations at germination, whereas those stored full imbibed show essentially none.[107] Furthermore, when dry seeds showing different levels of chromosomal aberrations were then imbibed (but maintained under non-germinating conditions), they showed no further increase in aberrations with time. However, although the loss of DNA integrity was halted, there was no significant evidence (as judged by the number of aberrant cells at subsequent germination) of a repair of the DNA lesions.[98]

The suppression of DNA synthesis may not be complete, however, for although by incorporation studies a lack of DNA replication can be verified, it is difficult to exclude the possibility of some DNA repair synthesis.[108] Other experiments in which dormant lettuce seeds have been exposed to gamma-irradiation in the dry or imbibed state, and then subsequently maintained imbibed for different periods, have shown that post-irradiation imbibition leads to fewer aberrations at germination.[107] However, it is still an open question whether this signifies a DNA repair system operative during the period of imbibed dormancy or rather, a DNA repair that occurred after dormancy was broken, i.e., during the early hours of germination. If the latter is the case, then both in dry and in imbibed seeds the rate of loss of DNA integrity could be a major controlling factor in the rate of cell senescence in a quiescent seed. In the imbibed seed, however, the general maintenance of all other cellular functions by continuous metabolic turnover would tend to sustain a fully effective cytoplasmic machinery and membrane compartmentalization not privileged to the seed stored dry.[102]

IV. CONCLUSION

Seeds, like all other living tissues, become senescent and die, but the manner of seed senescence is quite unusual, for it can happen at moisture contents of less than 5% in essentially dry cells, and it can happen in fully hydrated cells that are incapable of growth. The transition from the viable to the non-viable state in dry seeds is summarized in Fig. 6. Central features are the damage to membrane systems in all parts of the cell and the differential stability of the different enzyme proteins. The high stability

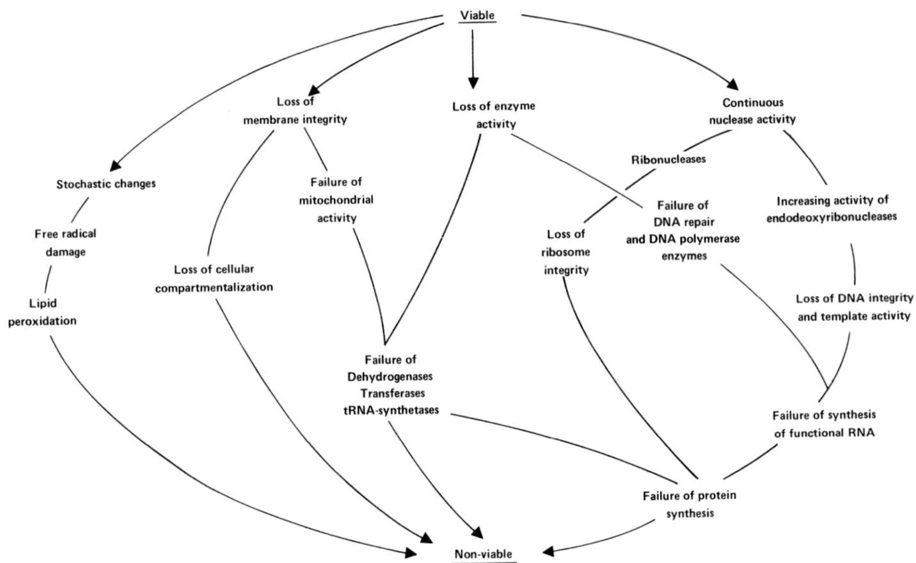

FIGURE 6. Scheme for progressive loss of cellular integrity during senescence of seeds stored dry.

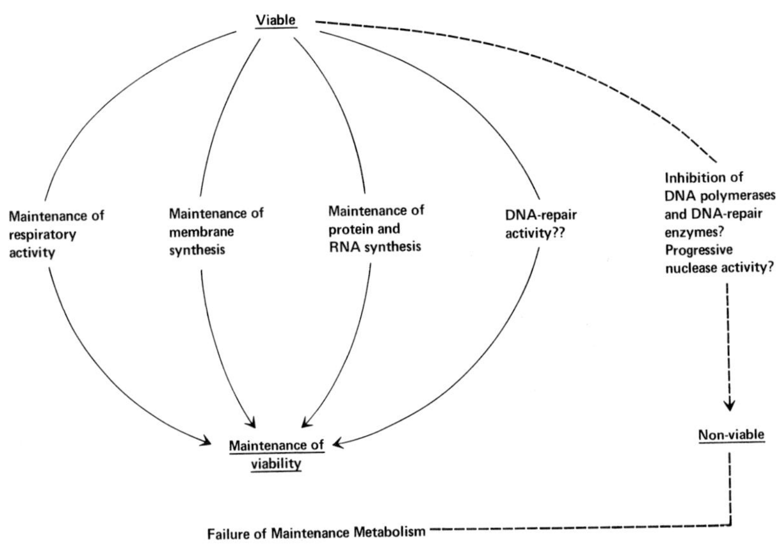

FIGURE 7. Scheme for maintenance of viability and progressive loss of cellular integrity during senescence of dormant seeds stored imbibed.

and likely activation of endo-deoxyribonucleases with ensuing loss of integrity of the DNA template precludes the synthesis of labile enzymes and repair systems at germination. The lesions are cumulative and interrelated, leading to complete loss of cellular organization.

In imbibed dormant seeds all metabolic processes are assumed to be operative (as outlined in Fig. 7) with the exception of those involved in cell enlargement and DNA replication. Since RNA (including mRNA) is transcribed, proteins are translated, and membrane turnover is continuous, the loss of DNA integrity may again be the determining factor in senescence. If DNA replicases and repair enzymes do not function or are not synthesized until dormancy is broken, and particularly if DNA lesions include the replicase or repair enzyme cistrons, a slow programmed senescence can be envisaged.

The understanding of senescence is a challenge to all biologists, but for the seed biologist the opportunities offered by these small structures, containing within them cells of a complete new individual, are the most provocative of all.

REFERENCES

1. **Levitt, J.,** *Responses of Plants to Environmental Stress,* Academic Press, New York, 1972.
2. **Hsiao, T. C.,** Rapid changes in levels of polyribosomes in *Zea mays* in response to water stress, *Plant Physiol.,* 46, 281—285, 1970.
3. **Kersting, J. F., Stickler, F. C., and Pauli, A. W.,** Grain sorghum caryopsis development I. Changes in dry weight, moisture content, and viability, *Agron. J.,* 53, 36—38, 1961.
4. **Roberts, E. H.,** *Viability of Seeds,* Chapman & Hall, London, 1972.
5. **Ewart, A. J.,** On the longevity of seeds, *Proc. R. Soc. Victoria,* 21, 1—210, 1908.
6. **Gwozdz, E. A. and Bewley, J. D.,** Plant desiccation and protein synthesis. An *in vitro* system from dry and hydrated mosses using endogenous and synthetic messenger ribonucleic acid, *Plant Physiol.,* 55, 340—345, 1975.
7. **Gaff, D. F. and Hallam, N. D.,** Resurrecting desiccated plants, in *Mechanisms of Regulation of Plant Growth,* Bieleski, R. L., Ferguson, A. R., and Cresswell, M. M., Eds., The Royal Society of New Zealand, Wellington, Bull. 12, 1974, 389—393.
8. **Brunori, A.,** Relationship between DNA synthesis and water content during ripening of *Vicia faba* seed, *Caryologia,* 20, 333—338, 1967.
9. **Scharpe, A. and van Parijs, R.,** The formation of polyploid cells in ripening cotyledons of *Pisum sativum* L. in relation to ribosome and protein synthesis, *J. Exp. Bot.,* 24, 216—222, 1973.
10. **Hallam, N. D.,** Embryogenesis and germination in rye *(Secale cereale* L.) I. Fine structure of the developing embryo, *Planta,* 104, 157—166, 1972.
11. **Osborne, D. J.,** Seeds of the past, *New Sci.,* 63, 252—254, 1974.
12. **Hallam, N. D. and Osborne, D. J.,** Fine Structure and Histochemistry of Ancient Viable and Nonviable Seeds, 8th Int. Congr. Electron Microscopy, Canberra, 1974, 2, 598—599.
13. **Osborne, D. J. and Payne, P. I.,** unpublished data.
14. **Porsild, A. E., Harington, C. R., and Mulligan, G. A.,** *Lupinus arcticus* Wats. grown from seeds of Pleistocene age, *Science,* 158, 113—114, 1967.
15. **Odum, S.,** Germination of ancient seeds. Floristical observations and experiments with archaeologically dated soil samples, *Dan. Bot. Ark.,* 24, 1—70, 1965.
16. **Ohga, I.,** On the age of the ancient fruit of the Indian Lotus which is kept in the peat bed in South Manchuria, *Bot. Mag. (Tokyo),* 41, 1—6, 1927.
17. **Libby, W. F.,** Radiocarbon dates II, *Science,* 114, 291—296, 1951.
18. **Goodwin, H. and Willis, E. H.,** The viability of Lotus seeds *(Nelumbium nucifera* Gaertn.), *New Phytol.,* 63, 410—412, 1964.
19. **Lerman, J. C. and Cigliano, E. M.,** New carbon-14 evidence for six hundred years old *Canna compacta* seed, *Nature (London),* 252, 568—570, 1971.
20. **Sivori, E., Nakayama, F., and Cigliano, E.,** Germination of Achira seed *(Canna* sp.) approximately 550 years old, *Nature (London),* 219, 1269—1270, 1968.
21. **Ramsbottom, J.,** Recent work on germination, *Nature (London),* 149, 658, 1942.
22. **Theophrastus,** Enquiry into plants, *Horticulture,* 11, 97 and 209, 1916 transl.
23. **Beal, W. J.,** The vitality of seeds, *Bot. Gaz. (Chicago),* 40, 140—143, 1905.
24. **Darlington, H.,** The 70 year period for Dr. Beal's seed viability experiment, *Am. J. Bot.,* 38, 379—381, 1951.
25. **Kivilaan, A. and Bandurski, R. S.,** The 90 year period for Dr. Beal's seed viability experiment, *Am. J. Bot.,* 60, 140—145, 1973.
26. **Duvel, J. W. T.,** The vitality of buried seeds, *U.S. Dep. Agric. Bur. Plant Ind. Bull.,* 83, 1905.

27. **Mackay, D. B. and Flood, R. J.,** Investigations in crop seed longevity II. The viability of cereal seed stored in permeable and impermeable containers, *J. Natl. Inst. Agric. Bot. (G.B.)*, 2, 378—403, 1968.
28. **Aksenov, S. I., Askochenskaya, N. A., and Petinov, N. S.,** The fractions of water in wheat seeds, *Fiziol. Rast.*, 16, 58—64, 1969.
29. **Askochenskaya, N. A. and Aksenov, S. I.,** Characteristics of the hydration of seeds of different composition, *Fiziol. Rast.*, 17, 95—100, 1970.
30. **Verzar, F.,** Intrinsic and extrinsic factors of molecular ageing, *Exp. Gerontol.*, 3, 69—75, 1968.
31. **Lipman, C. B.,** Normal viability of seeds and bacterial spores after exposure to temperatures near the absolute zero, *Plant Physiol.*, 11, 201—205, 1936.
32. **Strehler, B. L.,** *Time, Cells, and Aging,* Academic Press, New York, 1962.
33. **Toole, V. K. and Toole, E. H.,** Seed dormancy in relation to seed longevity, *Proc. Int. Seed Test. Assoc.*, 18, 325—328, 1953.
34. **Fykse, H.,** Investigations into the germination, dormancy, and longevity of wild oat seeds, *Meld. Nor. Landbrukshoegsk.*, 49, 1—120, 1970.
35. **Thurston, J. M.,** Weed studies: wild oats, *Rep. Rothamsted Exp. Stn.*, 1960, 102—103.
36. **Simon, U.,** The germinability of old seeds, *Z. Acker Pflanzenbau,* 106, 108—118, 1958.
37. **Cuming, A. C. and Osborne, D. J.,** Membrane turnover in imbibed and dormant embryo of the wild oat *Avena fatua* L. I. Protein turnover and membrane replacement, *Planta,* 139, 209—217, 1978.
38. **Rademacher, B. and Kiewnick, L.,** The effect of mineral and organic fertilizer on the viability and periodicity of germination of wild oat, *Avena fatua* L., seed, *Z. Acker Pflanzenbau,* 119, 369—385, 1964.
39. **van Staden, J., Gilliand, M. G., and Brown, N. A.,** Ultrastructure of dry, viable and non-viable *Protea compacta* embryos, *Z. Pflanzenphysiol.*, 76, 28—35, 1975.
40. **Hallam, N. D., Roberts, B. E., and Osborne, D. J.,** Embryogenesis and germination in rye *(Secale cereale* L.) III. Fine structure and biochemistry of the non-viable embryo, *Planta,* 110, 279—290, 1973.
41. **Abdul-Baki, A. A.,** Relationship of glucose metabolism to germinability and vigour in germinating barley and wheat seeds, *Crop Sci.*, 9, 732—737, 1969.
42. **Obendorf, R. L. and Marcus, A.,** Rapid increase in adenosine-5′-triphosphate during early wheat embryo germination, *Plant Physiol.*, 53, 770—781, 1974.
43. **Marcus, A., Feeley, J., and Volcani, T.,** Protein synthesis in imbibed seeds III. Kinetics of amino acid incorporation, ribosome activation and polysome formation, *Plant Physiol.*, 41, 1167—1172, 1966.
44. **Sen, S., Payne, P. I., and Osborne, D. J.,** Early ribonucleic acid synthesis during the germination of rye *(Secale cereale)* embryos and the relationship to early protein synthesis, *Biochem. J.*, 148, 381—387, 1975.
45. **Osborne, D. J., Roberts, B. E., Payne, P. I., and Sen, S.,** Protein synthesis and viability in rye embryos, in *Mechanisms of Regulation of Plant Growth,* Bieleski, R. L., Ferguson, A. R., and Cresswell, M. M., Eds., The Royal Society of New Zealand, Wellington, 1974, Bull. 12, 805—812.
46. **Sen, S. and Osborne, D. J.,** Germination of rye embryos following hydration-dehydration treatments: enhancement of protein and RNA synthesis and earlier induction of DNA replication, *J. Exp. Bot.*, 25, 1010—1019, 1974.
47. **Sen, S. and Osborne, D. J.,** Decline in ribonucleic acid and protein synthesis with loss of viability during the early hours of imbibition of rye *(Secale cereale* L) embryos, *Biochem J.*, 166, 33—38, 1977.
48. **Sen, S.,** Germination and Viability of Rye Embryos. Nucleic Acid and Protein Synthesis during Early Hours of Germination, Ph.D. thesis, Cambridge, U. K., 1975.
49. **Aspinall, D. and Paleg, L. G.,** The deterioration of wheat embryo and endosperm function with age, *J. Exp. Bot.*, 23, 925—935, 1971.
50. **Floris, C.,** Ageing in *Triticum durum* seeds: behaviour of embryos and endosperms from aged seeds as revealed by the embryo transplantation technique, *J. Exp. Bot.*, 21, 462—468, 1970.
51. **Abdul-Baki, A. A.,** Relationship of glucose metabolism to germination and vigour in barley and wheat seed, *Crop Sci.*, 9, 732—737, 1969.
52. **Roberts, B. E. and Osborne, D. J.,** Protein synthesis and loss of viability in rye embryos. The lability of transferase enzymes during senescence, *Biochem. J.*, 135, 405—410, 1973.
53. **Roberts, B. E. and Osborne, D. J.,** Protein synthesis and viability in rye grains, in *Seed Ecology,* Heydecker, W., Ed., University of Nottingham 19th Summer School in Agricultural Science 1972, Butterworth, London, 1973, 99—114.
54. **Szezotka, Z.,** Changes in the intensity of protein synthesis in the embryo axes of northern red oak *(Quercus borealis* Michx) and English oak *(Quercus robur* L.) during storage under controlled conditions, *Arbor. Kornickie,* XX, 291—297, 1975.

55. **Harman, G. E.and Grannett, A. L.**, Deterioration of stored pea seed: changes in germination, membrane permeabilty, and ultrastructure resulting from infection by *Aspergillus ruber* and from ageing, *Physiol. Plant Pathol.*, 2, 271—278, 1972.

56. **Parrish, D. J. and Leopold, A. C.**, On the mechanism of ageing in soybean seeds, *Plant Physiol.*, 61, 365—368, 1978.

57. **Villiers, T. A.**, Lysosomal activities of the vacuole in damaged and recovering plant cells, *Nature (London)* 233, 57—58, 1971.

58. **Berjak, P. and Villiers, T. A.**, Ageing in plant embryos I. The establishment of the sequence of development and senescence in the root cap during germination, *New Phytol.*, 69, 919—938, 1970.

59. **Berjak, P. and Villiers, T. A.**, Ageing in plant embryos. II. Age-induced damage and its repair during early germination, *New Phytol.*, 71, 135—144, 1972.

60. **Harman, G. E. and Mattrick, L. R.**, Association of lipid oxidation with seed ageing and death, *Nature (London)*, 260, 323—324, 1976.

61. **Abu-Shakra, S. S. and Ching, T. M.**, Mitochondrial activity in germinating new and old soybean seeds, *Crop Sci.*, 7, 115—118, 1967.

62. **Throneberry, G. O. and Smith, F. G.**, Relation of respiratory and enzymatic activity to corn seed viability, *Plant Physiol.*, 30, 337—343, 1955.

63. **Lakon, G.**, The topographical tetrazolium method for determining the germinating capacity of seeds, *Plant Physiol.*, 24, 389—394, 1949.

64. **Ching, T. M.**, Adenosine triphosphate content and seed vigor, *Plant Physiol.*, 51, 400—402, 1973.

65. **Barker, G. R. and Bray, C. M.**, Nucleic acid synthesis in germinating seeds, *Symp. Biol. Hung.*, 13, 61—68, 1972.

66. **Bray, C. M. and Chow, T-Y.**, Lesions in post-ribosomal supernatant fractions associated with loss of viability in pea *(Pisum arvense)* seed, *Biochim. Biophys. Acta*, 442, 1—13, 1976.

67. **Osborne, D. J., Dobrzanska, M., and Sen, S.**, Factors determining nucleic acid and protein synthesis in the early hours of germination, *Symp. Soc. Exp. Biol.*, 31, 177—194, 1977.

68. **Dobrzanska, M. and Osborne, D. J.**, unpublished data.

69. **Marcus, A. and Feeley, J.**, Activation of protein synthesis in the imbibition phase of seed germination, *Proc. Natl. Acad. Sci. U.S.A.*, 51, 1075—1079, 1964.

70. **Dure, L. S. and Waters, L. C.**, Long-lived messenger RNA: evidence from cotton seed germination, *Science*, 147, 410—412, 1965.

71. **Weeks, D. P. and Marcus, A.**, Preformed messenger of quiescent wheat embryos, *Biochim. Biophys. Acta*, 232, 671—684, 1971.

72. **Spiegel, S. and Marcus, A.**, Polyribosome formation in early wheat embryo germination independent of either transcription or polyadenylation, *Nature (London)*, 256, 228—230, 1975.

73. **Dobrzanska, M., Tomaszewski, M., Grzelezak, Z., Rejman, E., and Buchowicz, J.**, Cascade activation of genome transcription in wheat, *Nature (London)*, 244, 507—509, 1973.

74. **Roberts, B. E., Payne, P. I., and Osborne, D. J.**, Protein synthesis and the viability of rye grains. Loss of activity of protein-synthesising systems *in vitro* associated with a loss of viability, *Biochem. J.*, 131, 275—286, 1973.

75. **Bray, C. M. and Chow, T-Y.**, Lesions in the ribosomes of non-viable pea *(Pisum arvense)* embryonic axis tissue, *Biochim. Biophys. Acta*, 442, 14—23, 1976.

76. **Dell'Aquila, A., Zocchi, G., Lanzani, G. A., and De Leo, P.**, Different forms of EF1 and viability in wheat embryos, *Phytochemistry*, 15, 1607—1610, 1976.

77. **Chen, D., Sarid, S., and Katchalski, E.**, Studies on the nature of messenger RNA in germinating wheat embryos, *Proc. Natl. Acad. Sci. U.S.A.*, 60, 902—909, 1968.

78. **Bhat, S. P. and Padayatty, J. D.**, Presence of conserved messenger RNA in embryos, *Ind. J. Biochem. Biophys.*, 11, 47—50, 1974.

79. **Jachymczyk, W. J., Sieliwanowicz, B., and Chlebowicz, E.**, Activation of pre-existing messenger RNA in dry pea embryo axes, *Acta Biochim. Pol.*, 21, 137—143, 1974.

80. **Payne, P. I., Gordon, M. E., Dobrzanska, M., Parker, M. L., and Barlow, P. W.**, The long-lived messenger RNA of dry seeds: evidence for its existence, intracellular location, and role in germination, *Colloq. Int. C.N.R.S.*, 261, 487—499, 1976.

81. **Hecker, M.**, Untersuchungen uber stabile und wahrend der fruhen Quellungs Stadien neu synthetisierte poly(A)-RNA von *Agrostemma* samen, *Experientia*, 32, 50—53, 1976.

82. **Hammett, J. R. and Katterman, F. R.**, Storage and metabolism of poly(adenylic)-mRNA in germinating cotton seeds, *Biochemistry*, 14, 4375—4379, 1975.

83. **Payne, P. I., Gordon, M. E., Barlow, P. W., and Parker, M. L.**, The subcellular location of the long-lived messenger RNA of rape seed, in *Translation of Natural and Synthetic Polynucleotides*, Legocki, A. B., Ed., Poznan Agricultural University, Poland, 1977, 224—227.

84. **Gordon, M. E. and Payne, P. I.**, *In vitro* translation of the long-lived messenger ribonucleic acid of dry seeds, *Planta*, 130, 269—273, 1976.

85. **Gordon, M. E. and Payne, P. I.,** *In vitro* translation of poly-A rich RNA from dry and imbibing seeds of rape, in *Translation of Natural and Synthetic Polynucleotides,* Legocki, A. B., Ed., Poznan Agricultural University, Poland, 1977, 228—231.
86. **Fountain, D. W. and Bewley, J. D.,** Polyribosome formation and protein synthesis in imbibed but dormant lettuce seeds, *Plant Physiol.,* 52, 604—607, 1973.
87. **De Vries, H.,** *Die Mutationstheorie,* Vol 1, Veit and Co., Leipzig, 1901.
88. **Nilsson, N. H.,** Sind die induzierten mutanten nur selektive erscheinung? *Hereditas,* 15, 320—328, 1931.
89. **Navashin, M. S.,** Ageing of seeds is a cause of chromosome mutations, *Planta,* 20, 233—243, 1933.
90. **Cartledge, J. L., Barton, L. V., and Blakeslee, A. F.,** Heat and moisture as factors in the increased mutation rate from *Datura* seeds, *Proc. Am. Philos. Soc.,* 76, 663—685, 1936.
91. **Harrison, B. J.,** Seed deterioration in relation to storage conditions and its influence upon germination, chromosomal damage, and plant performance, *J. Natl. Inst. Agric. Bot. (G.B.),* 10, 644—663, 1966.
92. **Roberts, E. H. and Abdalla, F. H.,** The influence of temperature, moisture, and oxygen on period of seed viability in barley, broad bean, and peas, *Ann. Bot. (London),* 32, 97—117, 1968.
93. **Abdalla, F. H. and Roberts, E. H.,** Effects of temperature, moisture and oxygen on the induction of chromosome damage in seeds of barley, broad bean, and peas during storage, *Ann. Bot. (London),* 32, 119—126, 1968.
94. **Cheah, K. S. E. and Osborne, D. J.,** Analysis of nucleosomal deoxyribonucleic acid in a higher plant, *Biochem. J.,* 163, 141—144, 1977.
95. **Cheah, K. S. E. and Osborne, D. J.,** DNA lesions occur with loss of viability in embryos of ageing rye seed, *Nature (London),* 272, 593—599, 1978.
96. **Modak, S. P. and Bollum, F. J.,** Detection and measurement of single-strand breaks in nuclear DNA in fixed lens sections, *Exp. Cell Res.,* 75, 307—313, 1972.
97. **Nichols, C.,** The effects of age and irradiation on chromosomal aberrations in *Allium* seed, *Am. J. Bot.,* 29, 755—759, 1942.
98. **Osborne, d. J. and Sharon, R.,** unpublished data.
98a. **Veleminsky, J., Sadrazil, S., Pkorny, V., Gichner, T., and Svachulova, J.,** Repair of single-strand breads and fate of N-$_7$-methylguanine in DNA during the recovery from genetical damage induced by N-methyl-N-nitrosourea in barley seeds, *Mut. Res.,* 17, 49-58, 1973.
98b. **Soyfer, V. N., Krausse, G. V., Pokrovskaya, A. A., and Yakovleva, N. I.,** Repair of single-strand DNA breaks and recovery of chromosomal and chromatid aberrations after preatment of plant seeds with propyl-methanesulfonate *in vivo, Mut. Res.,* 42, 51-64, 1977.
98c. **Villiers, T. A. and Edgecumbe, D. J.,** On the cause of seed deterioration in dry storage, *Seed Sci. Technol.,* 3, 761-774, 1975.
98d. **Dell'Aquila, A., Savino, G., and De Leo, P.,** Metabolic changes induced by hydration-dehydration treatment in wheat embryo, *Plant Cell Physiol.,* 19, 349-354, 1978.
98e. **Yamaguchi, H., Naito, T., and Tatur, A.,** Decreased activity of DNA polymerase in seeds of barley during storage, *Jpn. J. Genet.,* 53, 133-135, 1978.
99. **Jenns, S. M. and Bryant, J. A.,** Correlation between deoxyribonuclease activity and DNA replication in the embryonic axes of germinating peas *(Pisum sativum* L). *Planta,* 138, 99—103, 1978.
100. **Buchowicz, J., Krazewska, E., and Eberhardt, J.,** Characterization of the early synthesized DNA in germinating *Triticum aestivum* embryos, *Phytochemistry,* 17, 1481—1484, 1978.
101. **Simmonds, J. and Simpson, G. M.,** Regulation of the Krebs cycle and pentose phosphate pathway activities in the control of dormancy of *Avena fatua* L., *Can. J. Bot.,* 50, 1041—1048, 1972.
102. **Chen, S. C. C. and Varner, J. E.,** Respiration and protein synthesis in dormant and after-ripened seeds of *Avena fatua, Plant Physiol.,* 46, 108—112, 1970.
103. **Hecker, M. and Bernhardt, D.,** Protein biosynthesen in dormanten und nachgereiften embryonen und samen von *Agrostemma githago, Phytochemistry,* 15, 1105—1109, 1976.
104. **Jarvis, B. C., Frankland, B., and Cherry, J. H.,** Increased nucleic acid synthesis in relation to the breaking of dormancy of hazel seed by gibberellic acid, *Planta,* 83, 257—266, 1968.
105. **Bradbeer, J. W. and Colman, B.,** Studies in seed dormancy: the metabolism of [2-^{14}C] acetate by chilled seeds of *Corylus avellana* L., *New Phytol.,* 66, 5—15, 1967.
106. **Cuming, A. C. and Osborne, D. J.,** Membrane turnover in imbibed and dormant embryos of the wild oat *Avena fatua* L. II. Phospholipid turnover and membrane replacement, *Planta,* 139, 219—226, 1978.
107. **Villiers, T. A.,** Seed aging: chromosome stability and extended viability of seeds stored fully imbibed, *Plant Physiol.,* 53, 875—878, 1974.
108. **Hecker, M. and Huth, A.,** Untersuchungen uber die DNA-synthese in quellenden dormanten und nachgereiften *Agrostemma githago-* Embryonen (II). *Biochem. Physiol. Pflanz.,* 167, 261—266, 1975.

109. **Turner, J. H.,** The viability of seeds, *Bull. Misc. Inf. R. Bot. Gard. Kew,* 6, 257—269, 1933.
110. **Becquerel, P.,** La longevite des graines macrobiotiques, *C. R. Acad. Sci. (Paris),* 199, 1662, 1934.
111. **Shull, G. H.,** The longevity of submerged seeds, *Plant World,* 17, 329—337, 1914.

Chapter 3

ULTRASTRUCTURAL CHANGES IN SEED DORMANCY AND SENESCENCE

T. A. Villiers

TABLE OF CONTENTS

I. INTRODUCTION

In addition to their primary role as the reproductive and dispersal units of higher plants, seeds are of interest to the plant physiologist because of the ability of the majority of species to withstand extended periods of suspended biochemical and physiological activity in an air-dry state. In the present context they are of special interest, providing means of studying the effect of aging on the same biological system in a natural state at three levels.

When in the air-dry state with a low moisture content, there appears to be little or no possibility of enzyme action or of normal physiological activities. A study of the air-dry seed should therefore provide information on the degree of stability of a highly-organized biological system, which is unable to turn over and replace its component parts. This is a normal stage in the life-cycle of the plant rather than an experimentally-induced condition. However, it should be remembered that although the ability to become dry is a natural property of many seeds, prolonged storage at controlled low moisture contents is an artificial state, usually maintained for human convenience.

Secondly, imbibed (but dormant) seeds can be shown to be capable of a wide range of biochemical activities and physiological processes. As there is no cell division in the majority of dormant embryos, it becomes possible to study the aging of a system in which enzyme activity is able to occur, and in which cellular components are probably able to be replaced, but in which there is no growth and therefore no possibility of replacing aberrant cells.

Thirdly, upon release from dormancy the germinating embryo undergoes rapid extension growth and commences active cell division. This gives an opportunity of observing the effects of cell division, and other processes involved in the commencement of growth, upon embryo tissues in various stages of deterioration through aging within the seed, either in the air-dry or dormant state.

For the present purpose it is necessary to be quite clear about what is meant by the term 'dormancy', and to distinguish between the seed in its air-dry state, and a non-germinating seed in an imbibed, but dormant state. Unfortunately, much confusion exists in the literature on this point, and seeds have frequently been described as 'dormant' when they are merely quiescent, in an air-dry condition, requiring only the addition of water to be able to recommence growth and development. An air-dry seed is therefore not necesarily dormant in the sense of possessing physiological or biochemical mechanisms preventing the resumption of growth.

For the purpose of this review a dormant seed is defined as one that is unable to germinate and produce a seedling when supplied with water and held at temperatures suitable for the normal growth of the species. Seeds in an air-dry state, which may or may not be dormant if supplied with water, will be referred to as air-dry or 'quiescent'.

In this review it is intended to survey changes that occur in seed tissues during prolonged air-dry storage, becoming manifest upon imbibition, and the degenerative structural changes that take place in imbibed seeds during prolonged dormancy states. The observable events occurring when aged, quiescent seeds, and long-dormant, imbibed seeds are allowed to germinate will be described and their possible significance discussed.

II. THE AIR-DRY STATE

In order to be able to store seeds for use at his own convenience, Man takes advantage of the ability of most seeds to lose water and exist in an air-dry state. When seeds are to be used as food sources, the main consideration is to prevent spoilage by microorganisms, and for this purpose it is important to be able to control the seed moisture content at a sufficiently low level to reduce fungal growth to a minimum. When seeds are to be used for crop production, however, the problem becomes one of preserving the seed, and particularly its enclosed embryo, in good structural and biochemical order, free from genetic deterioration and the loss of seedling vigor that can occur during prolonged dry storage, or storage in adverse conditions.

A so-called 'dry' seed in reality contains an appreciable amount of water adsorbed to the seed colloids, the system being in equilibrium with the relative humidity of the storage atmosphere. The amount of water present is of the order of 10% by weight of the seed tissues, but can vary widely according to the conditions of storage and the type of seed; in particular the seed food reserve materials. However, the tissues can become hydrated very rapidly when a seed is placed in water, and structural changes have been reported to be detectable in less than 10 sec after placing tissues in water.[1]

As normal fixation procedures for electron microscopy involve the use of aqueous media, it is obviously very difficult to observe the structure of the seed tissues as they exist in the air-dry state. The process of early imbition is a physical one occurring both in living and in non-living seeds and can normally be completed within the time taken for tissue fixation. Moreover, in aqueous fixation procedures, it is probable that water infuses into the cells before the fixative molecules enter, and so very early processes involving the rearrangement of cellular units could conceivably occur before fixation proper. However, it is usual to examine seed tissues fixed in this way and to assume that this represents the state of the cellular organization in the non-active condition, before any changes have taken place that would lead to germination of the embryo. The work of Buttrose[1] provides evidence of changes in cellular membranes occurring within the first few seconds of hydration, and although not part of the actual germination process, these may nevertheless have a significant bearing upon the problem of the retention of seed viability.

A. The Normal Seed

Attempts have been made, (e.g., Perner[2]) to fix air-dry seed tissues by prolonged exposure to osmium tetroxide vapor for extended periods. In general, it has been found that many membrane systems considered part of normal cellular structure, such as the mitochondria, the nuclear envelope with its system of pores, and the plasma membrane, can be seen to occur in the air-dry seed before the uptake of water, but the resolution has not been found to be sufficiently good to discern the fine detail of the structure of dry membranes.

With the reservation that changes in membrane conformation might take place during aqueous-medium fixation, some studies of the fine structure of 'dry' seed tissues have been made, e.g., by Paulson and Srivastava[3] on lettuce embryos, Swift and O'Brien[4] on wheat scutellum, and Webster and Leopold[5] on soybean cotyledons. This last report shows that in the air-dry seed the membranes may be discontinuous and that the plasma membrane is separated from the cell wall, whereas within 20 min from the beginning of hydration it becomes continuous and appears closely appressed to the cell wall.

It has been suggested that the phospholipid components of such discontinuous membranes might be re-deployed from small lipid bodies (which occur along the sites of the membranes) and used in the re-assembly or re-constitution of membranes during early imbibition. It is well known that when seeds are placed in water a loss of solute occurs by leakage from the seed tissues. This has been attributed to the disorganized state of the plasma membranes of the air-dry cells.[6] After a few minutes the rate of solute loss decreases sharply, when it is assumed that the plasma membranes of the embryo cells have regained their normal 'unit membrane' bi-lamellar structure and behave as osmotic membranes once more, retaining solutes within the cells.[7]

Evidence has also been presented that when the water content of a phospholipid-water system is reduced to a low level, the physical structure of the membrane changes. Phospholipids are assumed to exist in the normal hydrated state as a bilayer with the hydrocarbon 'tails' of each molecule directed inwards towards the central zone of the membrane between the two layers. The phospholipids become spatially rearranged when the water content is reduced below about 20% so that they now assume groups with the molecules in radial arrangements having the polar ends directed inwards and the hydro-carbon tails facing outwards.[8,9] This is likely to occur in seed membranes upon drying of the tissues and, should the phospholipid molecules have been subject to damage, e.g., by peroxidation, during the period of dry storage of the seeds, re-organization to form the normal hydrated bilayer might not be possible.

Organelles recognizable in the dry seed are the nucleus, mitochondria, plastids, protein bodies, and lipid droplets. The endoplasmic reticulum (ER) is ill-defined or absent, and Golgi bodies are rarely seen in dry or newly-imbibed seed tissues. Within 20 min the plasma membrane has become organized into a continuous, lamellar, unit membrane structure, the mitochondria have formed internal cristae, and recognizable ER systems have formed where none were previously detectable, both dispersed within the cytoplasm and closely surrounding the protein bodies. Ribosomes are attached to the newly-constituted ER in polysomic arrays, and, whereas they occurred as single units (monosomes) in the dry seeds, they aggregate to form free polysomes in the ground cytoplasm upon imbibition. The protein bodies lose internal crystalline structures very early in imbibition and become more electron transparent, possibly indicating the beginning of hydrolysis.

In summary, the changes upon imbibition of the seed appear to restore the integrity of the cellular structure such that the cells become physically capable of, for example, acting as osmotic units. Internal cellular compartmentation becomes re-established and the cells are prepared for the biochemical synthetic activities that lead to germination.

B. The Aged Seed

Unless special precautions are taken, when seeds are stored for long periods some degree of deterioration usually occurs. In the majority of species this is manifested firstly as an increase in the time taken for germination to begin following imbibition, together with an increase in the variability in time of emergence within a given seed lot. In addition, aging results in a general loss of growth vigor in the seedling produced. Ultimately there is a decrease in the total number of seeds germinating, which continues more or less rapidly according to the conditions of storage. One of the results of the aging process that has frequently been studied is an increase in the number of chromosome aberrations detectable in the first cell divisions of the germinating radicle.[10] An increase in the number of malformed plants and mutations in the populations produced from aged seeds has also often been recorded.[11]

In general, low seed moisture contents combined with low temperatures are essential for successful long-term storage.[11] From a purely practical point of view, it becomes a matter of maintaining the best conditions of storage for the time required that the economics of the situation will allow. However, it is possible to improve greatly on acceptable commercial methods either experimentally, or for special purposes such as seed stores in so-called 'gene banks'.

It remains to be demonstrated whether there are any differential effects of moisture content, or of temperature, upon separate systems in the seed, for example on the genetic rather than the structural component, and therefore one can only assume a general increase in deleterious effects as one raises the temperature and the moisture content of a seed sample.

Reports have appeared of work on seed aging carried out on seeds stored under very adverse conditions, i.e., at warm temperatures (ca. 35 to 40°C) and high seed moisture contents. Such treatment is usually referred to as 'accelerated aging'. It is by no means certain that this treatment will result in a mere extrapolation of the adverse effects accumulating within the more normal ranges of storage parameters, but from a practical point of view the differences between ordinary chronological aging and accelerated aging are accepted as being of degree rather than of kind. However, it is important to be aware of the actual conditions of storage of samples of seeds that are being compared.

Evidence that seeds aged in air-dry store undergo membrane damage has been presented by Berjak and Villiers.[12] In this study seeds of *Zea mays* were subjected to 40°C for 20 days at a moisture content of 14%. Cells of the radicle root cap, which would otherwise be expected to commence a normal sequence of production and differentiation from cell division to maturity, became precociously senescent. Whereas the lysosomal enzyme acid phosphatase was confined to small pro-vacuoles or protein body-like organelles in cells immediately adjacent to the senescent ones, in the precociously senescent cells the acid phosphatase reaction was dispersed within the cytoplasm, indicating that the usually intact membranes had not been effective in retaining the lytic enzymes.

Similarly, prolonged air-dry storage causes damage to the structure of mitochondria, which, upon imbibition of the seed tissues, present an 'exploded' appearance in which the outer membranes have given way allowing the mitochondrial matrix to flow outwards.[13] However, this damage is not caused by the air-dry state *per se*, as both mitochondria and plastids are damaged in this way in long-dormant embryo cells that have been continuously imbibed for the entire period of aging[14] (Figure 17). Connected with this structural damage to the mitochondria, it is known that aged seeds that have decreased vigor show decreased phosphorylative activity in mitochondrial preparations and have a more dense matrix,[15] and also show decreased activities of cytochrome oxidase, malic dehydrogenase, and alcohol dehydrogenase.[16]

In the normal, dry soybean seed the plasma membrane has been described as having discontinuous and disorganized areas interspersed among areas in which a more normal unit membrane structure can be seen.[5] In some areas no membrane could be seen at all, but in its place there appeared a series of small vesicles. During the very early stages of imbibition, within the first 20 min, the continuity of the plasma membrane had become re-established. This period corresponds to that required for the establishment of cellular integrity as determined by the decrease in solute leakage by diffusion, which occurs when seeds are first placed in water.[17] However, as seeds become aged in dry store the amount of solute lost at imbibition increases and the length of time during which leakage occurs also increases.[18] It is possible that during aging, deleterious changes have occurred in the components used to re-establish the continuity of the plasma membranes.

Visual evidence for damage to the plasma membranes of dry-stored seeds has been presented. Initially, damage is seen as a withdrawal of the plasma membrane from the cell wall[13,19] and is also seen in seeds rendered non-viable following fungal infection.[20] This is probably best interpreted as a failure of the protoplast to swell osmotically and become appressed to the cell wall during seed hydration, because the plasma membrane has not become properly reconstituted. Webster and Leopold[5] show that in dry seeds the cell wall and cytoplasm appear to be separated in places by a series of very small vesicles. Thus in the absence of protoplast turgidity, as the wall colloids swell upon hydration and the tissue volume increases, a readily-visible space will be left between the cytoplasm and the wall. Disintegration of the plasma membrane occurs as a late stage in cellular senescence of plant organs, e.g., leaf mesophyll cells.[21]

Within embryo cells there occur organelles classed as protein bodies, which are perhaps the most prominent organelles in seed tissues. They are surrounded by a single unit membrane and may contain crystalline inclusions and a dispersed granular material. They appear to store reserve substances in the seed, and upon germination these reserves become mobilized by hydrolysis. In relation to this, the protein bodies have been reported to contain a range of hydrolytic enzymes. The protein bodies swell upon seed germination and appear to act in the manner of lysosomes within which other cellular material is broken down. The protein bodies enlarge and coalesce, eventually forming the large central vacuole in the differentiated cell, with its contents separated from the general cytoplasm by the tonoplast. Rost[22] has presented a review of protein body occurrence and function.

This sequence of events, proceeding with time during imbibition and germination, can also be seen as a 'static' sequence if one surveys the cells of an embryo axis from the meristem zone towards the region of cellular elongation. It is probably the normal sequence of vacuole formation that occurs in a meristem, modified to have the function of storing reserve materials for use in germination and establishment of the seedling, and becoming arrested in its sequence during seed maturation.

Damage to membranes such as the plasma membrane has already been described, and it is evident that by similar means the tonoplast and the protein body membrane could become damaged during dry storage. Consequently, if upon imbibition repair of the bounding membrane is not possible because of changes to the components normally incorporated, or if there ensues a period of delay before repair can be affected, the cytoplasm could be exposed to the actions of hydrolytic enzymes normally segregated within the protein body.

Evidence has been presented[12,19] of general cytoplasmic lysis following dissolution of the protein body membranes in cells of the embryo axis, in which the lysosomal marker enzyme acid phosphatase was shown to be confined to the protein bodies until their membranes dispersed, after which the reaction appeared to be generally distributed in the cytoplasm among the disintegrating organelles. Thus the protein bodies present a special problem within the seed tissues, as once their efficiency as special compartments in the cell has failed, then general cell death is likely to occur. The lysosomal nature of the protein body has been further demonstrated in studies of aged seed tissues in which (upon germination) cellular organelles such as plastids and mitochondria were taken into the protein bodies and destroyed.[14,23]

A major change seen in the nuclei during the imbibition of long-stored, low-viability seeds is the condensation of chromatin accompanied by an intense staining reaction of the clumps of material formed.[19,24] In embryo cells of normal, unaged seeds the nucleoplasm stains homogeneously and shows a fine network presumed to be the heterochromatin (Figure 2). During extended air-dry storage, the heterochromatin comes to stain more intensely until (in embryos from seeds of low viability) the nuclei stain almost homogeneously black (Figure 3). Damage has been shown to occur in the chromatin of rye embryos during air-dry storage, resulting in a lowered integrity of the genetic material,[25] and is suggested to result in the loss of coordination of synthetic processes and to be the cause of the increase in chromosomal aberrations found in many kinds of seeds during aging.

There has been some controversy concerning the nature of the limiting surfaces of the lipid bodies, which in normal cells are not removed by lipid solvents applied before fixation.[26] In seed samples of low viability, however, embryos frequently contain cells in which the lipid bodies have become confluent, forming large masses of homogeneously-staining lipid material[27] (Figure 4). This has also been shown to occur in seeds of low viability infected with fungi, and had occurred during the time that the seeds were in store.[20] Such a change during the air-dry state would be possible, because even when the cytoplasm has lost much water and become non-mobile, the lipid presumably does not change its state, and could therefore flow and become confluent should either the structure or a retaining membrane or the forces acting at the lipid body surfaces undergo change.

When seeds presumed to be dead, being drawn from aged samples showing no germination, are fixed shortly after imbibition, the majority of organelles can usually be distinguished within the cells. However, few or no cells show signs of possessing a functional plasma membrane, and the boundary layers of the cytoplasm have retracted from the cell walls (Figure 5). Keeping such seeds imbibed and periodically sampling for electron microscopy shows a rapid degeneration of the cytoplasm in which the lipid bodies become confluent (where this has not already occurred), and eventually total dissolution of the internal cell structure ensues (Figure 6).

Seeds from samples of low vigor, and with only partially impaired germination, often show these degenerative changes in isolated cells (or groups of cells) among apparently normal tissue, and it is assumed that, when extensive, these relate to the necrotic spots or areas seen in some low-vigor germinating seedlings.

One of the effects of seed aging is to delay the onset of visible signs of germination. The cause of this delay is not known, but some evidence has been presented that one process that might occur during this period is a rehabilitation of at least some of the cellular structure.[12] Mitochondria and plastids were seen to be abnormal in structure in embryos of *Zea mays* subjected to accelerated aging and the ER rapidly developed to an abnormal extent, but by the time that protrusion of the radicle had begun, an apparently normal state of cellular organization had been regained. Aged embryos showing delayed germination also possessed enhanced rates of protein and RNA synthesis during the pre-germination period by comparison with normal unaged embryos. The rates of synthesis had returned to normal levels by the time that germination commenced, but it is not known whether the higher rate was related to the necessity for repair and replacement of cellular structures, or whether it was a reflection of a temporary derangement of the control exerted in the normal embryo cells.

It is evident that structural changes do occur in seeds during air-dry storage. However, as might be expected at such low water contents, these must be at the molecular rather than the visible structural level. When such changes require energy and enzyme activity for their reversal, they would be expected to be manifested as structural abnormalities upon re-hydration of the tissues.

III. THE IMBIBED DORMANT STATE

In considering the deterioration of seeds in storage it is necessary to study, in addition to the perhaps more familiar case of air-dry seeds, that of seeds in a natural environment, either fully, partially, or intermittently imbibed in the soil. Wild plants are finely adapted to the climatic regimes of their areas of origin by a variety of dormancy mechanisms, allowing seedling production only at favorable times of the year. Seeds may be dormant when they are produced on the parent plant, or may be rendered dormant should environmental conditions be unsuitable for immediate germination. The general changes occurring during the breaking of dormancy are not considered here, since they are not directly connected with senescence.[27a]

It was stated in the previous section that the overriding factor in seed longevity in dry storage is low seed moisture content. It can be shown that within the air-dry range, raising the moisture content has an increasingly deleterious effect on seed longevity, and it might therefore be expected that seeds in an imbibed state, with a moisture content of the order of 50%, would have only a very short storage life. However, it is a common ecological observation that seeds may lie buried in the soil and yet retain their viability for many years, and it is simple to demonstrate that this is also true for imbibed seeds stored in the laboratory in a dormant state.[10] Such imbibed seeds are able to remain alive much longer than air-dry seeds unless special precautions, such as artificial drying, are taken. In addition to increased longevity, it was found that the incidence of chromosome aberrations remained low in imbibed seeds compared with those stored air-dry.[10]

It is, of course, not possible to maintain seeds in an imbibed state unless they are naturally dormant, or, in the case of merely quiescent crop plant seeds, can be rendered dormant in some way, e.g., temperature sensitivity in fruits of *Lactuca sativa*. As experimental imbibed-stored seeds can be compared with those lying dormant in the soil, it is therefore of interest to examine the changes that can occur with time during imbibed storage. The situation also bears comparison with other senescence processes, and is also of more general interest because this is the intermediate of the states discussed in the Introduction, in which metabolism can take place, allowing turn-over at the molecular level without growth or cell division.

When seeds imbibe water, respiration becomes measurable, and its rate increases sharply. However, if the seeds become dormant and do not germinate, the rate falls to a low level. Such dormant seeds are capable of protein synthesis at rates comparable to those of non-dormant seeds of the same species during the pre-germination period.[28] Additionally, it has been shown that during dormancy, food inter-conversions and the production of organelles, and therefore membrane synthesis, are possible[29] (Figure 7). Thus during dormancy embryos are able to carry out metabolic processes and have the ability to turn over and replace membranes.

In the previous section it was shown that cytological changes that occur in embryo cells upon imbibition include reconstitution of the plasma membranes and the membranes of certain cytoplasmic organelles, such as the ER. When imbibition is followed by germination, as in many crop plants, other changes occur that are probably associated with the preparation for the period of rapid metabolism and growth in the production of a seedling. However, when a seed becomes dormant upon imbibition, e.g., a light-sensitive seed or one requiring chilling, certain differences have been noted, such as the apparent absence of polysome formation (Figure 9).

Depending on the state of maturity of the embryo in the seed as it matures on the parent plant, two distinct situations can arise when dormant seeds are imbibed. In seeds such as *Avena fatua*, the embryo may remain apparently unchanged, maintaining a slow metabolic rate,[28] but in some species changes concerned with the final stages of an interrupted development may begin. These changes may include mobilization and transport of food materials from the endosperm to the embryo, and differentiation or growth of the embryo such as occurs in *Heracleum sphondylium*[30] and *Fraxinus excelsior*.[31]

A. After-Ripening

In the European ash-tree *F. excelsior*, lipid bodies are originally a very prominent feature of the cells of the embryo axis, and are disposed as continuous layers around the protein bodies and at the periphery of each cell (Figure 1). These decrease in number dramatically during imbibed storage, the extractable lipid content of the embryos falling from 20% to only 4% of the dry weight by the time the maturation of the embryo is complete. During the same period, the ER becomes very extensive (Figure 7), attached ribosomes can be seen, and the plastids develop internal membrane sys-

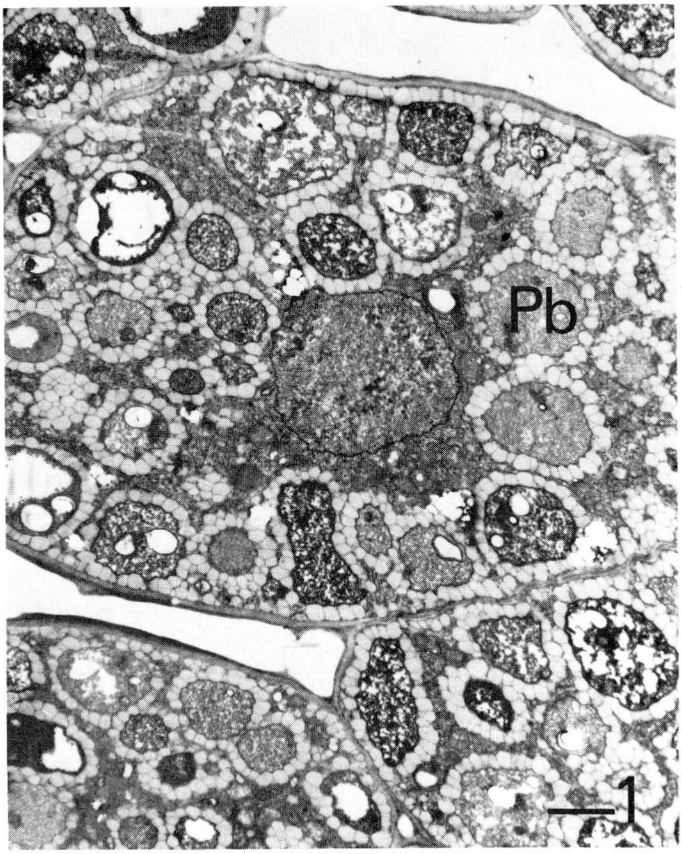

FIGURE 1. Radicle meristem cell from newly-imbibed embryo of *Fraxinus excelsior*, showing the distribution of lipid bodies around the periphery of the cell and closely appressed around each protein body (Pb). Ruled line corresponds to 1 μm.

tems and deposit starch. Active Golgi bodies are formed, apparently producing vesicles, and microbodies appear. Within the nuclei, the chromatin is stainable as a finely-dispersed network and the nucleoli become small and stain very densely compared with other nuclear components (Figure 8). The ribosomes that occur free in the cytoplasm do not appear to be arranged into polysomic aggregates (Figure 9), such as those seen in non-dormant embryos, although protein synthesis is possible, as the protein content of the embryos rises from 14% of the embryo weight upon first imbibition to 24% after 3 months of imbibed storage.

On receipt of the germination stimulus, certain changes become evident. In the case of low-temperature dormancy release, the changes occur slowly during an extended period of time, which may be weeks or even months. The main changes that appear to relate to metabolic activities during the chilling of *F. excelsior*[32] concern the structure of the nucleolus, which increases in volume twenty-fold, develops the normal structure of the nucleoli seen in active plant root cells,[33] and becomes active in RNA synthesis. In the case of light-controlled dormancy, the changes occur much more rapidly over a few hours or days, and again involve the enlargement and differentiation of the nucleoli (Figure 10) and also the aggregation of the free cytoplasmic ribosomes into polysome groups[34] (Figure 11). In addition, changes in other organelles give the appearance of a general activation of the tissues in preparation for growth.

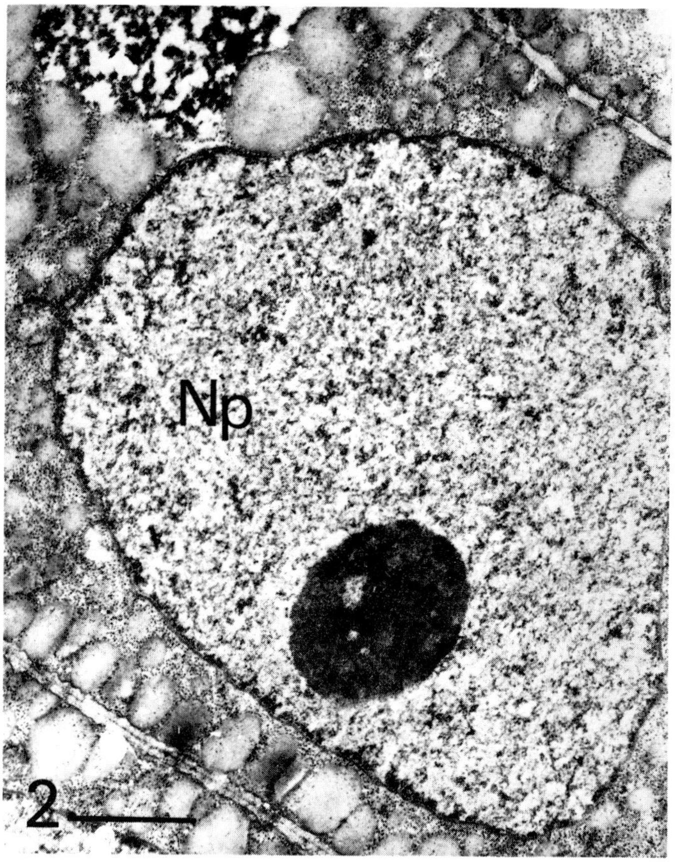

FIGURE 2. Nucleus from radicle meristem cell of lettuce embryo showing the granular appearance of the nucleoplasm (Np) in a viable embryo. Ruled line: 1 μm.

B. Dormant Seeds

Such changes must also occur in the majority of the seeds of wild plants where seasonal dormancy normally defers seedling production until after the coldness of winter in temperate regions, or the dryness of summer in, e.g., Mediterranean climatic regions. However, it also frequently occurs that in many species the period of dormancy is spread over many seasons, and populations of seeds of varying ages normally occur in the majority of soils. In any seed lot there usually occurs some variation in the depth of dormancy, there being an obvious biological advantage in spreading the germination of a crop of seeds over a period of time that may extend to more than one year. In addition, many weed seeds can be thrown into a state of secondary dormancy, only germinating after being brought to the soil surface by tillage, or the activities of the soil fauna. In these cases, long-term dormancy in an imbibed, or intermittently-imbibed, state presumably becomes deleterious after an extended period of time, and it is of interest to examine the state of embryo tissues subjected to prolonged dormancy.

It has been shown that, far from being deleterious, imbibed storage can maintain seeds in good order compared with dry-stored seeds, unless special precautions are taken that are very artificial, and in most cases expensive. This has been demonstrated and ascribed to the capacity of the hydrated tissues to turn over their components and

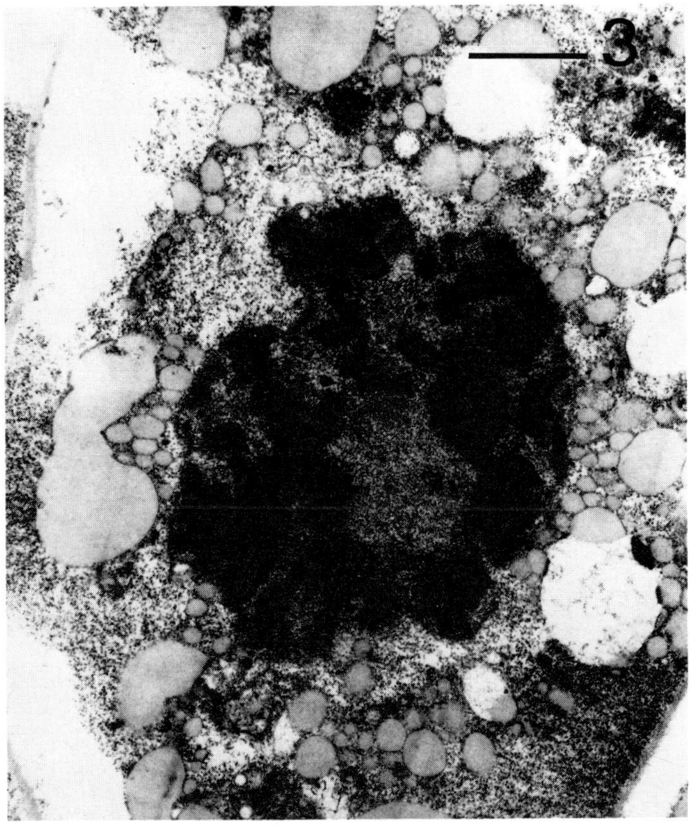

FIGURE 3. Nucleus with condensed chromatin from radicle meristem cell of lettuce embryo from a seed subjected to an accelerated aging regime of storage at 30°C with a moisture content of 13% for 4 weeks. The seed sample showed 15% germination. Ruled line: 1 μm.

maintain themselves in good structural order.[10] In imbibed storage it was shown that the germination percentage remained high, good seedling vigor was retained, and the incidence of chromosome aberrations was maintained at a very low level.

The accumulation of chromosome damage, which can be demonstrated as aberrations at the first cell divisions of the embryo radicle, has been suggested to be one of the causes of senescence,[35] and is an indication of the amount of genetic damage accumulated in non-dividing cells with time. Thus it appears that at least one important symptom of senescence is shown by aging seeds, and whereas genetic damage would accumulate slowly with time in non-dividing cells, such as the embryo cells of imbibed, dormant seeds, they would be expected to accumulate far more rapidly in air-dry seeds because of the inability of DNA repair mechanisms to operate.[10]

C. Long-Dormant Seeds

The cytological changes that take place in seeds that are imbibed, but enter a period of dormancy, have been described above. If dormancy is not broken and the seeds remain ungerminated for long periods, apparent pathological changes eventually occur. In embryos of *F. excelsior* maintained in an imbibed, dormant state for 6 years, lipid bodies had been removed, and the protein bodies had become electron transparent. ER with attached ribosomes surrounded groups of cytoplasmic organelles, apparently enclosing these in the manner of autophagic vacuoles, or cytolysomes[14] (Figures

FIGURE 4. Embryo radicle meristem cells from lettuce seed sample of low viability subjected to an accelerated aging regime. The plasma membrane (arrow) has collapsed and the lipid body contents have become confluent (L). Ruled line: 1 μm.

12 and 13), similar to those described for animal cells in starved or tissue-damaged subjects.[36] These become a prominent feature of long-dormant seed tissues and could be concerned either with the removal of damaged organelles, or the re-cycling of materials in a continuing turn-over of cell components. During prolongd dormancy, certain of the mitochondria became very enlarged, either iso-diametrically or by branched, serpentine growth (Figure 14), and plastids developed dense inclusions (Figure 15).

The ER became broken up into series of small vesicles (Figure 16), each of which, however, retained attached ribosomes. The outer membranes of some of the mitochondria and plastids appeared to collapse in places, allowing the enclosed matrix to spill outwards into the cytoplasm (Figure 17). Some nuclei became lobed, a symptom which has previously been ascribed to senescence[37] (Figure 18), and in some cells the two membranes constituting the nuclear envelope separated widely from each other and swelled into bladder-like extensions (Figure 19), which must have entailed changes to the structure of the nuclear pores. In the cells of the embryo axis, the cytoplasm eventually developed a noticeably empty appearance and, as time progressed, the incidence of membrane damage became higher, the Golgi bodies decreased in number and appeared to be inactive, and the protein bodies came to resemble small, empty cell vacuoles.

Recovery from such obviously aberrant states was still possible, however, even after

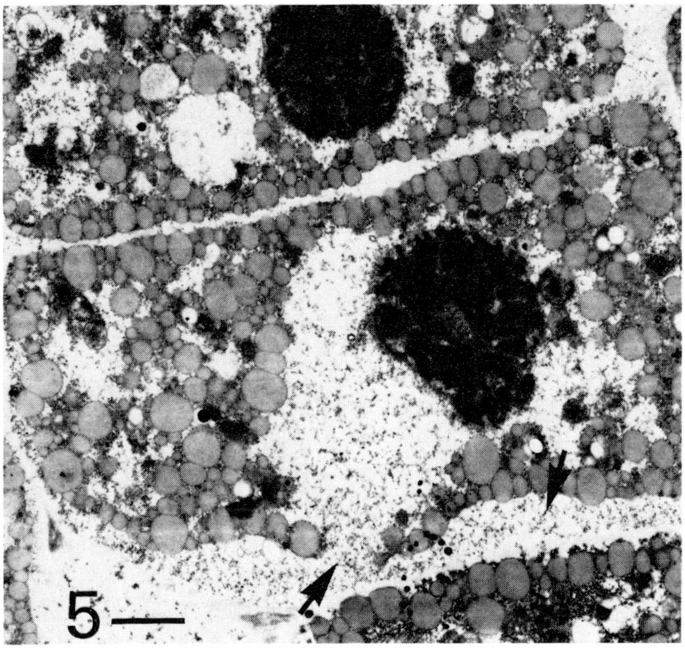

FIGURE 5. Embryo radicle meristem cells from lettuce seed subjected to an accelerated aging regime of storage at 30°C with a moisture content of 13% for 8 weeks. The sample was non-viable. The seed had been imbibed for only 2 hr before fixation. Many of the cytoplasmic organelles are recognizable, but the plasma membrane is not continuous and is retracted from the cell wall (arrows). Ruled line: 1 μm.

10 years of continuous imbibition in a dormant state. When stimulated to germinate, and the embryos sampled for electron microscopy over the period leading to germination, it was found that the cytological appearance of the cells could gradually revert to normal. It was presumed that in the case of *F. excelsior* the raw materials for these changes would be drawn from the endosperm tissue. The time taken for the embryos to begin germination after prolonged dormancy was always longer than in the case of fresh seeds. Golgi bodies became active in the production of vesicles and starch accumulated. Rough ER developed and normal mitochondria and plastids were seen. Before geotropic curvature of the embryos occurred, the nuclear envelope membranes appeared normal.

In the early stages of rehabilitation, the vacuole-like protein bodies could be found to be acting as lysosomes, and cytoplasmic vesicles and organelles such as mitochondria and plastids were seen apparently fixed in the act of moving into the vacuoles (Figure 20). This movement appeared to involve both the vacuole and the organelle, which extended towards each other in an amoeboid manner. At contact, the vacuole invaginated while the organelle appeared to continue to extend forward, passing into the vacuole. Near the area of contact with the vacuolar membrane, the matrix of the organelle being engulfed became electron transparent, and its membranes became swollen, bulging inwards into the vacuole. After entry into the vacuole the entire organelle became swollen and its internal membrane structure dispersed, and the vacuoles ultimately came to contain much disorganized membranous material (Figure 21). Both before and during this period of lysosomal activity, the vacuoles showed a positive reaction for acid phosphatase activity.

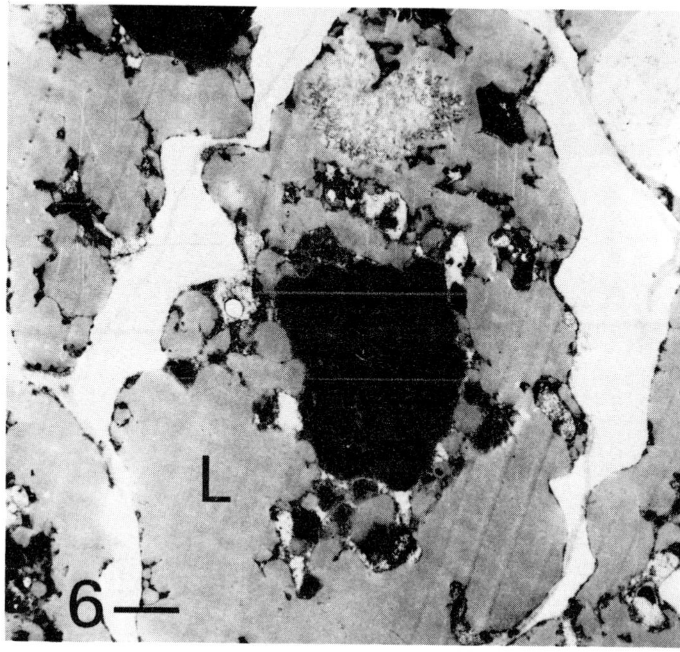

FIGURE 6. Embryo radicle meristem cell from the same sample shown in Figure 5, imbibed 72 hr before fixation. The lipid bodies have become confluent (L) and no cytoplasmic organelles are recognizable at this stage. Ruled line: 1 μm.

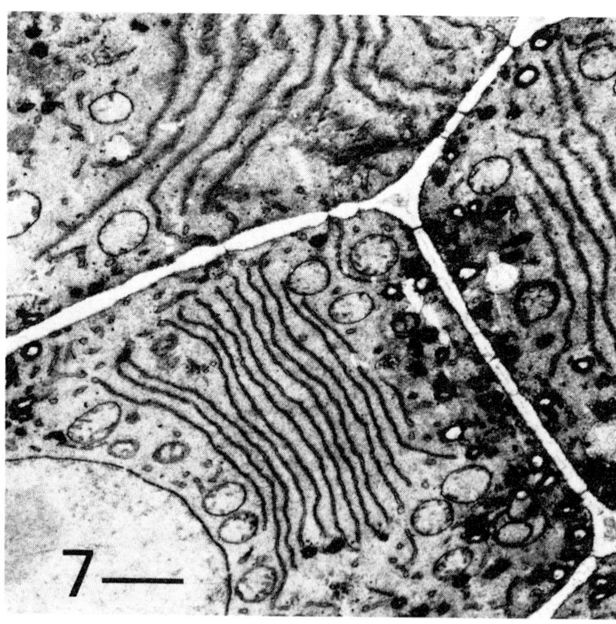

FIGURE 7. Radicle meristem cells from embryo of *Fraxinus excelsior* maintained fully imbibed at 20°C for 3 months. Extensive ranks of endoplasmic reticulum, previously rare, have now developed. Most of the lipid and protein bodies have disappeared. Ruled line: 1 μm.

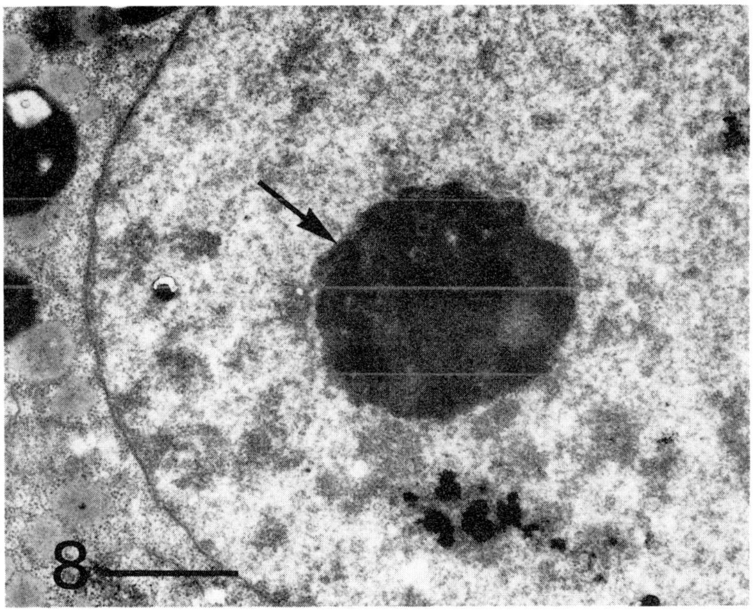

FIGURE 8. Part of nucleus showing small, dense nucleolus (arrow) from radicle meristem cell of dormant embryo of lettuce, variety Grand Rapids, maintained in the dark. Ruled line: 1 μm.

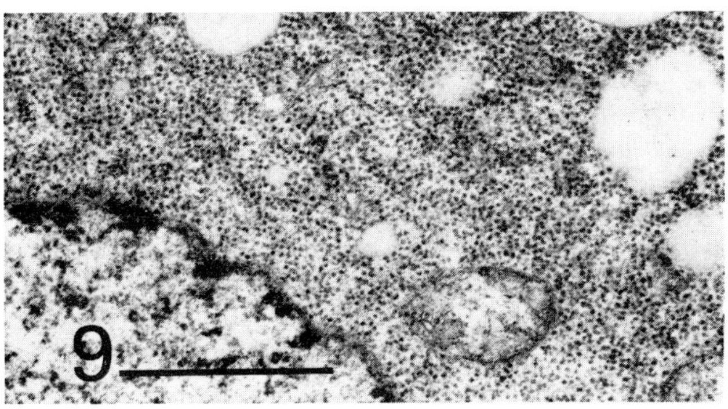

FIGURE 9. Ground cytoplasm showing ribosomes in radicle meristem cell of dormant embryo of lettuce, variety Grand Rapids, maintained in the dark. Ruled line: 1 μm.

In spite of this apparent removal of plastids and mitochondria by the vacuoles or protein bodies, the numbers of organelles in the embryo axis cells increased and in connection with this, stages of sub-division of plastids and mitochondria were frequently found. In some cases, owing to differences in content and density of staining between the sub-divisions, it appeared as though a damaged part of an organelle was being separated and removed (Figure 22).

IV. DISCUSSION

A study of the deterioration processes occurring in air-dry seeds is of importance for the practice of crop storage. Additionally, information on the senescence processes

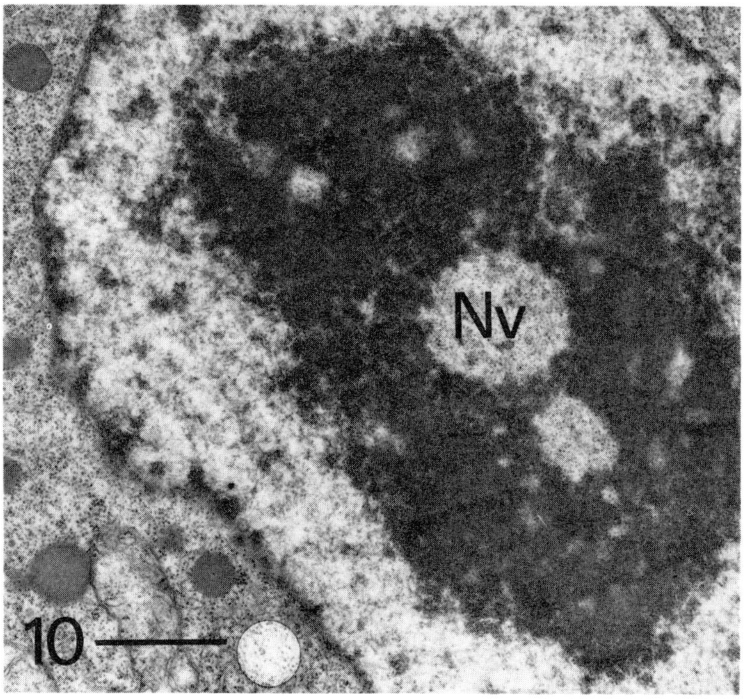

FIGURE 10. Part of nucleus showing very large, diffuse nucleolus with nucleolar vacuoles (Nv) from radicle meristem cell of seed from same sample as in Figure 8, fixed 8 hr after placing in the light to stimulate germination. Ruled line: 1 μm.

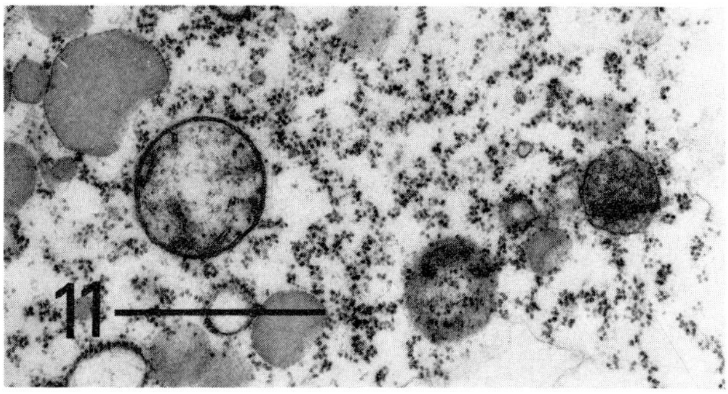

FIGURE 11. Ground cytoplasm showing ribosomes aggregated to form polysomes in radicle meristem cell of seed from same sample as in Figure 8, fixed 16 hr after placing in the light to stimulate germination. Ruled line: 1 μm.

occurring in imbibed, dormant seeds is of interest in plant ecology, for the understanding of the physiology of weeds of arable land, and possibly also for germ-plasm storage in seed banks. In some species of dormant, imbibed seeds cytological changes occur that are not deleterious, but are part of the differentiation process in embryo maturation which must take place before germination, while other changes may possibly be a result of metabolism under conditions of inadequate aeration. However, in long-term

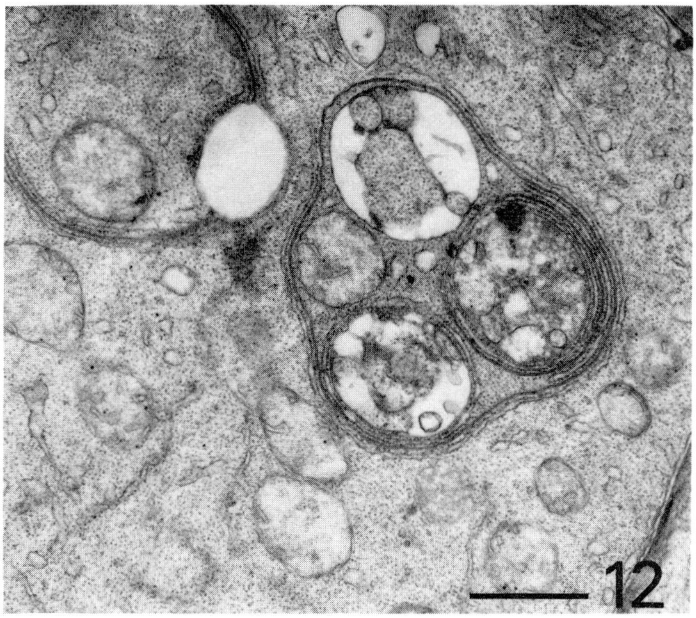

FIGURE 12. Associations of organelles within layers of endoplasmic reticulum. From a radicle meristem cell of an embryo of *Fraxinus excelsior* maintained imbibed in a dormant state by withholding the chilling stimulus. Ruled line: 1 μm.

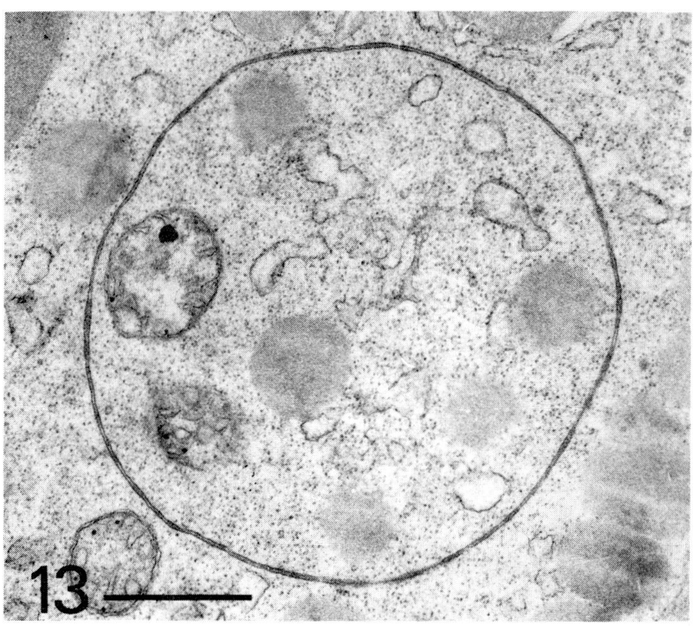

FIGURE 13. Cytolysome-like association in radicle meristem cell of embryo of *F. excelsior* maintained imbibed in a dormant state. Ruled line: 1 μm.

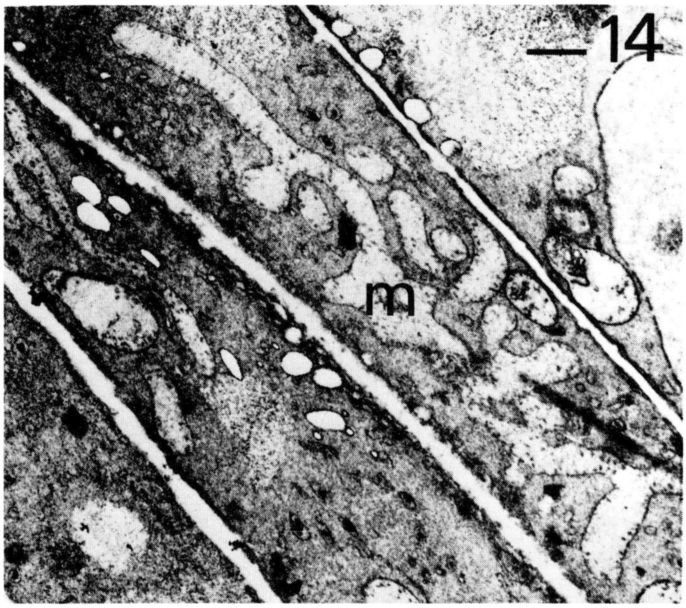

FIGURE 14. Greatly-enlarged mitochondria (M) in embryo of *F. excelsior* maintained imbibed in a dormant state. Ruled line: 1 μm.

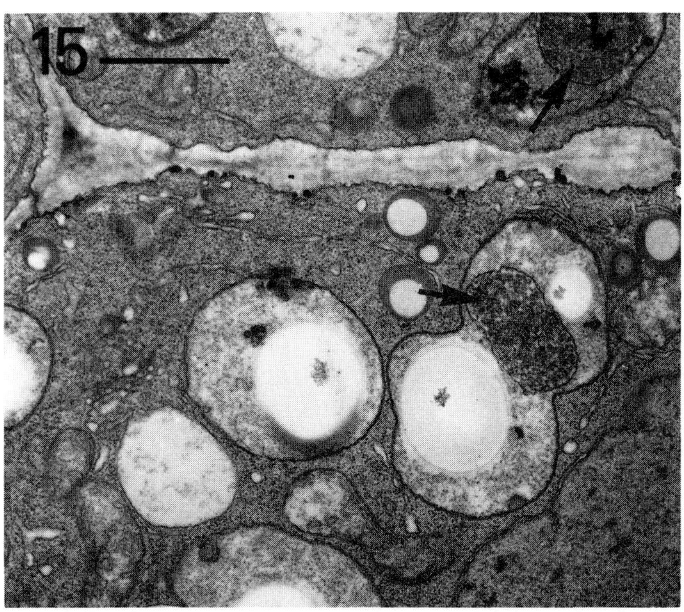

FIGURE 15. Dense inclusion body (arrows) in plastids of radicle meristem cells of *F. excelsior* maintained imbibed in a dormant state. Ruled line: 1 μm.

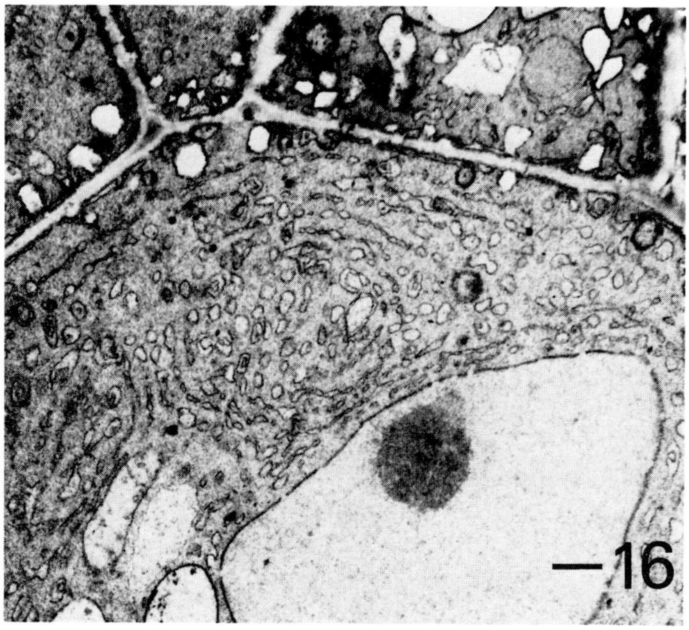

FIGURE 16. Endoplasmic reticulum broken into series of small vesicles in radicle meristem cell of *F. excelsior* maintained imbibed in a dormant state for 4 years. Ruled line: 1 μm.

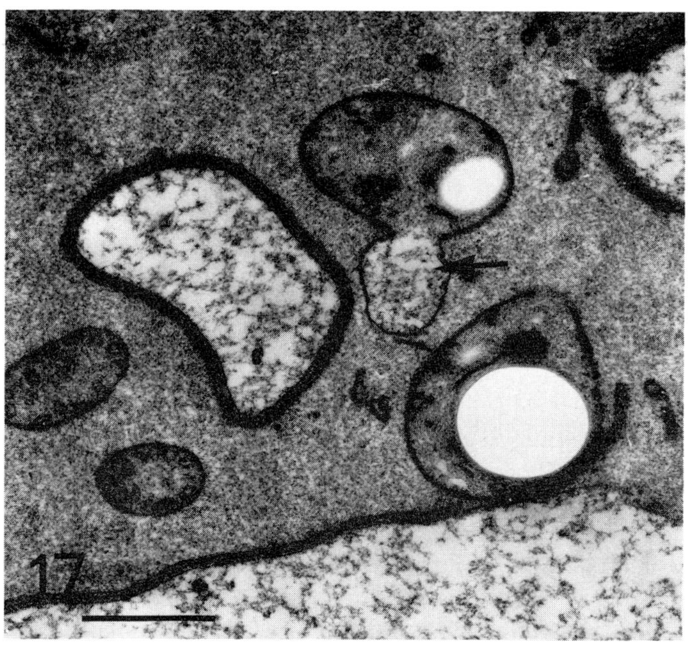

FIGURE 17. Plastid with burst membranes (arrow) in radicle meristem cell of *F. excelsior* maintained imbibed for 5 years. Ruled line: 1 μm.

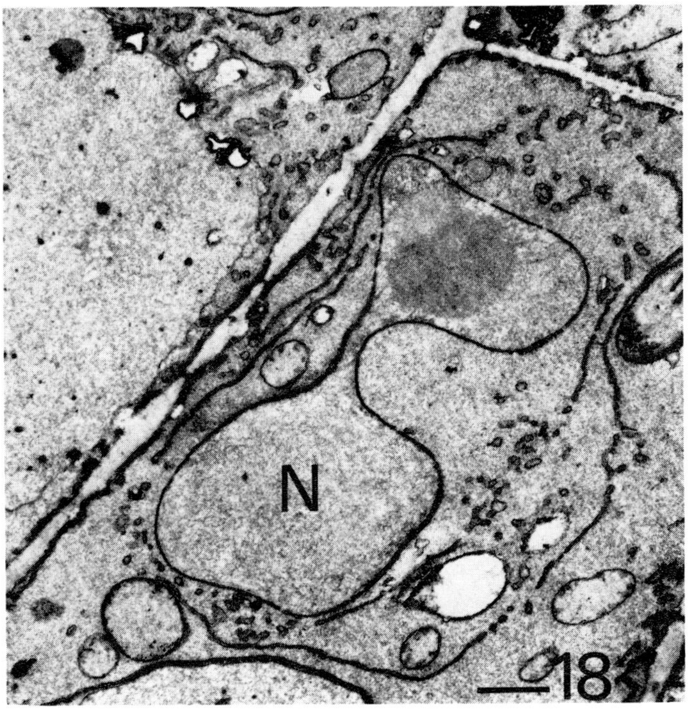

FIGURE 18. Lobing of nucleus (N) in radicle meristem cell of *F. excelsior* maintained imbibed for 4 years. Ruled line: 1 μm.

dormancy, the changes that occur appear to be pathological and will be discussed as a senescence process and compared with the aging of air-dry quiescent seeds and the senescence of growing plant organs.

A comparison of the cytological damage shows many differences in the process between air-dry and imbibed-stored seeds. The rate at which damage accumulates in air-dry seeds is directly related to the conditions of storage, especially the seed moisture content and the temperature. In imbibed-stored seeds the water content is not easily variable, and is of the order of 50% of the fresh weight of the entire seed, but while the temperature can be varied it is important for the theory that has been advanced in explanation of imbibed storage[10] that the temperature should allow metabolic activity while not breaking the state of dormancy necessary for continued storage.

In the dry-stored seed most of the changes that occur appear to be at the molecular level and do not show any effect until the seed is imbibed. One visible change that can occur in the air-dry state is most easily seen in lipid-storing seeds, where the lipid bodies break down and allow the oils to become confluent in large masses within the cells. In the imbibed-stored seed this is not seen to occur, but these organelles disappear, presumably because the lipid becomes partly metabolized and partly converted into other materials, such as starch.

Other changes are not manifest until the time of seed imbibition and include deterioration of the nucleus, where the chromatin becomes highly condensed as the seed loses viability. The rate of this deterioration is related to storage conditions, and it is probable that the optimum conditions of moisture content, temperature, and atmosphere are those that provide the greatest physical stability to the macromolecules. Probably an earlier manifestation of such changes in the chromatin is the incidence of chromosome aberrations, which increase steadily with time in seeds. The rate of accu-

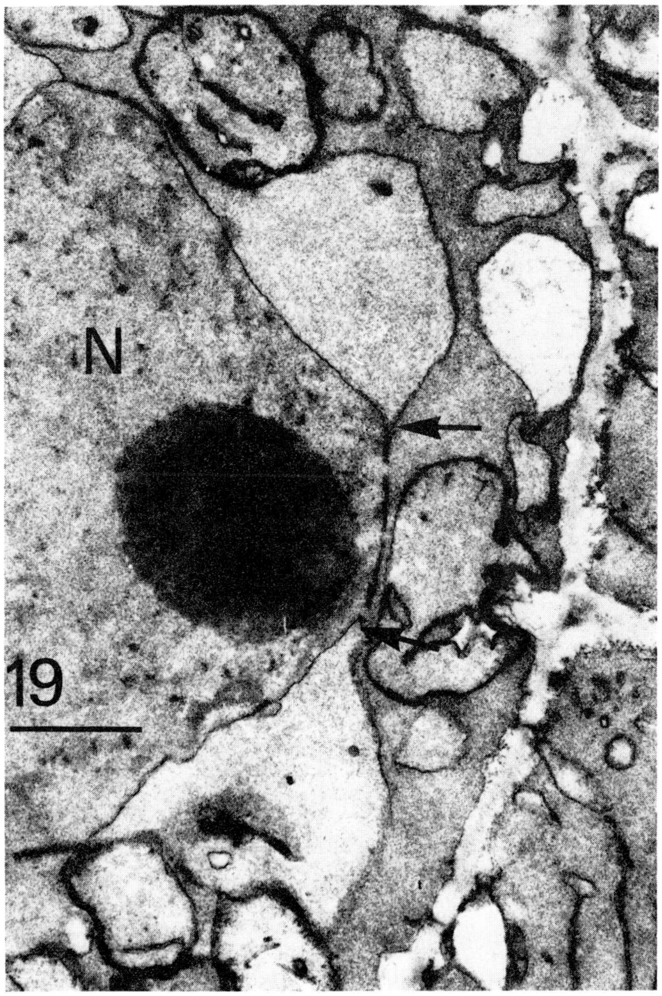

FIGURE 19. Nucleus (N) in radicle meristem cell of embryo of *F. excelsior* maintained imbibed for 5 years. The membranes of the nuclear envelope have separated and become dilated (arrows). Ruled line: 1 μm.

mulation of chromosome damage is greatly reduced in imbibed-stored seeds, and the condensation of chromatin described above did not occur over a period of 8 years' imbibed storage in the laboratory. Evidently the state of hydration of the tissues either stabilizes the chromatin, or damage is able to be repaired at the time of occurrence, rather than being allowed to accumulate.

Structural changes that probably occur in the configuration of the phospholipids of the cellular membranes such as the plasma membrane, the tonoplast, and the ER during desiccation of the seed have been described above. A re-organization of phospholipid molecules constituting the bi-lamellar unit membrane probably occurs when the seed moisture content falls below about 20%. Further, it has been suggested that cell membranes in desiccated tissues lose a proportion of their lipid content, which may appear in the cytoplasm in the vicinity of the membranous structures as numerous small lipid droplets in desiccated roots[38] and in seeds[5] and disappear again upon rehydration of the tissues.

In this rearranged state, the seed membranes would no longer exist as such, and this

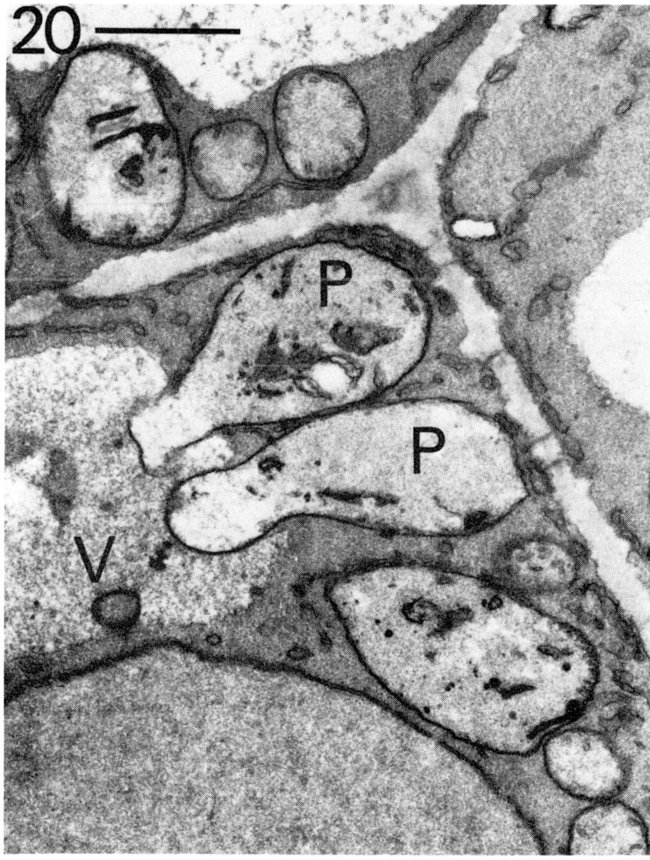

FIGURE 20. Plastids (P) apparently entering a vacuole (V) in radicle meristem cell of embryo of *F. excelsior* maintained imbibed for 6 years. Embryo fixed 2 days after stimulating to germinate by excision from the seed. Ruled line: 1 μm.

has been used to explain the leakage of solutes that occurs when seeds are first placed in water. Deleterious changes in the molecules might make the reconstitution of the normal, functioning membrane difficult or incomplete, and this would contribute to the extension of the period during which leakage occurs when aged seeds are re-imbibed. Damage to the membrane components could include lipid oxidation, which is known to occur in aged seeds, and the beneficial effect on seed longevity of the exclusion of oxygen could have its effect on this process. Koostra and Harrington[39] have shown that a decline in extractable polar lipids occurs with loss of viability in seeds stored air-dry, and ultrastructural studies have been described above that show that collapse of the plasma membrane or the failure of the protoplasts to regain turgidity is commonly seen in aged seeds, whereas this was not seen in seed stored continuously hydrated. Leakage of electrolytes and loss of weight from senescing leaf tissues has been reported and correlated with membrane phospholipid decrease,[40] and this appears to be similar to the leakage occurring from aged seeds.

Freeze-etch studies[1] have suggested that the membranes of the protein bodies undergo disorganization similar to that of the plasma membrane upon drying of the cells. Freeze-etching, in which rapid freezing of the tissues would be expected to prevent changes in configuration of the membranes of cytoplasmic organelles, has shown

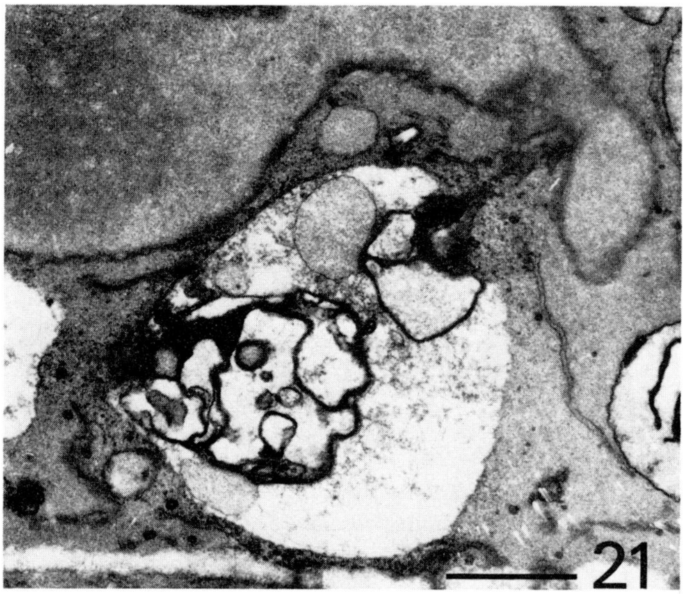

FIGURE 21. Organelles enclosed within vacuole in embryo of *F. excelsior* dormant for 6 years. Fixed 3 days after stimulating to germinate by excision. Ruled line: 1 μm.

that when tissues are prepared from the air-dry state by fracturing in pure glycerol, clear bounding membranes are not seen, but if allowed to hydrate even for less than 10 sec, the protein body membranes present a normal continuous appearance. That lipid may be reduced or absent from the protein body membranes was suggested, because the fracture path was not deviated in tissue that had been extracted with chloroform. Thus the argument previously applied to the deterioration of plasma membranes in dry storage also applies in this case, that deleterious changes to the re-arranged components could prevent or delay the reconstitution of the normal protein body membrane structure upon imbibition.

Again, although common in dry-stored seeds upon hydration, dispersion of the protein body membranes was not seen in imbibed-stored seeds. As the protein bodies in seeds are probably modified developing vacuoles, containing hydrolytic enzymes and having a function similar to that of lysosomes, it would be fatal to the cell to allow dispersion of these membranes too early in a senescence program, and the imbibed seed could therefore replace or turn over its membrane components, retaining the integrity of these organelles during dormancy.

ER is scarce or absent from dry seeds, but becomes very extensive in long-imbibed seeds. This development of the ER is unlikely to be necessary for synthesis, as a slowly-metabolizing system such as a dormant embryo presumably synthesizes protein at a relatively slow rate. Similar ER figures have been reported in plant tissues subject to anoxia,[29] and it may be that this over-development is a result of poor aeration rather than being concerned with the dormancy state. ER vesiculation, which occurs in long-imbibed seeds, is also seen in other types of cellular damage, such as chilling injury.[41]

Golgi bodies are rarely seen in dry seeds, and those present in imbibed-stored seeds become inactive and disperse after long periods of storage. Mitochondria and plastids are little changed in appearance in dry seeds, but on hydration they show aberrant membrane structures similar in appearance to those seen in very long-imbibed seeds.

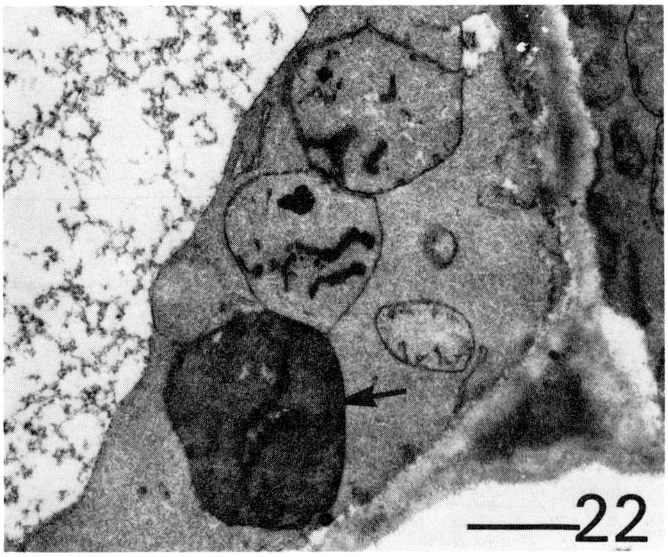

FIGURE 22. Sub-dividing plastid, with one dense, possibly non-functional, compartment (arrow) in embryo of *F. excelsior* dormant for 6 yr, fixed 5 days after stimulating to germinate by excision. Ruled line: 1 μm.

The cytolysomes occurring in imbibed-stored seeds have no counterpart in dry-stored seeds owing to the impossibility of ER formation or membrane re-arrangement in the dry state.

The degeneration of structure seen in seeds maintained imbibed is probably similar to that occurring in seeds in natural situations in the soil. The changes can be divided into two general kinds, the first being concerned with the genetic material and the molecules responsible for the transmission of genetic information. Unless special conditions are maintained, the DNA has been shown to become damaged far more quickly in the air-dry state, and this would logically seem to be connected with the inability of DNA repair enzymes to remain active in dry cells; correspondingly it would be expected that damage would also accumulate in other macromolecules, which could not be replaced, such as RNA and enzymes.

The second category of damage appears to be to the structural fabric of the cells in the form of impaired membranes and is concerned with the integrity of the protoplast as a unit, and the internal compartmentation of the cytoplasm. Neither type of damage is apparent in the air-dry state, and dry seeds of ancient origin can still be demonstrated to possess a cytologically recognizable structure.[13] However, upon hydration structural and genetic damage become demonstrable, and may even become amplified by the release of lytic enzymes from their normal compartments, or by the faulty functioning of key enzymes such as polymerases and the repair enzymes.

Unlike air-dry seeds, imbibed dormant seeds are more similar to normal plant tissues, and therefore some comparison with senescence processes in other organs is of interest. Much information has been obtained on the degeneration of chloroplasts, which are usually the first organelles to undergo changes in senescence — while the remaining cellular structures, such as the mitochondria, are still apparently normal and functional.[21] At about the time of degeneration of the chloroplasts a reduction in the number of free cytoplasmic monosomes has been recorded. The polysomes are more persistent, and attached ribosomes can be seen on vesicles at very late stages of cellular dissolution. Vesiculation of the ER appears to be typical of senescing plant cells, occurring in cotyledons and wheat leaf cells, as well as in aging dormant seeds. Swelling of the perinuclear 'space' between the membranes of the nuclear envelope has been noted in sieve elements[42] and in orange fruit[43] in addition to long-dormant seeds.[14] Mitochondria appear to be resistant, sometimes being the last complete organelles to be seen at the final stages of cell dissolution, when the tonoplast and finally the plasma membrane break down and cell lysis is completed.

From a study of the senescing leaf of *Phaseolus,* Barton[44] concluded that senescence appeared to be a well-ordered process initially affecting mainly the chloroplasts, and suggested that a specific enzyme system might be involved rather than a general deteriorative process. Martin and Thimann[45] demonstrated that active protein synthesis occurred at the start of leaf senescence and suggested the formation of a proteolytic enzyme system with L-serine at its active centre.

It is known that proteolysis occurs in attached senescing leaves and is accompanied by a polar basipetal transport of amino acids out of the leaf.[46] In the economy of the growing plant, this would obviously be of great advantage in the re-deployment of resources, and therefore in the senescence of leaves in the growing plant the process would not be expected to be one of general deterioration, but would proceed in an orderly programmed manner. The first visible signs of degeneration occur within the plastids, which represent a large proportion of the protein within a leaf, while the remainder of the cellular structure shows little change.* General damage to systems involved in transport, such as the mitochondria, would interfere with the organized withdrawal of such materials. Butler and Simon[21] present an argument for the case for genetically-controlled, i.e., 'programmed', senescence in plants, at the cellular, organ, and whole plant levels. The cytological events occurring in each type of senescence have great similarities, and it is likely that a similar sequence is followed in each case.

The deterioration of a seed before germination is unlikely to be programmed, as the embryo tissues, with the possible exception of the cotyledons in some cases, are intended for the most active period of metabolism and growth in the whole life of the plant, viz., seedling production. Thus the dry seed undergoes a general deterioration more or less rapidly according to the degree of stability afforded by the storage conditions, with the full consequences of the damage only becoming evident at seed hydration, and in the imbibed seed the various systems might be expected to degenerate more randomly than in a plant organ following a programmed sequence of events. McKersie and Thompson[47] have suggested that functionally distinct membranes are differentially sensitive in the senescence process. Following this argument it is possible that the mitochondria might be readily subject to damage, because of their high content of polyunsaturated lipids[48] and the presence of free radicals in the respiratory chain,[49] unless they are specially stabilized, repaired, or replaced. Varner[50] has suggested that mitochondria might be a basic factor in the onset of senescence, because if they should become damaged the production of ATP would fall, leading to loss of general membrane integrity.

A striking feature of the senescence of seeds is their ability to recover from considerable damage at the visible level. Plant organs (such as leaves) can reverse chlorophyll loss and this is accompanied by re-formation of the internal membrane structures. It has not been recorded that they are able to reverse structural damage to other organelles such as mitochondria, although this appears to be possible in imbibed seeds. As

* Proteolysis, however, is biochemically detectable many hours before chlorophyll decreases, at least in the oat leaf (Chap. 5).

their cells can become rehabilitated even after considerable damage has occurred to such essential organelles, presumably the point of no return in seed senescence would appear to be the rupture of the tonoplast, or of the plasma membrane.

In their survey of the ultrastructure of plant senescence Butler and Simon[21] concluded that although it is difficult to generalize about the causes of plant senescence, protein supply is probably likely to be a key factor. They point out that in view of the low specificity of many hydrolytic enzymes, their activation or loss from lysosome-like structures would quickly lead to total dissolution rather than produce the sequential process of the changes observed, especially in early senescence; they suggest that breakdown of the tonoplast may be the last stage, which is irreversible. In this respect it is notable that although plasma membrane and tonoplast damage can readily be induced in dry seeds, it is not seen in imbibed seeds until very late stages of tissue degeneration.

The deterioration of the plasma membrane in dry seeds is significant because the rapid re-establishment of osmotic integrity in the seed tissues is of vital importance in the early stages of imbibition, when leakage of solutes can occur. The internal compartmentation of the cells is also of fundamental importance, especially in the case of lysosome-like protein bodies, and a failure in the continuity of this type of membrane would lead to a loss of the control of metabolism and cellular organization.

It is therefore concluded that a fundamental cause of senescence in dry seeds is the loss of structural integrity of the cellular membranes, together with the deterioration of macromolecules such as the nucleic acids and certain key enzymes. Damage to the chromatin during air-dry storage can readily be demonstrated, but cell division does not begin until after protrusion of the radicle, and it has been shown that even when the DNA has been inactivated by irradiation, lettuce seeds can still germinate and produce seedlings without cell division.[51] However, a consideration of the imbibed, dormant seed, where senescence still occurs when metabolism is able to proceed, but growth is suspended in a static, non-dividing cell system, emphasizes the importance and necessity of turn-over at the DNA and cellular level, and of the process of growth, for the continued existence of life.

REFERENCES

1. **Buttrose, M. S.,** Rapid water uptake and structural changes in imbibing seed tissues, *Protoplasma,* 77, 111—122, 1973.
2. **Perner, E.,** Elektronenmikroskopische Untersuchungen an Zellen von Embryonen im Zustand völliger Samenruhe. I. Mitteilung. Die zellular Strukturordnung in der Radicula lufttrockner Samen von *Pisum sativum, Planta,* 65, 334—357, 1965.
3. **Paulson, R. E. and Srivastava, L. M.,** The fine structure of the embryo of *Lactuca sativa.* I. Dry embryo, *Can. J. Bot.,* 46, 1437—1445, 1968.
4. **Swift, J. G. and O'Brien, T. P.,** The fine structure of the wheat scutellum during germination, *Aust. J. Biol. Sci.,* 25, 469—486, 1972.
5. **Webster, B. D. and Leopold, A. C.,** The ultrastructure of dry and imbibed cotyledons of soybean, *Am. J. Bot.,* 64, 1286—1293, 1977.
6. **Simon, E. W.,** Phospholipids and plant membrane permeability, *New Phytol.,* 73, 377—420, 1974.
7. **Parrish, D. J. and Leopold, A. C.,** Transient changes during soybean imbibition, *Plant Physiol.,* 59, 1111—1115, 1977.
8. **Luzzati, V. and Husson, F.,** The structure of the liquid-crystalline phases of lipid-water systems, *J. Cell Biol.,* 12, 207—219, 1962.
9. **Finean, J. B.,** Biophysical contributions to membrane structure, *Q. Rev. Biophys.,* 2, 1—23, 1969.
10. **Villiers, T. A.,** Seed ageing: chromosome stability and extended viability of seeds stored fully imbibed, *Plant Physiol.,* 53, 875—878, 1974.

11. **Roberts, E. H.,** *Viability of Seeds,* Chapman and Hall, London, 1972.
12. **Berjak, P. and Villiers, T. A.,** Ageing in plant embryos. III. Acceleration of senescence following artificial ageing treatment, *New Phytol.,* 71, 513—518, 1972.
13. **Hallam, N. D., Roberts, B. E., and Osborne, D. J.,** Embryogenesis and germination in rye *(Secale cereale* L). III. Fine structure and biochemistry of the non-viable embryo, *Planta (Berlin),* 110, 279—290, 1973.
14. **Villiers, T. A.,** Cytological studies in dormancy. II. Pathological ageing changes during prolonged dormancy and recovery upon dormancy release, *New Phytol.,* 71, 145—152, 1972.
15. **Abu-Shakra, S. S. and Ching, T. M.,** Mitochondrial activity in germinating new and old soybean seeds, *Crop Sci.,* 7, 115—118, 1967.
16. **Throneberry, G. O. and Smith, F. G.,** Relation of respiratory and enzymatic activity to corn seed viability, *Plant Physiol.,* 30, 337—343, 1955.
17. **Simon, E. W. and Harun, R. M. R.,** Leakage during seed imbibition, *J. Exp. Bot.,* 23, 1076—1085, 1972.
18. **Abdul-Baki, A. A. and Anderson, J. D.,** Viability and the leaching of sugars from germinating barley, *Crop Sci.,* 10, 31—34, 1970.
19. **Villiers, T. A.,** Ageing and the longevity of seeds in field conditions, in *Seed Ecology,* Heydecker, W., Ed., Butterworth, London, 1973, chap. 15.
20. **Anderson, J. D., Baker, J. E., and Worthington, E. K.,** Ultrastructural changes of embryos in wheat infected with storage fungi, *Plant Physiol.,* 46, 857—859, 1970.
21. **Butler, R. D. and Simon, E. W.,** Ultrastructural aspects of senescence in plants, in *Advances in Gerontological Research,* Vol. 3, Strehler, B. L., Ed., Academic Press, New York, 1971, 73—129.
22. **Rost, T. L.,** The ultrastructure and physiology of protein bodies and lipids from hydrated dormant and non-dormant embryos of *Setaria lutescens* (Graminae), *Am. J. Bot.,* 59, 607—616, 1972.
23. **Villiers, T. A.,** Lysosomal activities of the vacuole in damaged and recovering plant cells, *Nature (London),* 233, 57—58, 1971.
24. **Van Staden, J., Gilliland, M. G., and Brown, N. A. C.,** Ultrastructure of dry viable and non-viable *Protea compacta* embryos, *Z. Pflanzenphysiol.,* 76, 28—35, 1975.
25. **Cheah, K. S. E. and Osborne, D. J.,** DNA lesions occur with loss of viability in embryos of ageing rye seeds, *Nature (London),* 272, 593—598, 1978.
26. **Ory, R. L., Yatsu, L. Y., and Kircher, H. W.,** Association of lipase activity with the sphaerosomes of *Ricinus communis, Arch. Biochem. Biophys.,* 123, 255—264, 1968.
27. **Villiers, T. A.,** Genetic maintenance of seeds in imbibed storage, in *Crop Genetic Resources for Today and Tomorrow,* Frankel, O. H. and Hawkes, J. G., Eds., University Press, Cambridge, 1975, chap. 23.
28. **Chen, S. S. C. and Varner, J. E.,** Respiration and protein synthesis in dormant and after-ripened seeds of *Avena fatua, Plant Physiol.,* 46, 108—112, 1970.
29. **Villiers, T. A.,** Cytological studies in dormancy. I. Embryo maturation during dormancy in *Fraxinus excelsior, New Phytol.,* 70, 751—760, 1971.
30. **Stokes, P.,** A physiological study of embryo development in *Heracleum sphondylium.* I. Effect of temperature on embryo development, *Ann. Bot. (London),* 16, 441—445, 1952.
31. **Villiers, T. A. and Wareing, P. F.,** Dormancy in fruits of *Fraxinus excelsior, J. Exp. Bot.,* 15, 359—367, 1964.
32. **Villiers, T. A.,** Cytological studies in dormancy. III. Changes during low-temperature dormancy release, *New Phytol.,* 71, 153—160, 1972.
33. **Hyde, B. B.,** Changes in nucleolar ultrastructure associated with differentiation in the root tip, *J. Ultrastruct. Res.,* 18, 25—29, 1967.
34. **Villiers, T. A.,** Light-controlled dormancy in *Fraxinus americana, Proc. Electron Microsc. Soc. Am.,* 4, 37—38, 1974.
35. **Curtis, H. J.,** A composite theory of aging, *Gerontologist,* 6, 143—149, 1966.
36. **Wattiaux, R.,** Biochemistry and function of lysosomes, in *Handbook of Molecular Cytology,* Lima-de-Faria, A., Ed., North-Holland, Amsterdam, 1969, 1159—1178.
37. **Shaw M. and Manocha, M. S.,** Fine structure in detached, senescing wheat leaves, *Can. J. Bot.,* 43, 747—755, 1965.
38. **Nir, I., Poljakoff-Mayber, A., and Klein, S.,** The effect of water stress on mitochondria of root cells, *Plant Physiol.,* 45, 173—177, 1970.
39. **Koostra, P. T. and Harrington, J. F.,** Biochemical effects of age on membrane lipids of *Cucumis sativus* L. seed, *Proc. Int. Seed Test. Assoc.,* 34, 329—340, 1969.
40. **Ferguson, C. H. R. and Simon, E. W.,** Membrane lipids in senescing green tissues, *J. Exp. Bot.,* 24, 307—316, 1973.
41. **Niki, T., Yoshida, S., and Sakai, A.,** Studies on chilling injury in plant cells. I. Ultrastructural changes associated with chilling injury in callus tissues of *Cornus stolonifera, Plant Cell Physiol.,* 19, 139—148, 1978.

42. **Esau, K.,** Changes in the nucleus and the endoplasmic reticulum during differentiation of a sieve element in *Mimosa pudica* L., *Ann. Bot. (London)*, 36, 703—710, 1972.
43. **Platt-Alloia, K. A. and Thomson, W. W.,** An ultrastructural study of two forms of chilling-induced injury to the rind of grapefruit *(Citrus paradisi* Macfed.), *Cryobiology*, 13, 95—106, 1976.
44. **Barton, R.,** Fine structure of mesophyll cells in senescing leaves of *Phaseolus, Planta (Berlin)*, 71, 314—325, 1966.
45. **Martin, C. and Thimann, K. V.,** The role of protein synthesis in the senescence of leaves. I. The formation of protease, *Plant Physiol.*, 49, 64—71, 1972.
46. **Thimann, K. V., Tetley, R. M., and Thanh, T. V.,** The metabolism of oat leaves during senescence. II. Senescence in leaves attached to the plant, *Plant Physiol.*, 54, 859—862, 1974.
47. **McKersie, B. D., and Thompson, J. E.,** Cytoplasmic membrane senescence in bean cotyledons, *Phytochemistry*, 14, 1485—1491, 1975.
48. **Tappel, A. L.,** Will antioxidant nutrients slow aging processes?, *Geriatrics*, 23, 97—105, 1968.
49. **Harman, D.,** The biologic clock: the mitochondria?, *J. Am. Geriatr. Soc.*, 20, 145—147, 1972.
50. **Varner, J. E.,** Biochemistry of senescence, *Annu. Rev. Plant Physiol.*, 12, 245—264, 1961.
51. **Haber, A. H. and Luippold, H. J.,** Separation of mechanisms initiating cell division and cell expansion in lettuce seed germination, *Plant Physiol.*, 35, 168—173, 1960.

Chapter 4

SENESCENCE IN FUNGI

Karl Esser and Paul Tudzynski

TABLE OF CONTENTS

I. INTRODUCTION

Senescence or aging is a basic biological phenomenon that concerns all living beings. Throughout the different levels of biological organization, senescence expresses itself in different phenotypes, but in each case, via irreversible alterations of the metabolism, the *goal* of senescence is attained, namely *cellular death.*

Not only the manifestation, but also the symptoms of aging are very different in different organisms. They range from an inability to propagate, (i.e., inhibition of cell division in either the vegetative or the sexual phase) to a gradual decrease of cellular functions in multicellular organisms. This versatile expression of senescence reveals that the phenomenon is a *syndrome composed of symptoms* that differ to a great extent from organism to organism, as may be seen from the following examples.

1. In *mammals,* the in vivo diagnosis of senescence is rather difficult. It seems to be caused by degenerative processes of the nervous system and the immune system.[1] However, as shown by Hayflick,[2,3] cultures of mammalian cells have only a restricted life span. Depending on their origins, after 40 to 50 divisions the growth rate is retarded in these cells and eventually, after a stoppage of mitotic divisions, they die.[4] Only so-called transformed cells (such as cancer cells) retain a more or less indefinite capacity for cell division. Transformation may be induced either by specific viruses or by treatment with carcinogenic substances.[5]
2. In *lower animals,* aging shows up either as a diminution of the mitotic division rate, (e.g., amoebae[6]; ciliata[7]) or as a general decrease of the metabolism in tissues or organs of multicellular organisms like nematodes[8] or insects.[9]
3. In *higher plants* senescence concerns either the whole plant or parts of the plant. Its onset may be influenced by plant hormones, (e.g., cytokinins or abscisic acid[10]).
4. Senescence in *filamentous fungi,* which will be the topic of this review, is characterized by the following symptoms: decrease of growth rate due to diminution of mitotic divisions, morphological alteration of the hyphae, eventually stoppage of growth, and finally cellular death.[11]

Many of these organisms have been subjected to physiological and sometimes genetical analysis.[10,12,13] There is at present no generally accepted interpretation or explanation of the senescence syndrome, but several hypotheses, based on data from single organisms, have been put forward. However, their general value is debatable.[14] In principle these conceptions comprise *two opposing theories:*

1. *Senescence* is understood as an *"accident",* caused by an accumulation of stochastic errors occurring during the whole life span of an organism at different metabolic levels like DNA-replication, translation, or membrane synthesis.[15,16]
2. *Senescence* is understood as a last and *final step* of cell or tissue *differentiation* coded, like all the previous steps, by the genetic information.[17,18,19]

Up to now, in spite of many experimental efforts, a decision between these two opposing theories has not been possible. This failure may be partially due to difficulties in the handling of animal cells in culture, to the complexity of symptoms in these organisms, and to their relative inaccessibility for genetic analysis. *Fungi* do not have most of these disadvantages and, because of their ease in handling under laboratory conditions and suitability for genetic analysis, they are ideal organisms to be used as models for the study of senescence on both the physiological and the genetical levels.

II. SURVEY OF SENESCENCE IN FUNGI

After the first description, the syndrome has been analyzed very thoroughly by Rizet and collaborator Marcou. Their data[20,24-31] are summarized in a published version of Marcou's thesis,[32] which will be frequently referred to. After short contributions by Holliday[33] and Smith and Rubenstein[34,35] we have in our laboratory started several years ago to work on the senescence syndrome of *Podospora*. For the reader's convenience we shall not present the data in chronological order.

III. THE *PODOSPORA* CASE

The life cycle of *Podospora anserina* (family: Sordariaceae) is similar to that of *Neurospora crassa,* the well-known object of genetics, but there are general differences:

1. *Podospora* exhibits colonial growth and the mycelia become green-black — due to the action of phenoloxidases, which catalyze an abundant formation of melanin pigments (Figures 1c, d). Its growth rate of 7 mm/day under optimal conditions (26°C) is about 8% of that of *Neurospora*. However, for the completion of the sexual cycle only 10 to 12 days are needed; *N. crassa* takes about 4 weeks, because the ascospores undergo a dormant phase.
2. The mycelia produce no macroconidia, and the microconidia do not germinate under normal conditions, but act as male gametes and are therefore called spermatia.
3. The asci contain two binucleate spores, due to a peculiar mechanism of nuclear distribution and spore formation during ascus development. With specific experimental methods they can as well be submitted to tetrad analysis as the ascospores of *Neurospora*. (For details concerning life cycle and genetics of *P. anserina,* see References 11 and 21).

In *Podospora anserina,* many wild strains have been isolated from nature (manure of herbivores) in various regions in France and Germany.[22,23] After continuous vegetative propagation* all strains become more or less rapidly senescent, i.e., the growth rate decreases rapidly (see also Figure 4), the marginal hyphae show abnormal branching and apical swellings (Figure 1 — compare a with b); the formation of aerial hyphae is reduced and the melanin formation increases (compare Figure 1c with d); eventually the mycelial growth stops and hyphal death occurs. When grown in race tubes under controlled conditions (25°C), it is easy to determine the growth rates, the length of mycelium, and the times for the onset and end of senescence (Figure 2).

After the first description, the syndrome has been analyzed very thoroughly by Rizet and collaborator Marcou. Their data[20,24-31] are summarized in a published version of Marcou's thesis,[32] which will be frequently referred to. After short contributions by Holliday[33] and Smith and Rubenstein[34,35] we have in our laboratory started several years ago to work on the senescence syndrome of *Podospora*. For the reader's convenience we shall not present the data in chronological order.

A. Environmental Control of Senescence

The onset of senescence depends to some degree on environmental conditions: permanent illumination, high temperatures (above 30°C), and frequent transplantation of small mycelial inocula shorten the life span considerably.

* The *Podospora* mycelia are grown on complete medium (corn meal agar[21]).

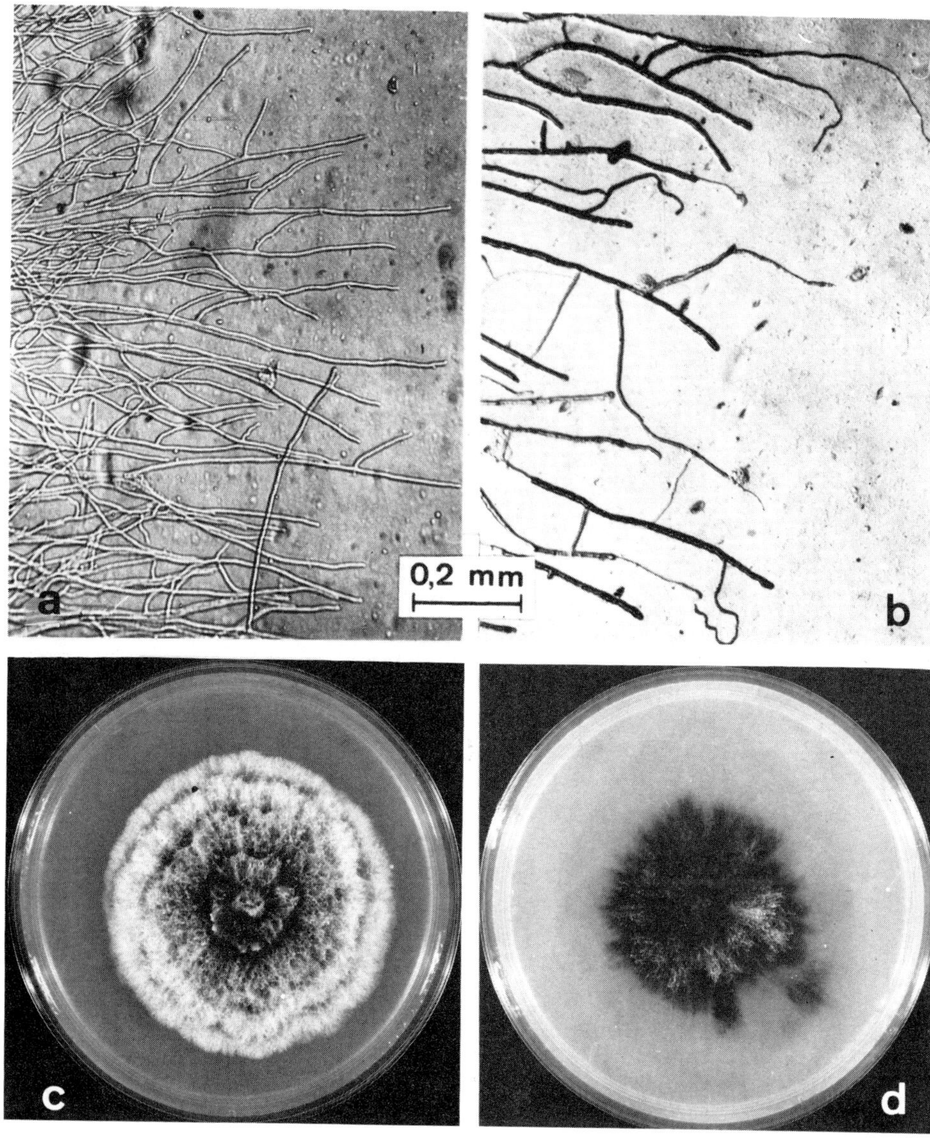

FIGURE 1. Juvenile and senescent mycelia of *Podospora anserina*. (a) Juvenile hyphal tips; (b) senescent hyphal tips; (c) juvenile colony; (d) senescent colony. (From Esser, K. and Tudzynski, P., *Nature (London)*, 265, 454-456, 1977. With permission.)

If the mycelial metabolism is drastically reduced (storage at 3°C or under paraffin oil or liquid nitrogen), the onset of senescence may be prevented for years. In addition, senescent or partially senescent mycelia may be regenerated under these conditions. These phenomena have been given the explanation that under these "hard" metabolic conditions the non-senescent parts of the mycelia have a selective advantage, because the senescent parts die off.[32]

B. Genetic Control of Senescence

Under constant environmental conditions it becomes evident that the onset of senescence is genetically controlled by chromosomal and also by extrachromosomal genetic information.

FIGURE 2. *Podospora anserina,* mycelia grown in race tubes. Above: mycelium still juvenile (age 7 days); below: mycelium shortly after onset of senescence (age 25 days). Note the difference between the marginal edges and compare with Figure 1. Further details in text.

1. Chromosomal Factors

Already the first quantitative analysis of the life span of various geographic races showed that the onset of senescence varied to a great extent (Table 2). Since the specific breeding system of pseudocompatibility in *Podospora* causes geographic races to exhibit great diversity in their genetic information,[36] it was not surprising that these differences in aging could be attributed to a heterogeneity of chromosomal genes, especially since even single genes like the mating type alleles (+ and −) and the gene *incoloris* (colorless mycelium, reduction in fruit body formation) were found to influence senescence.[32]

Recent investigations have led to a more comprehensive understanding of the correlation between genotype and onset of senescence.[37,38] It was found that some genes — which have a pleiotropic effect on the morphology (mycelial growth and color, shape of hyphae, fruit body formation, etc.) when combined in double mutants exhibit a synergistic action in drastically postponing the onset of senescence. In contrast, when grown as single mutants, only a rather small effect is observed (Figure 3). As may be also seen from Figure 3 this *prevention of senescence by synergistic action of pairs of morphogenetic genes* is not correlated with the position of these genes within the genome, because some of the effective gene pairs are linked, others not.

As an example, the growth curve for the double mutant *incoloris-vivax* is compared in Figure 4 with those of the single mutants and the wild strain. It may be seen that the wild strain decreases after about 20 cm of growth, i.e., after 25 days. The mutants grow somewhat longer (see also Figure 3), whereas the double mutant is still growing after 1400 days, by which time it has grown over a distance of 9 m. The physiological action of these life-prolonging genes is not yet understood.

2. Extrachromosomal Factors

From the beginning of the studies on aging, it was observed that the syndrome of senescence, once it has appeared, is *maternally inherited.* This was proved because senescent mycelia do not lose the capacity for sexual propagation. As may be seen from Figure 5, the test system is rather simple: if a homokaryotic juvenile mycelium is fertilized with the spermatia of a senescent strain of opposite mating type, all fruit bodies contain only spores giving rise to juvenile mycelia. (In this connection it should be mentioned that spermatia do not contain much cytoplasm, but only one large nucleus.) In contrast, in the reciprocal combination (♀ senescent × ♂ juvenile), there are

Linkage Groups	II	III				IV	V	VI	VII	
Genes	z	viv	n	i	ir	gr	la	ta	cr	ca
z	21±2	—	—	—	—	—	—	—	38±5	—
viv		66±8	**>400**	**>1400**	**>490**	**>510**	**>150**	—	**>225**	**>570**
n			29±1	**>930**	18±1	—	—	—	**>850**	37±3
i				42±2	**>790**	—	—	—	**147±8**	47±4
ir					25±1	26±1	—	—	52±6	31±3
gr						39±1	—	55±7	**301±42**	—
la							33±1	—	67±6	—
ta								66±3	58±8	—
cr									44±2	**438±59**
ca										39±3

FIGURE 3. Effect of pleiotropic morphogenetic genes on the life span of *Podospora anserina*. The lifetime of single mutants (diagonal) and double mutants is given in days with standard error. The combinations exhibiting a life-prolonging effect are shaded. The double mutants characterized by > were still growing when the manuscript was sent to the editor. The combinations marked — have not yet been tested. Ten of the double mutants were produced and tested by Frau I. Grubbert.

two types of perithecia, which contain exclusively either "senescent" or "juvenile" spores. The frequency of the two types of fruit bodies depends on the stage of aging and the genotype. It varies between 10 and 80%. Since there are no mendelian segregations in single asci, the conclusion is justified that *senescent strains contain in their cytoplasm genetic elements that cause aging when transferred to a juvenile cell.*[32] It was also possible to show that the senescent elements can be transferred vegetatively. After fusions between senescent and juvenile hyphae, the latter become rapidly senescent. Furthermore, one could follow with the aid of microsurgery the migration of the senescent elements within the juvenile hypha (nuclear transfer and migration was excluded). Therefore the *extrachromosomal genetic elements* must be *identical with an infective principle causing senescence when present in the cytoplasm.*[32]

Many fruitless efforts have been made to identify the infective principle. Neither mycoviruses nor RNA were found to be responsible for aging, as shown in our laboratory.[39] The same holds true for mycoplasmas.[40] However, studies of the physiological control of senescence, which will be described in the next section, give at least circumstantial evidence for the nature of the causative agent of senescence.

C. Physiological Control of Senescence

The present knowledge of the physiological basis of senescence depends mainly on the analysis of the action of metabolic inhibitors.

In preliminary experiments, it was shown by Holliday[33] that the life span was reduced when the mycelia were grown on media containing analogues of amino acids. His statement that these data confirm the "accident-hypothesis" seems somewhat premature, because an untimely death does not necessarily require a specific interference with normal aging.

On the contrary, the results of experiments that lead to a *postponement of the point of senescence* are much easier to understand and to explain. It was found that the addition to the medium of inhibitors of mitochondrial metabolism in low concentra-

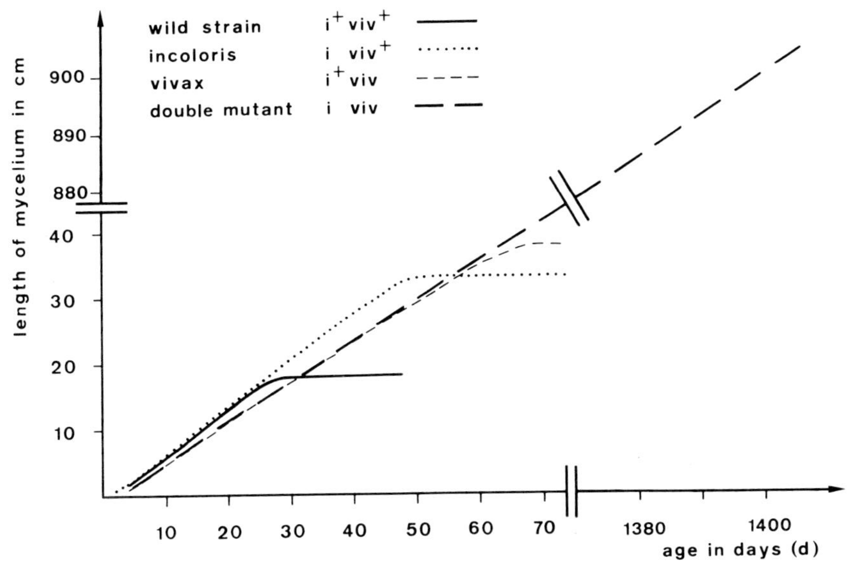

FIGURE 4. Growth curves of the wild strain, the morphogenetic mutants (*vivax* and *incoloris* and its double mutant. (Adapted from Esser, K. and Keller, W., *Mol. Gen. Genet.*, 144, 107—110, 1976. With permission.)

tions kept juvenile mycelia alive up to 200 days,[38,41,42] even though they had no great effect on the growth rate. Especially effective are intercalating *DNA dyes with a specific affinity for m+-DNA* and *"ribosomal inhibitors"*, which interfere with the mitochondrial protein synthesis on the level of translation (Table 3).

A more detailed analysis of these observations reveals that the life-prolonging effect of these drugs depends on the age of the mycelia. As may be seen from Figure 6, there is a point of no return, i.e., mycelia fail to react when grown too long on unsupplemented medium. This point is different for the two types of inhibitors: 10 days before the onset of senescence for the ribosomal inhibitors, but simultaneous with the macroscopically visible onset of senescence for the DNA inhibitors.

This means that there are two thresholds in the life span of *Podospora*. The first is a "physiological threshold", characterizing the end of the juvenile phase and the *beginning of a presenescent phase,* because in this phase there must have occurred some physiological alteration that can no longer be stopped or postponed by ribosomal inhibitors. The *second threshold* also characterizes a physiological point of no return, but at the same time reveals the senescent *morphological phenotype.* In this connection it is interesting to note that Marcou[32] had already postulated a presenescent phase ("phase d'incubation") from her growth experiments.

In addition it should be stressed that *no permanent recovery from senescence* may be obtained *by the action of these inhibitors.* In each case, after transfer from supplemented medium to unsupplemented medium, the mycelia die more or less rapidly. Only a treatment with ethidium-bromide may cause a temporary rejuvenation of presenescent mycelia.[38] A temporary rejuvenation was also observed by Smith and Rubenstein[35] after mycelial growth on cycloheximide or cyanide. However, this effect was correlated with a drastic reduction of the growth rate, and therefore may be caused by reduction of general metabolism, as described above, resulting from alteration of the environmental conditions (see Section III.A.).

These physiological findings lead to the conclusion that the *mitochondria and especially the mt-DNA and the mitochondrial protein synthesis are somehow involved in*

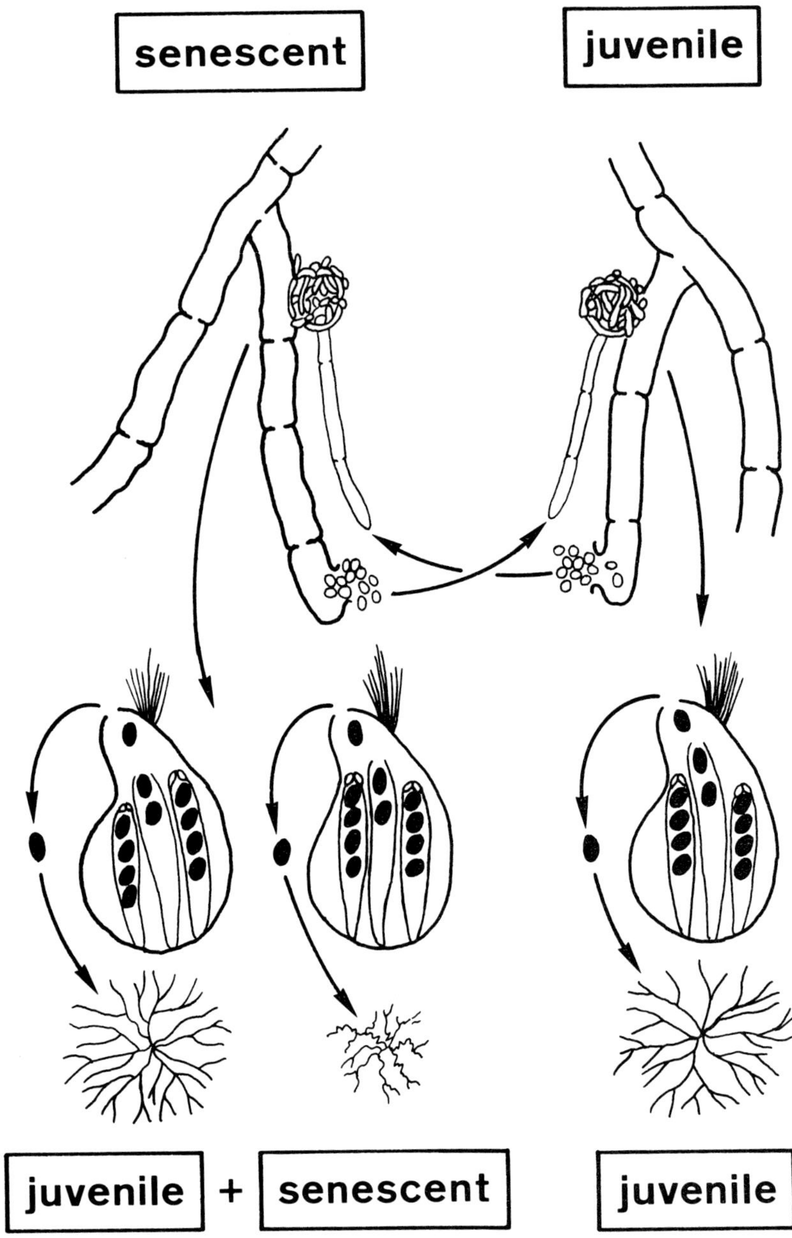

FIGURE 5. Scheme for testing the maternal inheritance of senescence in *Podospora anserina*. Two mycelia of different mating types are reciprocally crossed by transfer of spermatia. The offspring shows no Mendelian segregation, but a reciprocal difference. For further explanation see text.

the propagation and expression of the infective principle causing the onset of senescence. Before any hypotheses are offered, however, another experimental track will be delineated.

D. Nature of the Infective Principle

Mitochondria fulfill all criteria for an infective principle. They contain their own specific genetic information; they are capable of autoduplication; they are motile in

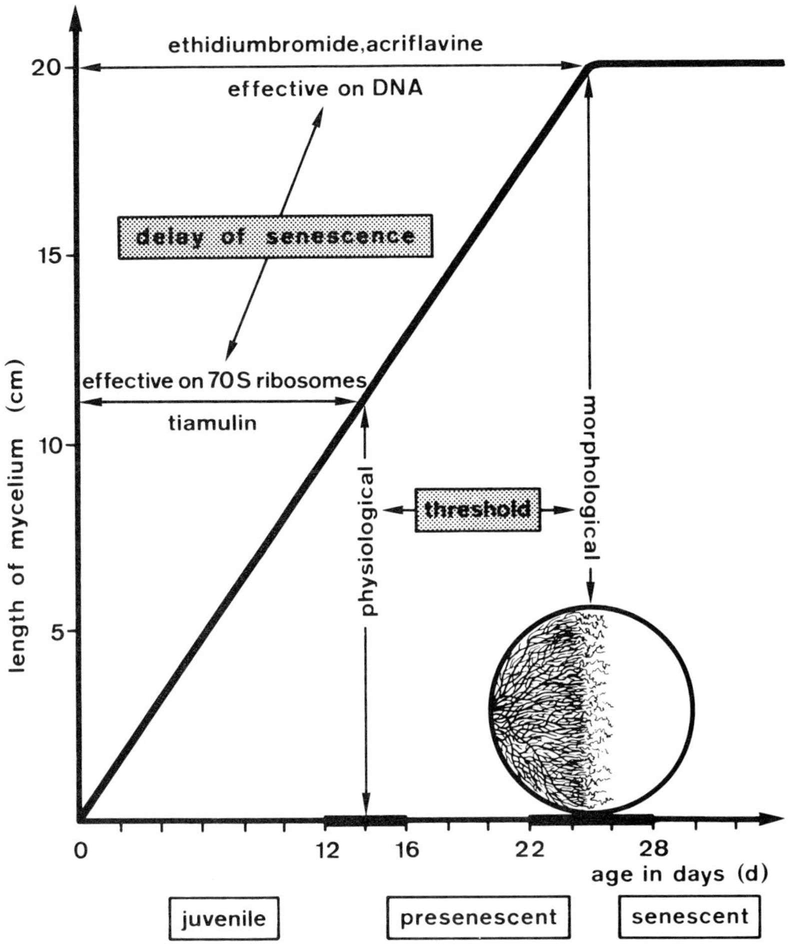

FIGURE 6. Growth curve for *Podospora anserina* mycelium indicating threshold periods for the effects of two classes of mitochondrial inhibitors. (From Esser, K. and Tudzynski, P., *Viruses and Plasmids in Fungi,* in press. With permission.)

the cytoplasm and may migrate via hyphal anastomoses (without nuclear exchange) into other hyphae; and last (but not least) they have an essential influence on the general metabolism of the cell. However, it seems improbable that the mitochondria themselves are identical with the infective principle, because Marcou and Raynal[44] found that there was no correlation between the transfer of senescence and mitochondrial marker genes. Also experiments to transform protoplasts from juvenile mycelia by infection with mitochondria from senescent mycelia gave no unequivocal results.[38] However, similar experiments in which the bulk of total DNA from senescent mycelia was used showed up to 20% transformation of juvenile protoplasts, which regenerated to senescent mycelia.[38]

A tentative conclusion from the latter experiment is that extrachromosomal DNA, connected somehow with the mitochondria, is the infective principle, rather than whole mitochondria. This working hypothesis is also strengthened by the fact that the dyes intercalating with DNA are effective in postponing senescence — even in the presenescent phase, where the ribosomal inhibitors fail. Another result may be interpreted along these lines: there is plasmid-like DNA present in the senescent mycelia.[45] This extrachromosomal DNA could be identified also by biophysical methods, as seen in the electron microscope (Figure 7).

At the present stage of research, it has not yet become evident whether these peculiar DNA molecules are brought about by cell degradation resulting from the aging procedure, or whether they are causative for aging and thus identical with the infective principle.

In summing up the results concerning the phenomenon of aging in *Podospora*, one may say: in this fungus senescence depends on the *presence of an infective principle* in the cytoplasm probably identical with extrachromosomal DNA. The expression of this principle requires a *specific configuration of nuclear genes*. It became further evident that senescence is not like a disease such as bacterial infection, which can be cured, but an *inevitable procedure*, which can *only be delayed by alteration of the environmental conditions*, unless there are drastic alterations of chromosomal genes.

IV. DATA ON SENESCENCE IN OTHER FUNGI

According to their phenotype the senescence syndromes in fungi fall into three groups (see also Table 1):

1. Senescence occurs in fully active hyphae spontaneously and more or less regularly.
2. Senescence occurs only after induction (unbalanced growth).
3. Symptoms of aging occur only in old parts of the mycelia, dependent on environmental conditions.

We have dealt with *Podospora anserina* in detail as a representative of the *first group*. The other fungi listed in this category exhibit a similar phenotype: retardation of hyphal growth and cellular death or "vegetative death" as it was called in *Aspergillus glaucus*. In this fungus the life span is rather indeterminate, it varies in isogenic cultures, inoculated in parallel, between 5 months and 2 years. However, "vegetative death" is infectious and can be transferred via hyphal anastomosis; it is also suppressive in heterokaryons between juvenile and senescent mycelia. Unfortunately, senescent mycelia do not retain the capacity for sexual propagation and thus do not allow a more detailed genetic analysis.

The mutant "natural death" in *Neurospora crassa* seems to have a similar phenotype as "vegetative death".

A comparable system was described in Caten for *Aspergillus amstelodami* (a member of the *A. glaucus* group). First called "vegetative death", it was later renamed "ragged", because the symptoms and occurrence of aging are similar to those in *A. glaucus*, but the "ragged" mycelia do not die after having obtained a senescent phenotype. Physiological data have produced hints that "ragged" might be caused by a mitochondrial mutation, because this strain has an altered spectrum of cytochromes.

The aging observed in the diploid plasmodial stage of the slime mould *Didymium iridis* depends, as in *Podospora*, on the genotype; an additive polygenic nuclear system determines the life span. Heterokaryon experiments indicate that in this organism senescence is caused either by an extrachromosomal element or is under positive nuclear control.

For some other fungi listed in group 1, like *Podospora setosa*, *Ascobolus stercorarius*, *Pestalozzia annulata*, and *Helminthosporium victoriae*, it was also shown that senescence can be transferred by an infective principle. However, there are no detailed studies to offer further evidence on this syndrome. But there are two more cases listed in Group 1 in which infectivity of senescence has not (or not yet?) been found:

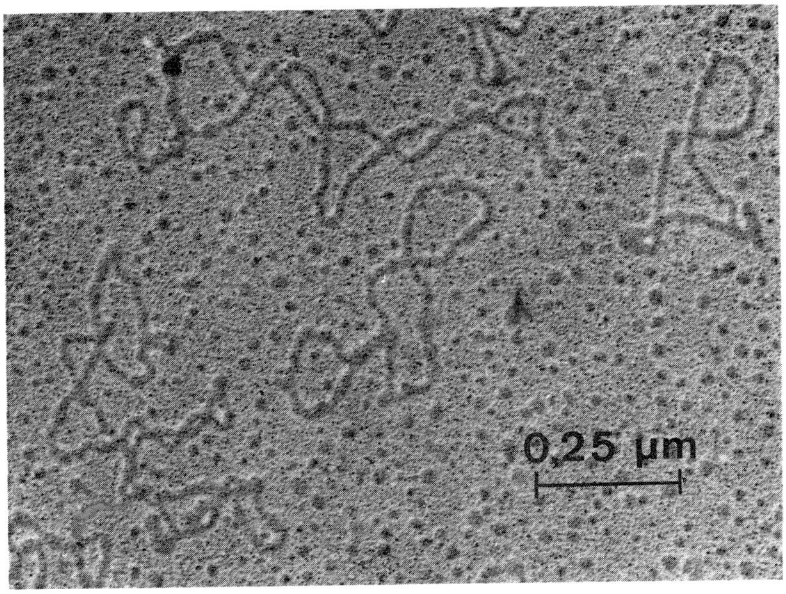

FIGURE 7. Plasmid-like DNA isolated from senescent mycelia from *Podospora anserina*.

1. In *Neurospora crassa,* the mutant "stopper" also exhibits symptoms of aging like drastic reduction of growth, but after some delay hyphal growth starts again. This mutant never dies; in contrast, it shows a kind of rhythmic growth expressed as a circular zonation comparable to the mutant *zonata* or *Podospora anserina.*[71]
2. In the slime mould *Physarum polycephalum,* after several plasmodial transfers, the cultures decrease in growth rate, become necrotic and die. The life span depends both on the genotype of the isolate (between some weeks and some months) and also on environmental conditions (temperature and light).

The symptoms of senescence included in group 2 of Table 1 are very little understood. They have in common that aging and cellular death only occur when induced by environmental conditions like nutrition or temperature. These symptoms are sometimes referred to as "unbalanced growth".

In *Neurospora crassa,* senescence is brought about by feeding with amino acid analogues. Some auxotrophic mutants of *Neurospora,* yeasts and *Aspergillus* die rather rapidly after transfer to minimal medium, whereas on supplemented medium aging has not been observed. Quite recently it was shown by Shimada et al.[72] that the death of a mutant of *Saccharomyces cerevisiae,* auxotrophic for biotin, is prevented by addition of a DNA inhibitor (hydroxyurea). This shows that evidently a synthesis of DNA is required during the procedure of this type of aging. It is hard to interpret this type of aging, which in its manifestation suggests the phenomenon of leakiness, found in most fungi during the screening for auxotrophs, and also in leaves (Chap. 5, Sec. 2E). The stoppage of growth of *Rhizoctonia* and *Sclerotium* in liquid medium has been very thoroughly analyzed *(group 3).* Since aging in these fungi never occurs when grown on a solid medium, since only the central parts of the mycelia (in liquid cultures) show aging symptoms, and since the peripheral hyphae, in spite of having stopped growth, never die, it seems inappropriate to integrate these phenomena into the concept of senescence or aging in the sense defined and discussed in this paper. The diver-

TABLE 1

Survey of senescence syndromes found in fungi[a]

Organism	Symptoms	Ref.
1. Spontaneous		
Podospora anserina		20
Podospora setosa		24
Aspergillus glaucus		46, 47
Aspergillus amstelodami	Stoppage of growth, death of hyphal tips, cytoplasmically inherited, infectious	48, 49
Ascobolus stercorarius		32
Pestalozzia annulata		50
Helminthosporium Victoriae		51
Neurospora crassa		52
mutant "natural death"		
mutant "stopper"	Reversible death of the marginal hyphae	
Physarum polycephalum	Death of plasmodium, genetically controlled	54
Didymium iridis		55
2. Induced		
Neurospora crassa		
wild strain	Amino acid analogues 33°C	56
leucin mutant		57
temperature sensitive mutant	Minimal medium 37°C	58
inositol mutant		59
Kloeckera apiculata		
inositol mutant		
Saccharomyces carlsbergensis		
inositol mutant	Stoppage of growth after a short linear phase of growth, Cellular death Minimal medium	60
S. cerevisiae		
biotin mutant		61
fatty acid synthetase mutant		62
pantothenate mutant		63
Aspergillus nidulans		
biotin mutant		64
3. Others		
Sclerotium bataticola	Stoppage of growth in liquid medium, No true symptoms of senescence, Aging only in central parts of mycelium.	65
Rhizoctonia solani		66
		67
		68
		69
Phytophtora palmivora	Degeneration of sporangia	70

[a] For further explanations see text.

gence of some physiological parameters (DNA content, enzyme activities) are not necessarily the expression of senescence. They can just as well be the manifestation of different developmental stages, comparable, for instance, with resting cells.

TABLE 2

Correlation between onset of senescence and genotype in various strains of *Podospora anserina*[a]

Strain		Onset of senescence in days
A	Geographical races	10
B		27
C		15
D		32
E		106
F		9
G		60
H		26
N		33
S		43
Mutant incoloris (race S)		Growth longer than 60 days (compared to 43 days for wild strain)
+/− Mating alleles		Show in some races variation of life span

[a] Data compiled from Marcou, D., *Ann. Sci. Nat. Bot. Biol. Veg̃.*, 12, 653—764, 1961.

V. CONCLUSIONS

A retrospective on the various phenomena that are found in fungi and classified under the heading of aging or senescence allows the following statements:

Senescence or aging manifests itself always as a syndrome consisting of a series of symptoms of rather great versatility. This explains the difficulties in interpretation of these syndromes, as becomes evident not only in the previous paragraph, but also in the other chapters of this book. From this it follows that one must be aware that senescence has to be accepted as a concept, which comprises a number of different mechanisms that all have one thing in common — they lead finally to cellular death. This concept shows a remarkable similarity to that of cancer, where (somewhat inversely) a proliferation of cells is induced by various mechanisms.

This leads to the question — to what extent the data obtained with fungi may contribute to a better understanding of aging in general. Before we look into this matter, it must be remembered that most of the fungal data (except the *Podospora* case) represent more a description of aging than an analysis; thus any conclusions drawn are charged with much speculation. Since this is equally true for the two opposing theories on senescence (which were mentioned in the introduction) the following remarks have to be evaluated under this perspective.

1. Are the data in fungi relevant for an understanding of aging in cells of higher organisms, e.g., mammalian cells? In other words: can we pool the data and apply conclusions from experimental results obtained with fungi to data obtained with cultures of animal cells, and vice versa? The answer is rather positive because there seems to be some similarity between senescence in *Podospora* and in fibroblast cultures: (a) in both, the capacity for cell division is limited and is under genetic control,[73] (b) also in fibroblasts the life span consists of three phases, the last of which is senescence.[2,74] (c) the participation of extrachromosomal genetic information has been proposed for both: a plasmid-like DNA in *Podospora*,[45] and viruses in fibroblasts,[74-76] (d) an accumulation of a pigment (lipofuscin) was found in aging fibroblasts. The same pigment seems also to be present in aging hyphae of *Podospora.*[77]

Thus it should be quite interesting to coordinate future research between *Podospora*

TABLE 3

Examples of the action of some inhibitors of the mitochondrial metabolism when added to the growth medium, on the onset of senescence in *Podospora*

Inhibitor	Site of action	Effective concentration (Mol/ℓ)	Growth rate (mm/day)	Length of linear growth (days)	Mycelial morphology	Remarks
Control	—	—	7.5 ± 0.2	23 ± 1	Normal	
Ethidiumbromide	DNA (mitochondrial or similar)	1.0×10^{-4}	7.5 ± 0.1	130	Normal	Lag phase 3-5 days
Acriflavin	DNA (mitochondrial or similar)	1.0×10^{-4}	7.0 ± 0.2	85 ± 10	Normal	Lag phase 4-5 days
Kanamycin	Prokaryotic ribosome, 30S	3.0×10^{-3}	7.2 ± 0.2	66 ± 9	Normal	—
Streptomycin	Prokaryotic ribosome, 30S	$1\text{-}3 \times 10^{-3}$	7.0 ± 0.2	40-100[a]	—	
Tiamulin[b]	Prokaryotic ribosome, 50S	3.3×10^{-4}	5.2 ± 0.1	286 ± 26	Normal	—
Puromycin	Prokaryotic ribosome, 50S	1.0×10^{-3}	5.8 ± 0.1	39 ± 1	Normal	—

[a] Effect depends greatly on the quality of the inhibitor charge.
[b] (2-diethyl-aminoethyl)-thioacetoxy-mutilin-hydrogen fumarate[79]

Adapted from Esser, K. and Tudzynski, P., Viruses and Plasmids in Fungi, in press. With permission.

and fibroblasts. Experimenters should consider the technical advantages of both organisms for the one or the other kind of experiment, e.g. the advantage of *Podospora* for all kinds of genetic experiments.

2. Do the data obtained with fungi support one or the other of the two opposing models of aging, the accumulation-of-error-theory, or the differentiation-theory? The data obtained on senescence syndromes that are induced (Table 1, group 2) evidently support the error-theory. On the other hand, the observations made on those senescence syndromes that occur spontaneously (Table 1 group 1), especially in *Podospora* and the slime moulds, point to a genetic control of aging and thereby seem to support the differentiation-theory. These two modes of interpretation are not necessarily contradictive, because one may presume that when senescence occurs spontaneously the induction of aging could be the last genetically controlled step in the long sequence of differentiation, comparable to a self-destroying mechanism. It should also then be possible, in organisms having a longer life span than *Podospora,* to cause this induction experimentally. The results of this kind of induction are absolutely similar in phenotype to those of an "error catastrophe". It is therefore very well possible that further experiments should lead the two theories to become amalgamated into one.

Thus it is almost trivial to say that the aging syndromes in fungi need a more thorough analysis, but certainly this will then help to understand the most important syndrome of senescence — the aging of mammalian cells, a phenomenon that concerns more or less all of us.

REFERENCES

1. **Curtis, H. J.,** The nature of the ageing process, in *The Biological Basis of Medicine,* Vol. 1, Bitter, E. E. and Bitter, N., Eds., Academic Press, London, 1968, 521—542.
2. **Hayflick, L.,** The establishment of a line of human amnion cells in continuous cultivation, *Exp. Cell Res.,* 23, 14—25, 1961.
3. **Hayflick, L.,** The limited in vitro life time of human diploid cell strains, *Exp. Cell Res.,* 37, 614—620, 1965.
4. **Good, P. I.,** Aging in mammalian cell populations: a review, *Mech. Ageing Dev.,* 4, 339—348, 1976.
5. **Mauersberger, B.,** Die tierische Zelle in Dauerkultur, in *Aktuelle Probleme der Zellzüchtung,* Mauersberger, C. B., Ed., Gustav Fischer, Jena, 1971, 131—202.
6. **Danielli, J. F. and Muggleton, A.,** Some alternative states of amoeba, with special reference to life span, *Gerontologia,* 3, 76—85, 1959.
7. **Sonneborn, T. M.,** The relation of autogamie to senescence and rejuvenescence in *Paramecium aurelia, J. Protozool.,* 1, 38—53, 1954.
8. **Kisiel, M. J., Castillo, J. M., Zuckerman, L. S., Zuckerman, B. M., and Himmelhoch, S.,** Studies on aging in *Turbatrix aceti, Mech. Ageing Dev.,* 4, 81—88, 1975.
9. **Levenbook, L. and Krishna, I.,** Effect of aging on amino acid turnover and rate of protein synthesis in the blowfly *Phormia regina, J. Insect Physiol.,* 17, 9—17, 1971.
10. **Platt, D.,** *Biologie des Alterns,* Quelle & Meyer, Heidelberg, 1976.
11. **Esser, K. and Kuenen, R.,** *Genetics of Fungi,* Springer, Berlin, 1967.
12. **Curtis, H. J.,** *Biological Mechanisms of Aging,* Charles C Thomas, Springfield, Ill., 1966.
13. **Comfort, A.,** The position of aging studies, *Mech. Ageing Dev.,* 3, 1—10, 1974.
14. **Hayflick, L.,** Current theories of biological aging, *Fed. Proc.,* 34, 9—13, 1975.
15. **Orgel, L. E.,** The maintenance of the accuracy of protein synthesis and its relevance to aging, *Proc. Natl. Acad. Sci. U.S.A.,* 49, 517—521, 1963.
16. **Curtis, H. J.,** Genetic factors in aging, *Adv. Genet.,* 16, 305—324, 1971.
17. **Hayflick, L.,** The biology of human aging, *Am. J. Med. Sci.,* 265, 433—445, 1973.
18. **Strehler, B. L., Hisch, G., Gussek, D., Johnson, R., and Bick, M.,** Codon-restriction theory of aging and development, *J. Theor. Biol.,* 33, 429—443, 1971.
19. **Bullough, W. S.,** Ageing of mammals, *Nature (London),* 229, 608, 1971.

20. **Rizet, G.**, Sur l'impossibilité d'obtenir la multiplication vegetative ininterrompue et illimitede l'ascomycète *Podospora anserina, C. R. Acad. Sci. (Paris),* 237, 838—840, 1953.
21. **Esser, K.**, Podospora anserina, in *Handbook of Genetics,* Vol. I, King, R. C., Ed., Plenum Press, New York, 1974, 531—551.
22. **Esser, K.**, Breeding systems in fungi and their significance for genetic recombination, *Mol. Gen. Genet.,* 110, 86—100, 1971.
23. **Rizet, G. and Esser, K.**, Sur des phénomènes d'incompatibilité entre souches d'origines différentes chez *Podospora anserina, C. R. Acad. Sci.,* 237, 760—761, 1953.
24. **Rizet, G.**, Sur la longévité des souches de *Podospora anserina, C. R. Acad. Sci. (Paris),* 237, 1106—1109, 1953.
25. **Rizet, G.**, Les modifications qui conduisent à la sénescence chez *Podospora anserina* sont-elles de nature cytoplasmique? *C. R. Acad. Sci. (Paris),* 244, 663—665, 1957.
26. **Rizet, G. and Marcou, D.**, Longévité et sénescence chez l'ascomycète *Podospora anserina,* C.R. 8th Congr. Int. Bot. Sect., Paris, 1954, Vol. 10, 121—128.
27. **Marcou, D.**, Sur la longévité des souches de *Podospora anserina* cultivées a diverses températures, *C. R. Acad. Sci. (Paris),* 239, 895—897, 1954.
28. **Marcou, D.**, Sur la rajeunissement par le froid des souches de *Podospora anserina, C. R. Acad. Sci. (Paris),* 239, 1153— 1155, 1954.
29. **Marcou, D.**, Rajeunissement et arrêt de croissance chez *Podospora anserina, C. R. Acad. Sci. (Paris),* 244, 661—663, 1957.
30. **Marcou, D.**, Sur le déterminisme de la sénescence observée chez l'ascomycetè *Podospora anserina,* Proc. 10th Int. Congr. Genet., Montreal, 1958, 179.
31. **Marcou, D. and Schecroun, J.**, La sénescence chez *Podospora* pourrait être duê â des particules cytoplasmiques infectantes, *C. R. Acad. Sci. (Paris),* 248, 280—283, 1959.
32. **Marcou, D.**, Notion de longévité et nature cytoplasmique du déterminant de la sénescence chez quelques champignons, *Ann. Sci. Nat. Biol. Bot. Veg.,* 12, 653—764, 1961.
33. **Holliday, R.**, Errors in protein synthesis and clonal senescence in fungi, *Nature (London),* 221, 1224—1228, 1969.
34. **Smith, J. R. and Rubenstein, I.**, The development of senescence in *Podospora anserina, J. Gen. Microbiol.,* 76, 283—296, 1973.
35. **Smith, J. R. and Rubenstein, I.**, Cytoplasmic inheritance of the timing of senescence in *Podospora anserina, J. Gen. Microbiol.,* 76, 297—304, 1973.
36. **Esser, K.**, Die Incompätibilitatsbeziehungen zwischen geographischen Rassen von *Podospora anserina* (Ces.) Rehm. I. Genetische Analyse der Semi-Incompatibilität, *Z. Indukt. Abstamm. Vererbungsl.,* 87, 595—624, 1956.
37. **Esser, K. and Keller, W.**, Genes inhibiting senescence in the ascomycete *Podospora anserina, Mol. Gen. Genet.,* 144, 107—110, 1976.
38. **Tudzynski, P.**, Untersuchungen zur Genetik und Physiologie der Seneszenz des Ascomyceten *Podospora anserina,* Dissertation Abt. für Biologie, Ruhr-Universität, Bochum, 1978.
39. **Bozarth, R. F.**, Physicochemcal properties of double-stranded RNA myco-viruses, in *Viruses and Plasmids in Fungi,* Lemke, P. A., Ed., Marcel Dekker, New York, in press.
40. **Tudzynski, P.**, unpublished data.
41. **Esser, K. and Tudzynski, P.**, The prevention of senescence in the ascomycete *Podospora anserina* by the antibiotic Tiamulin, *Nature (London),* 265, 454—456, 1977.
42. **Tudzynski, P. and Esser, K.**, Inhibitors of mitochondrial function prevent senescence in the ascomycete *Podospora anserina, Mol. Gen. Genet.,* 153, 111—113, 1977.
43. **Esser, K. and Tudzynski, P.**, Genetic control and expression of senescence in *Podospora anserina,* in *Viruses and Plasmids in Fungi,* Lemke, P. A., Ed., Marcel Dekker, New York, in press.
44. **Marcou, D. and Raynal, A.**, Cytoplasmic determinism of longevity and senescence in the fungus *Podospora anserina,* in *IUBS Colloquium Molecular and cellular models of ageing,* (Abstr.), Paris, 1978.
45. **Stahl, U., Lemke, P., Tudzynski, P., Kück, U., and Esser, K.**, Evidence for plasmid-like DNA in a filamentous fungus: the ascomycete *Podospora anserina, Mol. Gen. Genet.,* 162, 341—343, 1978.
46. **Jinks, J. L.**, Naturally occurring cytoplasmic changes in fungi, *C. R. Trav. Lab. Carlsberg Ser. Physiol.,* 26, 183—203, 1956.
47. **Jinks, J. L.**, Lethal suppressive cytoplasm in aged clones of *Aspergillus glaucus, J. Gen. Microbiol.,* 21, 397—409, 1959.
48. **Caten, C. E.**, Vegetative incompatibility and cytoplasmic infection in fungi, *J. Gen. Microbiol.,* 72, 221—229, 1972.
49. **Handley, L. and Caten, C. E.**, Vegetative death: a mitochondrial mutation in *Aspergillus amstelodami, Heredity,* 31, 136, 1975.
50. **Chevaugeon, J. and Digbeu, S.**, Un second facteur cytoplasmique infectant chez le *Pestalozzia annulata, C. R. Acad. Sci.,* 251, 3043—3061, 1960.

51. **Lindberg, G. D.**, A transmissible disease of *Helminthosporium victoriae, Phytopathology*, 49, 29—52, 1959.
52. **Sheng, T. C.**, A gene that causes natural death in *Neurospora crassa, Genetics*, 36, 199—212, 1951.
53. **Bertrand, H., McDougall, K. J., and Pittenger, T. H.**, Somatic cell variation during uninterrupted growth of *Neurospora crassa* in continuous growth tubes, *J. Gen. Microbiol.*, 50, 337—350, 1968.
54. **Poulter, R.**, Senescence in the myxomycete *Physarum polycephalum*, Ph.D. Dissertation, University of Leicester, 1969.
55. **Clark, J.**, personal communication.
56. **Munkres, K. D.**, Ageing of *Neurospora crassa*. IV. Induction of senescence in wild type by dietary amino acid analogues and reversal by antioxidants and membrane stabilizers, *Mech. Ageing Dev.*, 5, 171—191, 1976.
57. **Printz, O. B. and Groß, S. R.**, An apparent relationship between mistranslation and an altered leucyl-t-RNA synthetase in a conditional lethal mutant of *Neurospora crassa, Genetics*, 55, 451—467, 1967.
58. **Inoue, H. and Ishikawa, T.**, Death resulting from unbalanced growth in a temperature-sensitive mutant of *Neurospora crassa, Arch. Microbiol.*, 104, 1—6, 1975.
59. **Sullivan, J. L. and Debusk, A. G.**, Inositol-less death in *Neurospora crassa* and cellular aging, *Nature (London), New Biol.*, 243, 72—74, 1973.
60. **Ridgeway, C. J. and Douglas, H. C.**, Unbalanced growth of yeast due to inositol deficiency, *J. Bacteriol.*, 76, 163—166, 1958.
61. **Kuraishi, H., Takamura, Y., Mizunaga, T., and Uemura, T.**, Factors influencing death of biotin deficient yeast cells, *J. Gen. Appl. Microbiol.*, 17, 29—42, 1971.
62. **Henry, S. A.**, Death resulting from fatty acid starvation in yeast, *J. Bacteriol.*, 116, 1293—1303, 1973.
63. **Shimada, S., Kuraishi, H., and Aida, K.**, Unbalanced growth and death of yeast due to pantothenate deficiency, *J. Gen. Appl. Microbiol.*, 18, 383—397, 1972.
64. **Pontecorvo, G., Roper, J. A., Hemmons, L. M., McDonald, K. D., and Bufton, A. W. J.**, The genetics of *Aspergillus nidulans, Adv. Genet.*, 5, 141—185, 1953.
65. **Gottlieb, D., Molitoris, H. P., and Van Etten, J. L.**, Changes in fungi with age. III. Incorporation of amino acids into cells of *Rhizoctonia solani* and *Sclerotium bataticola, Arch. Microbiol.*, 61, 394—398, 1968.
66. **Molitoris, H. P., Van Etten, J. L., Gottlieb, D.**, Altersbedingte Änderungen der Zusammensetzung und des Stoffwechsels bei Pilzen, Proc. 2nd Sci. Symp. 7th Int. Congr. Mushroom Sic., 1968, XIII, 59—67.
67. **Obrig, T. G. and Gottlieb, D.**, In vitro protein synthesis and aging in *Rhizoctonia solani, J. Bacteriol.*, 101, 755—762, 1970.
68. **Gottlieb, D. and Raghn, K.**, Methylation and aging in *Rhizoctonia solani, Mycopathologia*, 60, 57—61, 1976.
69. **Gottlieb, D. and Sharma, V. D.**, Changes in ribonuclease with age in the fungus *Rhizoctonia solani, Mech. Ageing Dev.*, 5, 25—32, 1976.
70. **Hemmes, O. E. and Hohl, H. R.**, Ultrastructural changes and a second mode of flagellar degeneration during aging of *Phytophtora palmivora* sporangia, *J. Cell Sci.*, 19, 563—577, 1975.
71. **Esser, K.**, The influence of pH on rhythmic mycelial growth in *Podospora anserina, Mycologia*, 61, 1008—1011, 1969.
72. **Shimada, S., Kuraishi, H., and Ko, A.**, The role of DNA-synthesis in death caused by starvation for biotin in the yeast, *Saccharomyces cerevisiae, J. Gen. Appl. Microbiol.*, 24, 89—102, 1978.
73. **Hayflick, L. and Moorhead, P. S.**, The serial cultivation of human diploid cell strains, *Exp. Cell Res.*, 25, 585—592, 1961.
74. **Oldstone, M. B. A. and Dixon, F. J.**, Aging and chronic virus infection: is there a relationship? *Fed. Proc.*, 33, 2057—2059, 1974.
75. **Wallace, D. J.**, Aging and RNA — tumor viruses, *Clin. Res.*, 22, 198—200, 1974.
76. **Bayreuther, K., Buck, T., Kontermann, K., Mollenhauer, J., and Rodemann, H. P.**, A unifying concept of the molecular mechanisms of the cellular aging and the cellular neoplastic transformation of dividing cells, in *IUBS Colloquium Molecular and Cellular Models of Aging*, (Abstr.), Paris, 1978.
77. **Munkres, K. D.**, personal communication.
78. **Riedl, K.**, Studies on Pleuromutilin and some of its derivatives, *J. Antibiot.*, 29, 132—139, 1976.

Chapter 5

THE SENESCENCE OF LEAVES

Kenneth V. Thimann

TABLE OF CONTENTS

I. INTRODUCTION

From the experimentalist's viewpoint, the study of the senescence of leaves has one great advantage over the process in seeds, namely that it is under direct control. With seeds one can only wait for senescence, much as in studying animals, or else one can use material that has previously been stored until germination failure signals its full or partial senescence. With leaves, on the other hand, one can rapidly induce senescence in a number of ways, and there is as yet no reason to think that such induced senescence differs in any way from the natural process. In partial recompense, however, leaves suffer from the complication of bearing the photosynthetic apparatus; yet, as we shall see, this is not wholly a disadvantage since it shows promise of leading us to a general concept of the nature of the senescence process in tissues. Some aspects of the topic have recently been reviewed elsewhere.[162]

II. METHODOLOGY AND CHOICE OF MATERIALS

After the reality of the existence of auxin had been established, a major contributory reason for the relatively rapid development of the understanding of its physiology was the fact that many laboratories used the same experimental plants, namely the oat coleoptile and the etiolated pea seedling. Results from different institutions could thus be quickly compared, confirmed, or disproved. No such uniformity characterizes the work on the senescence of leaves. Mature leaves, cotyledons, seedling leaves, intact seedlings, dicots and monocots, herbs and trees, sun and shade plants have all been used, and in general it is the exception rather than the rule to find two laboratories using the same plant material. Even the same material is treated in ways that differ in important physiological respects — attached to the plant, detached and either held in moist atmosphere or floating on solutions, or with petioles dipping in the solutions, in light or in darkness, etc. One is tempted to predict that really fundamental understanding of the subject will not come until both the plant material and the environmental conditions are widely agreed upon and strictly limited. However, such pessimism is not really justified since, in fact, a good deal has been discovered despite these haphazard methodologies.

It is not suggested that the senescence syndrome is necessarily the same in all plant species — indeed, the hormonal relations certainly differ from one species to another — but rather that when one system is thoroughly worked out and even partly understood, it should be much easier to address the variations on the theme that are presented by leaves of other types and more varied external conditions.

Because of the irregularities of leaf shape, work on the leaves of dicotyledons (especially on leaves of large size) has often been done with discs, usually cut from the center of the leaf and as far as possible without major veins. Such discs are then surrounded by wounded cells. But wounds stimulate cell division in many tissues. Wound effects have been studied, though sporadically, since the days of Jost and Haberlandt. Wounds have varied effects, including the breakdown of phospholipids, and therefore of membranes — a type of breakdown that extends well beyond the cells immediately adjacent.[161] In work on auxin transport, it has been found that wound effects develop within about 4 hr, and the data could best be interpreted as meaning that the effects concern not only the cells directly adjacent, but spread to others beyond them.[55] In a leaf disc of 1 cm diameter, wound effects to a depth of 500 μm would involve almost 20% of the total area, and such effects could markedly interfere with the changes due to senescence proper. As an example, making 1 cm leaf discs from tobacco leaves increases the peroxidase activity fifteenfold, and largely prevents the normal decrease in catalase.[122] Peroxidations in the tissue would thus be greatly increased. For these reasons the continued use of leaf discs may be less than desirable.

This chapter will deal primarily with detached leaves. To study the senescence of leaves that are still attached to the parent plant is to invite complications from: (1) metabolite transfer in and out of the leaves from the other organs of the plant, (2) cytokinin formation in the roots, auxin production in adjacent younger leaves, or other hormonal influences, and (3) such unknown materials as the postulated "senescence factor" of hormonal type, which may be postulated to control senescence in the whole plant (see Chapter 10). Differences specifically related to the influence of detachment will be taken up where appropriate.

III. THE GENERAL METABOLIC CHANGES IN LEAF SENESCENCE

A. The Loss of Chlorophyll

The obvious character of leaf senescence is yellowing. "Golden" fields of ripe wheat, and the autumn colors of the northern hardwood forests, have long been subjects for poets and essayists. Chlorophyll disappearance has been the principal criterion of senescence for the largest number of workers, (e.g., Molisch,[106] Leopold et al.,[84] Whyte and Luckwill,[176] Goldthwaite and Laetsch,[57] Fletcher,[49] Goldney and Van Steveninck,[53] Back and Richmond.[10] Exactly why the chlorophyll fades has not been made clear. Since it disappears in the dark, photo-oxidation is ruled out. Chemical instability, in the absence of new formation, might be suggested, but it is excluded because chlorophyll in solution does not decompose or lose its color nearly as rapidly as it does in darkened leaves. Mixtures of four hydrolytic enzymes can cause chloroplasts to fade in the dark, though not quite as rapidly as in senescing leaves.[26] This enzyme mixture does cause protein loss about as rapidly as in the leaf. The pH exerts an important modifying effect. Chloroplasts are very sensitive to acidity, and it may be that a proton pump functions in the dark to lower the pH surrounding the thylakoids and thus supplement the action of enzymes. Additionally, respiratory CO_2 could add some acidity, especially if (cf. Section 6) stomata are closed in the dark.

Along with chlorophyll disappearance goes the process of photosynthesis. Declines in photosynthetic rate were noted as early as 1935 by Singh and Lal[147] and more recently by Freeland,[51] Richardson,[133] Sorokin and Krauss,[153] Smillie and Krotkov,[150] Smillie,[151] and Woolhouse.[182] In Richardson's data[133] the main change in tree leaves is in the saturating level of light intensity, which rises from 7000 lux in the young leaf to about 21000 lux in the fully expanded leaf, and falls back to 8000 lux with increasing age; three species of *Acer* and two of *Quercus* gave comparable changes.

Nevertheless, chlorophyll content and photosynthetic rate by no means follow one another closely. For instance, in the very slowly senescing leaves of poplar (attached to the tree), CO_2 assimilation and phosphorylation are at their maximal rates when the leaves are about 10 days old, but the chlorophyll content continues to increase until about the age of 50 days.[63] Even in the much more rapid senescence of seedling leaves the photosynthetic decline is also not ascribable to chlorophyll loss, because the enzyme ribulose-biphosphate carboxylase (RuBPCase, also referred to as Fraction I protein) is rapidly broken down in senescing leaves.[73,16,127,126,124] In barley leaves, after 5 days in the dark, RuBPCase accounts for 90% of the protein lost, though this may be an extreme case. The sensitivity of this enzyme doubtless explains the several earlier observations that photosynthetic rate decreases *before* appreciable loss of chlorophyll.[147,141,62] There are contradictory data, though, for in isolated wheat leaves the fraction I and RuBPCase activity decrease faster than total protein,[16] while in intact wheat seedlings the Fraction I protein is at first more stable than the other leaf proteins, then it decreases sharply after the 11th day.[82]

The fact that leaves turn *yellow* (rather than white) seems to show that the carotenoids are more stable than the chlorophylls, though quantitative data to support this do not seem to have been collected. Fields of grass that have first yellowed have a tendency to turn brown or even grayish after additional days in the sun, showing that the carotenoids may undergo bleaching later. Curiously enough, when etiolated oat leaves senesce in the dark they lose their carotenoids at about the same rate as the green leaves of the same age lose their chlorophyll[159] (see Figure 4). The fate of carotenoids in leaf senescence evidently needs considerable further study, though it may not be a matter of high priority.

B. Proteolysis

That proteins break down during leaf senescence has been realized from some of the earliest studies. Vickery et al.[173] in following the senescence of detached tobacco leaves, found that proteolysis soon began and continued for at least 7 days. Yemm[186,187] made comparable observations on detached barley leaves and noted that the respiratory quotient fell after the first day to around 0.75, which suggested that the amino acids liberated by proteolysis were now the main respiratory substrate. A later study by Vickery et al. of whole detached rhubarb leaves[174] also showed extensive protein breakdown. Since then virtually all workers, using widely different plant species and different experimental conditions, have noted protein hydrolysis (Wood and Cruikshank,[181] Racusen and Aronoff,[131] Osborne,[114] Shaw et al.,[142] Simon,[146] Wollgiehn,[180] Ecklund and Moore,[42] Atkin and Srivastava,[9] Fletcher,[49] Shibaoka and Thimann,[144] Millikan and Ghosh,[98] Parisco and O'Leary,[121] Martin and Thimann,[94] and Tetley and Thimann[159]). In a few cases (e.g., Goldthwaite and Laetsch,[56] Osborne,[118] Kuraishi,[78] Mizrahi et al.,[104] and Tung and Brady[169a]), proteolysis has actually been used as a measure of senescence. The active process of incorporation of $^{14}CO_2$ into amino acids is more sensitive to senescence than is the breakdown of proteins, and in bean leaves such incorporation begins to decrease even before sugar formation decreases.[129] This is doubtless connected with the breakdown of RuPBCase as noted above, together with the proteolysis of some other enzymes. In a mutant of *Festuca pratensis* (meadow fescue), the chlorophyll is relatively stable during senescence, but the leaf proteins are broken down at the normal rate.[167]

Among the proteins undergoing hydrolysis are included not only RuBPCase, but also RNAase, which interrelates the process with nucleic acid metabolism. The sequence of breakdown is suggestive with reference to the loss of photosynthetic activity, for in seedling oat leaves proteolysis is detectable after 6 hr in darkness, while no loss of chlorophyll occurs until the second day[94] (cf. Brady et al.[16] and Section B below). Isolated oat chloroplasts retain their protein much better than when in the leaf,[25,26] which indicates that the proteases are mainly cytoplasmic in origin.

The liberated amino acids are secondarily converted to asparagine and/or glutamine and partly deaminated to liberate ammonia. (The smoke of many varieties of cigars carries a strong smell of ammonia. The parts played by leaf enzymes and by bacteria in curing tobacco leaves have not been rigidly separated, however.) In spinach leaves, while several enzymes decrease in the dark by 25 to 50%, glutamate dehydrogenase increases by nearly 400% and falls off again on transferring the leaves back to the light.[67a] This will promote deamination.

If the leaf is attached to the plant, the amino acids are transported out to the stem, probably as amides, and thus made available to other organs. When the first leaf of an intact barley seedling begins to lose nitrogen (which it does after the 8th day) some of the amino acids go into the second leaf and some into the roots.[33] Similarly in the intact oat seedling, the amino acids are transported out by a basipetally polar system comparable to that transporting auxin — though much slower — and they accumulate in the roots, where the addition of kinetin can cause them to be resynthesized into protein.[165] There is evidence for a proton-amino acid exchange system in this type of transport.[46] Possibly the special properties of this transport out of the leaf may be the reason why, in *Hibiscus* leaves, removal of the petiole has such a strong effect, for it delays senescence by about 10 days.[102] If the plants are flowering or fruiting, the amino acids move to these reproductive organs, as in fruiting cotton,[32,27] while if they are growing rapidly the amino acids move into the apices and young leaves.[177,65,146]

The probable metabolic pathway of such movement into fruits has been elucidated by Storey and Beevers.[156,157] They found that the level of glutamine synthetase (GS) during the senescence of bean leaves is high enough to convert all the free amino acids liberated by hydrolysis into glutamine. At this time the bean pod contains both GS and glutam*ate* synthetase, while the seeds (or rather the cotyledons) have only the latter. Thus the enzyme data are consistent with the scheme:

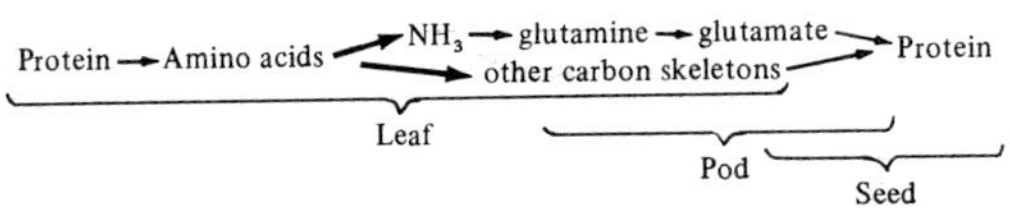

The well-known formation of asparagine in the etiolated lupine or pea seedling is not strictly comparable with this, since in that case the amino acids come directly from the cotyledons into the growing etiolated axis. A more comparable case is that of the leaves of the runner bean *(Phaseolus coccineus)*, in which proteolysis begins every night and is followed by protein resynthesis from the accumulated asparagine each day — a kind of reversible, though transient, senescence.[23] Another aspect of this continuing turnover is shown in the incorporation of amino acids into protein in isolated leaves, for this process goes on in oat leaves simultaneously with protein hydrolysis.[144]

Proteases are of course always present in leaves, and their amounts increase during senescence, as was first reported by Anderson and Rowan in tobacco.[6] The role of the proteases, and what little is known of their enzymatic and chemical nature, will be taken up in the following chapter.

C. Changes in Nucleic Acids

Much of the work on senescence in animals has been dominated by the idea that there is a steady accumulation of errors in transcription, leading to decreases in the biological effectiveness of the several forms of RNA. At first this concept was very much in the minds of plant workers (e.g., Woolhouse,[182] and many of the papers in the 1967 S.E.B. symposium), but with the increasing attention paid to instances of very rapid senescence attention has largely shifted away from it. Obviously for leaves that senesce in 72 hr, and that had essentially ceased cell division before that, some other explanation is needed. The general behavior of tissue cultures presents a supporting argument, though it does not yet apply to leaves, for they have not been brought into continued culture (though Tran Thanh Van[168] has come close to it). This argument derives from the simple fact that tissues from *annual* plants, that would have undergone senescence at the end of the season, can be brought into culture and kept alive and growing for years. When isolated leaves form roots, they similarly survive almost indefinitely. Even pollen, which normally loses its ability to germinate with great rapidity, has been so cultured for a decade or more.[169] Such cases make it "unlikely that genetic damage can account for much of the aging of a higher plant",[22] although, as seen in Chapter 2, it may well account for the gradual loss of the ability to germinate in aging seeds.

In any event, whether causative or not, there are marked changes in the nucleic acids of senescing leaves (cf. pp. 22-27 of the review by Oota.[111]). A general decline in total RNA was first noted by Smillie and Krotkov[150] in pea leaves, in which both RNA and DNA decrease steadily from very early stages on. Similar decreases, specifically during senescence, were reported by Böttger and Wollgiehn in tobacco leaves, and later, using varied plant species, by Osborne,[114,116,118] Shaw et al.,[142] Simon,[146] Wollgiehn,[180] Fletcher,[49] Millikan and Ghosh,[98] Brady et al.,[16] and Gorsheva.[58] The decrease in DNA is typically smaller than that in RNA. Wollgiehn found with tobacco leaves that after 3 days the RNA had decreased by 16% and the DNA by 3%; Dyer and Osborne[41] with *Xanthium* leaves found that after 4 days the RNA had decreased by 55% and the DNA by 39%.

While total RNA decreases, the synthesis of some RNA continues. Both Wollgiehn[180] and Osborne[116,118] showed that incorporation of adenine or orotic acid into the RNA of chloroplasts, mitochondria, ribosomes, and cytoplasm is still continuing, though slowly, even when the leaves are fully yellow. Brady et al.[16] noted that cytoplasmic r-RNA continues to be synthesized for a time in senescing wheat leaves. Synthesis and breakdown of protein are similarly continuing together, as noted above. Curiously enough, the synthesis of RNA, as shown by the incorporation of 8-^{14}C-adenine, actually accelerates when radish leaf discs senesce on water;[30] in 6 days the rate almost doubles.

Loss of RNA can be due to decreases in the rates of synthesis or to increases in the activity of RNase. The latter increases can be large, but our knowledge of them is imperfect. In senescing corn leaves the RNase level became high enough that it was considered a very rich source of the enzyme.[178] In oat and barley leaves, however, the total RNase rises at first and then falls again to its initial level on about the 4th day.[154,144] However, the situation is made more difficult by the presence of more than one RNase in most leaves. For instance, only one of the four nucleases in oat seedlings was found to increase during senescence.[48] Similarly, senescing barley leaves showed no marked change in total RNA in 4 days, but when the chromatin was isolated from these leaves both the DNAase and RNAase associated with it had increased some tenfold.[155,154] The decrease in RNA is of course accompanied by disappearance of ribosomes, which indeed seems to start as soon as growth stops.[16]

Some suggestive findings are associated with *regreening*. This is the change that occurs when mature plants are decapitated; the senescing lower leaves (or in young plants the still attached cotyledons) then tend to recover from senescence and begin to synthesize chlorophyll, protein, and RNA. For example, in soybeans, decapitation at the age of 16 or 17 days causes the senescing cotyledons to recover rapidly, even though 90% of the nucleic acids and 80% of the proteins have been lost.[77] Decapitation on or after the 18th day causes no such reversal. Nucleic acids recover the soonest and the order of recovery of the different types is shown in Figure 1. The proteins broken down during senescence include the enzyme RNase along with others, and that enzyme shows very little recovery during regreening. For this reason it would be worth while to study the subsequent second senescence of the regreened cotyledons, for it would be uncomplicated by the presence of RNase. In *Perilla frutescens*, decapitation and removal of all axillary buds similary arrests senescence and yellowing of the lower leaves, regreening being detectable within 12 to 24 hr.[21] The cytoplasmic RNA here recovers more rapidly than the RNA of the chloroplasts. Beever and Woolhouse[12] subsequently reported that induction of flowering (*Perilla* is a short-day plant) had an effect very similar to that of decapitation and debudding; indeed, since in both treatments the apex no longer bears a vegetative bud they might well be considered similar.

The question whether these nucleic acid changes are causative of senescence or merely a part of a syndrome whose ultimate causes lie elsewhere, will be deferred to Section 6.

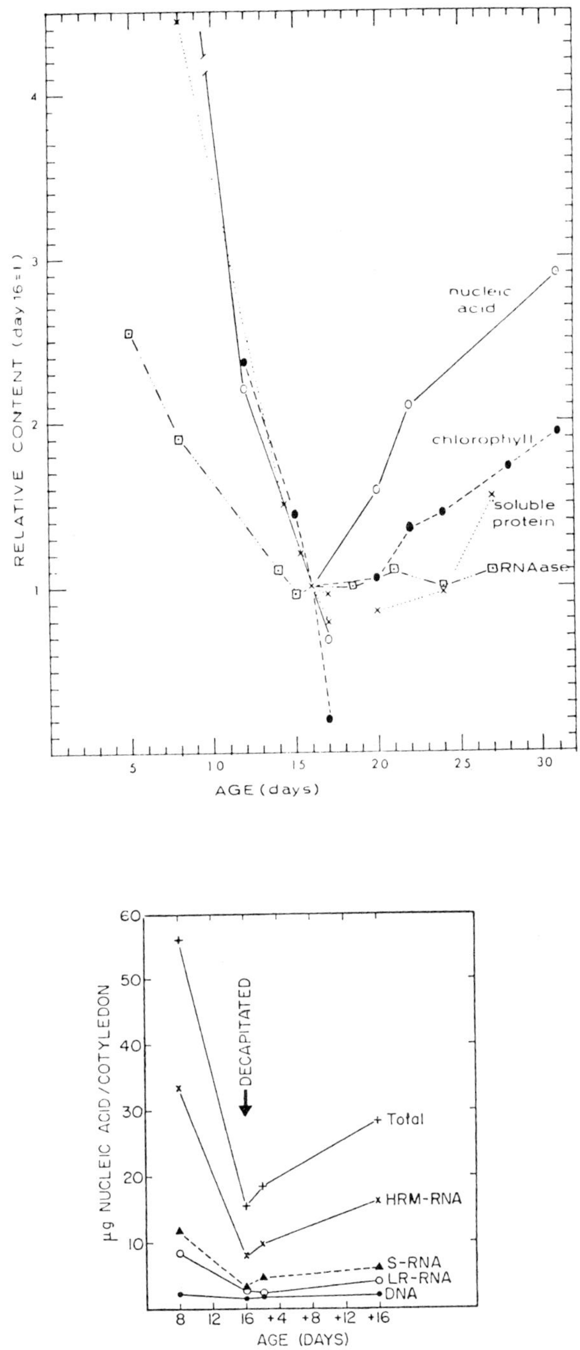

FIGURE 1. Above, decrease of major constituents of soybean cotyledons during senescence and their partial return after the plant was decapitated on the 16th day (Values on day 16 plotted as 1). Below, detail of the nucleic acid fractions in a similar experiment. Absolute values per cotyledon (based on 10 cotyledons). HRM-RNA, heavy ribosomal messenger RNA; LR-RNA, light ribosomal RNA. (From Krul, W. A., *Plant Physiol.*, 54, 36—40, 1974. With permission.)

D. Respiration

One might expect that if senescence is viewed as a general decline in function, the respiration rate would steadily decline. In fact such a decrease does generally occur, but only quite late in the senescence process, and at first there is a sharp increase. The form and timing of this increase is suggestively like the climacteric rise in the respiration of ripening fruit (see Figure 2). First noted by Hover and Gustafson in 1926, and then by Krotkov[76] with wheat seedlings in 1939 (Figure 2), this respiratory increase was studied in more detail by Yemm[188] and later workers. Its incidence and magnitude vary with the material and the circumstances. In the relatively long-lived *Perilla* leaves (attached to the plant) the respiratory rate is almost constant till the 56th day, then rises to double its value just before abscission.[62] In tomato leaf segments the rise is preceded by a steep fall for the first 6 days[96] (see Figure 2). On oat leaves (below) it starts after about 24 hr and reaches 2.5 times the initial rate in about 2 more days; then it decreases rapidly. Some workers have not noticed any such rise, but this may be because the rate has not been followed frequently enough.

It was Yemm who first noticed that the R.Q., which in green tissue usually lies close to 1.0, falls to about 0.74 in senescence, a value which on classical animal-physiological grounds would indicate the oxidation of lipids, but in leaves more likely means that the liberated amino acids are becoming the major respiratory substrates. Alternatively it might indicate some process of withholding of CO_2, like that in succulents. No marked increase in acidity occurs, however, in these leaves. In any case, such a low value does not support the idea that the rise in rate could be due to the soluble sugars, liberated by hydrolysis of polysaccharides, acting as respiratory substrate. As it happens, in oat seedlings the concentration of reducing sugars reached by this hydrolysis during senescence is not high enough to account for the very greatly increased respiratory rate. With these facts in mind, and with daily measurements of both the soluble sugars and free amino acids, Tetley and Thimann[159] concluded that although the increased rate could be to some extent accounted for by the provision of increased substrate (mainly amino acids), it was to a much larger extent due to *partial uncoupling* of respiration from phosphorylation. In support of this notion, they found that increases of nearly this magnitude could be obtained with freshly cut leaves by adding dinitrophenol at optimal pH and concentration. At the height of senescence the rate showed a much smaller increase on adding the DNP. An earlier study with tobacco leaves[88] had demonstrated some decrease in the incorporation of tritiated phosphate into ATP, but since there was at the same time greater incorporation into sugar phosphates it was difficult to draw a clear conclusion. Later work has not borne out this idea, since the rise in the respiratory rate is accompanied by a comparable rise in the ATP content of the leaf tissue.[59a] This matter will be referred to again in connection with the action of kinetin.

E. Changes in Fine Structure

Most workers agree that among the first microscopically visible changes in leaf senescence are those in the chloroplasts (see the review of Butler and Simon[20]). Starch disappears early, causing the bulk of carbohydrate to appear as reducing sugars, and the chloroplasts to lose their characteristic starch deposits. In some cases, especially wheat leaves [143] and leaf discs of *Raphanus* and *Brassica*,[37,31] the chloroplasts swell up in the first day or two, but this did not occur in parsley leaves.[105] The tips of the endoplasmic reticulum sometimes swell also. Swelling of organelles typically denotes loss of energy (cf. Section 6). The thylakoids become less dense and osmiophilic granules appear close to them.[11a,19,37,105] These granules doubtless derive from the lipids of the adjacent thylakoids. In cucumber cotyledons the free ribosomes disappear about

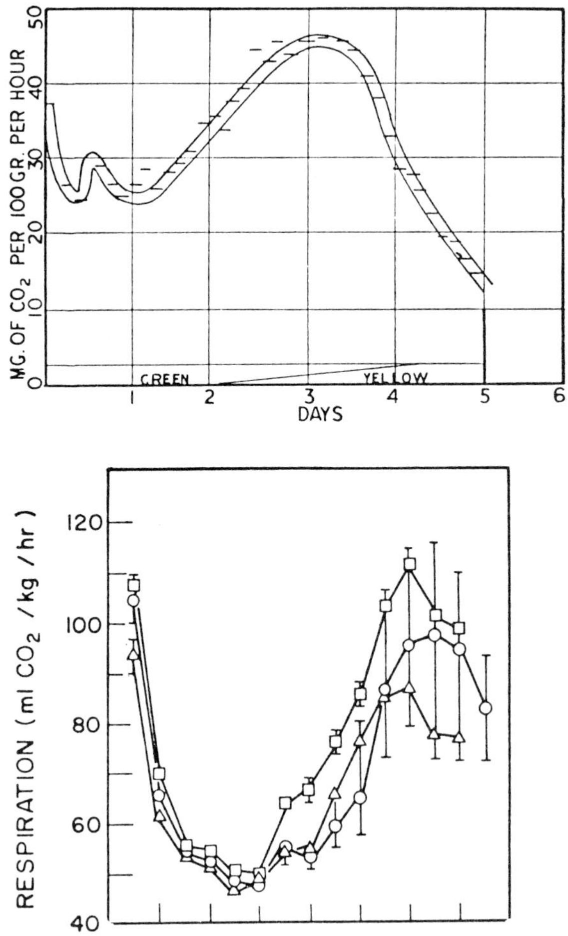

FIGURE 2. Two examples of the respiratory rise of detached leaves in darkness. Above, wheat leaves, from Krotkov;[76] below, tomato leaf discs, from McGlasson et al.[96] Circles, normal cultivar ("Rutgers"); squares and triangles, the non-ripening cultivars, *rin* and *nor,* respectively.

as fast as the first changes in the chloroplasts, and then the polysomes disappear a little later[19] (cf. Chapter 3). The mitochondria show comparable changes, their cristae first becoming distorted, then either disappearing,[143] or at least shrinking.[19]

In later stages, the cells of the same leaf evidently senesce at different rates, for in some the whole plastid envelope disappears, while in others this membrane is still present even after 12 days.[31] Eventually the tonoplast and plasmalemma both disappear — the former first. If the vacuole contains hydrolases these would be expected to attack the cytoplasm. Evidence that the vacuole does indeed contain hydrolytic enzymes has been largely developed by Matile,[95] who proposed that plant vacuoles may be functionally cognate with animal lysosomes. Some hydrolases are normally attached to membranes, particularly the inner side of the tonoplast,[123] and probably enter the vacuole from there. Others may come in from the cisternae of the rough endoplasmic reticulum. Vacuoles of *Ricinus* endosperm now are found to contain acid protease, carboxypeptidase, phytase, β-glucosidase, and RNase, among others,[110] thus bearing out Matile's views in a striking way.

It is evident that such weakening and dissolution of membranes must of necessity produce large changes in the permeability of cells of senescing tissue. Indeed, permeability changes and "leakage" of solutes have been reported by several workers, first by Sacher[137] with bean pods and subsequently with leaf discs from *Mesembryanthemum* and *Rhoeo*;[138] the latter showed visible leakage of anthocyanin from the vacuole. Senescing bean leaves also develop increased permeability and leakage,[34] but when the plants were decapitated to inhibit further senescence the permeability barriers were reinstated, which indicates rapid resynthesis of membrane material comparable to the process described for seeds in Chapter 3. Draper and Simon[39] ascribed the leakage to the breakdown of lipids in the membranes; in this connection they observed that free fatty acids move out of the chloroplasts into other parts of the leaf. The changes probably involve numerous constituents, however, since several detergents that are known to disrupt membranes did not seem to affect senescence.[94] On the whole the data, especially the rather late occurrence of some of the effects noted above, suggest that the changes in fine structure and in membrane integrity may be *results* rather than *causes* of senescence.

IV. THE MANY INFLUENCES OF HORMONES ON LEAF SENESCENCE

A. Auxins

The first reports that hormones might participate in the control of leaf senescence were those of Sacher,[137,138] who found α-naphthalene-acetic acid to retard senescence of leaf discs and bean pods. Shortly thereafter Osborne and Hallaway[115,117] reported that 2,4-D butyl ester (a synthetic auxin largely used as a herbicide) retarded the proteolysis of detached leaves of *Prunus serrulata* (one of the flowering cherries). Auxins have not, however, been generally found very potent in retarding leaf senescence. In the oat leaf the auxin (IAA was used) retarded senescence only at concentrations around 500 times that of kinetin.[144] In wheat seedlings naphthalene-acetic acid has a small effect, but only at 100 mg/ℓ, i.e., over 100 times the effective concentration of BAP, and one which would soon be toxic.[179] Dove[38] reported a clear auxin effect on tomato leaflets, and one or two others have noted minor effects in passing. A complication is that auxin, at levels only slightly above the physiological, stimulates the production of ethylene, which in some tissues promotes senescence; however, its influence in leaves is not strong, and often indeed it seems to show no effect at all (see Section E below). Thus the influence of auxin, though slight, is probably to be ascribed to the auxin itself.

B. Cytokinins

The discovery of the cytokinins in 1957 brought to light much more potent agents. Richmond and Lang in that year[135] reported that *Xanthium* leaves with their petioles in water (in daylight) lost 60% of the protein in 12 days, while with their petioles in kinetin (5 ppm) they lost only 15% in the same time. Loss of chlorophyll was less marked, and was variable. Interest was heightened when Mothes and Engelbrecht[107,45] showed that if a small area of a basal tobacco leaf was treated with kinetin solution only that area remained green, the rest of the leaf yellowing normally. When a ^{14}C-amino acid was applied to another area its radioactivity became concentrated in the kinetin-treated area (Figure 3). At first this accumulation of the ^{14}C was ascribed to activation of protein synthesis by the kinetin, but then ^{14}C-α-aminobutyric acid, which does not become incorporated into protein, was also found to accumulate in the kinetin-treated area. At about the same time Osborne[114] found that the incorporation of labeled orotic acid into RNA, and of labeled leucine into protein, both decrease during senescence, and kinetin reinstates the original rate of incorporation, i.e., of synthesis. Formal explanation of the accumulation phenomenon has not yet been given, but a strong suggestion will be offered in Section VI. C.

A flood of studies followed these first papers, almost all concurring in the effectiveness of kinetin or another cytokinin in preventing, or at least greatly delaying, senescence. Some of the major conclusions from these studies will be grouped together here.

Firstly, all cytokinins that have been tested have this property; they delay proteolysis and chlorophyll loss in the dark. Kinetin and benzylaminopurine (BAP) have been most used, while zeatin (the one cytokinin known to be present in flowering plants) is usually less effective. The relative activities of the cytokinins vary with the plant species; BAP is somewhat more active than kinetin or zeatin on legume leaves[100] and about 5 times more active than kinetin on oat leaves,[159] but on mature cereal leaves kinetin is more effective than BAP.[100] For whole wheat seedlings held in the dark, zeatin has to be at 0.1 mM to delay development of solute leakage.[179] A cytokinin of very high activity on tissue cultures, namely 2-isopentenyladenine (2-iP) has only 1% of the activity of kinetin on the senescence of oat leaves,[159] and also has very low activity on radish leaf discs.[61] Kinetin, zeatin, and 2-iP were placed in that order of (decreasing) activity on tobacco and *Brassica* leaf discs by Letham.[85] Diphenylurea, a weak cytokinin on tissue cultures, is almost inactive on oat leaves, but quite active on radish leaf discs, and three of its derivatives, 3,4-dichlorophenylurea, N-3-fluorophenyl-N-phenylurea, and N-4-acetamido-N-phenylurea are about as active as kinetin in that test.[18] Even benzimidazole, an extremely weak cytokinin, largely prevents proteolysis at 30 ppm in oat leaves[159] and at 50 ppm in detached wheat leaves;[125] in Mishra and Misra's tests[100] it was about as effective as kinetin. Thus there is little quantitative parallel between the action of cytokinins on cell division and on leaf senescence.

Secondly, the action on leaves is generally paralleled by comparable effects in delaying the senescence of flowers, as described in Chapter 7.

Thirdly, there are a few plants whose leaves do not respond to cytokinins, and these respond instead to gibberellins (see Section C below).

The fourth area of agreement is that in leaves attached to the plant the effects of the cytokinins are usually less marked. Such leaves senesce more slowly than when detached, so that the differences from controls are smaller. Clear effects of cytokinins on attached leaves have been shown, however, by Fletcher[49] and Adedipe et al.[3] using BAP on bean leaves, and by Thimann et al.[165] with oat leaves; Schreyer and Noodén[140] obtained only a moderate delay with soybean leaves. Weak effects were reported by Mothes[107] and Muller and Leopold.[108]

A likely explanation of this general phenomenon is that attached leaves receive a supply of endogenous cytokinin from their roots, for Kende[75] and Itai and Vaadia[67] showed real cytokinin activity in the roots and in the stem exudate of *Helianthus*, and numerous later workers have confirmed this in other plants. The supply of cytokinin traveling upwards seems to be shared among the various shoot organs, for removal of one can lead to increases in the cytokinin content of others. Thus, Beever and Woolhouse[12] showed that debudding greatly increases the cytokinin concentration in the exudate of *Perilla*. Some of their data are summarized in Table 1. Fruit removal can have an effect similar to that of debudding (see Chapter 8).

The cytokinin is not only formed in roots, but is transported upwards from them to the leaves through the xylem, along with some amino acids. It can be shown to become distributed among the various sinks, especially the fruits. Some care is needed in relying wholly upon this explanation, however, for some cytokinin activity can be extracted from leaves too, and strong (but indirect) evidence points to its being synthesized in buds in the light.[83] The export from leaves, as well as the synthesis there, may

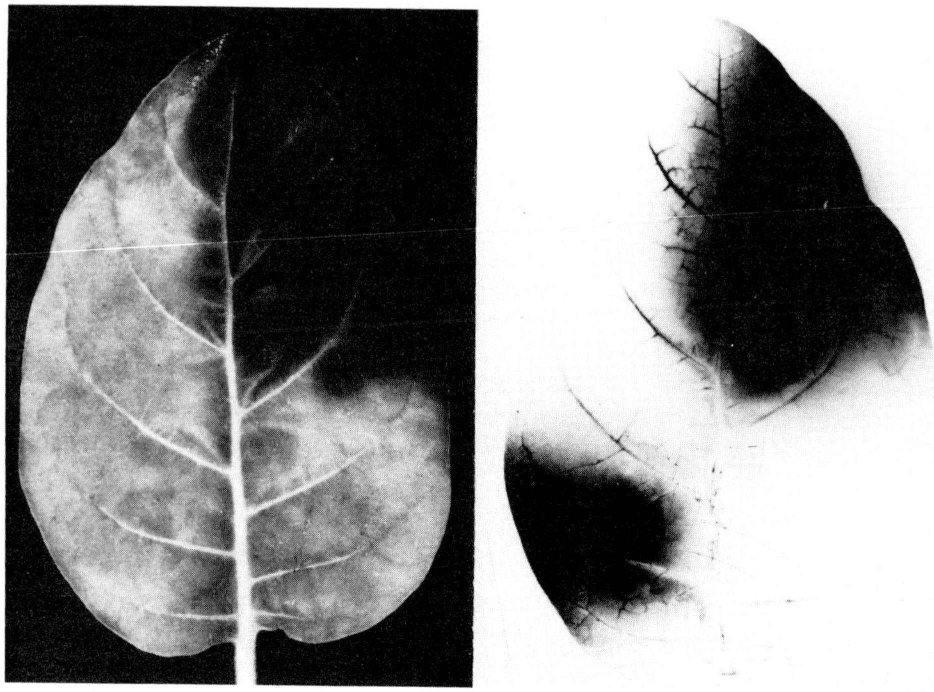

FIGURE 3. Tobacco leaf treated with kinetin on its upper right half and with ^{14}C-α-aminobutyric acid on the lower left half. Above, photograph (with green filter), showing that the kinetin area remains green; below, radioautograph, showing that ^{14}C has migrated to the treated area. (Cf. Figure 9). (From Mothes et al.,[107] original photo supplied by Prof. Mothes; reproduced from Reference 161a. With permission.)

be quite limited, for Malik and Berrie[90] concluded from experiments on partial and total defoliation that leaves do not much influence the course of senescence in the whole plant.

The fifth important consideration, which is agreed upon by several groups of workers, is that there is a strong effect of light, delaying the typical senescence syndrome in a manner superficially similar to the effect of cytokinins. On this account an additional effect of cytokinin in presence of light is often difficult to demonstrate. Clear examples of delay by light are given by Engelbrecht,[45] Hopkinson,[65] Goldthwaite and Laetsch,[56] Haber et al.,[60] Mishra and Pradhan,[101] and Thimann and Satler.[163] The length of day has an influence; leaves of *Hibiscus rosa-sinensis,* which senesce in the dark in 20 days, survive to 35 days in 8 hr photoperiods and to 45 days in 16 hr photoperiods.[102] Thus the effect of light is both quantitative and additive. In three cases, however, light has been reported to increase the action of cytokinin — in tobacco leaf discs,[111] in oat leaves, [163] and in detached rice leaves.[148] In the latter case, red light and benzimidazole each had a weak effect and they were simply additive. In detached oat leaves what kinetin does is to prolong the action of the white light,[163] an action that is discussed further below.

Light also inhibits the hydrolysis of RNA in leaves of potato,[28,93] and tobacco.[158] Red light was reported to delay senescence in rice leaves, and far red to antagonize this effect,[101] but so far such phytochrome-like effects have not been reported in other leaves, and generally blue and red lights are about equally effective.[60] The mechanism of the action of light will be taken up in detail in Section 6.

Kinetin exerts a striking effect on respiration. Instead of the climacteric-like rise and fall of the respiratory rate of oat leaves in the first few days, the respiration in kinetin

TABLE 1

Cytokinin concentration (as nanograms kinetin equivalents per m*l* sap) by bioassay, in the xylem sap of *Perilla frutescens*

Days	Plants intact	Plants decapitated and disbudded
0	14	—
12	21	44
22	16	68
38	20	168

From data of Beever, J. E. and Woolhouse, H. W., Mech. of Regulation of Plant Growth, 1974, 681—686. With permission.

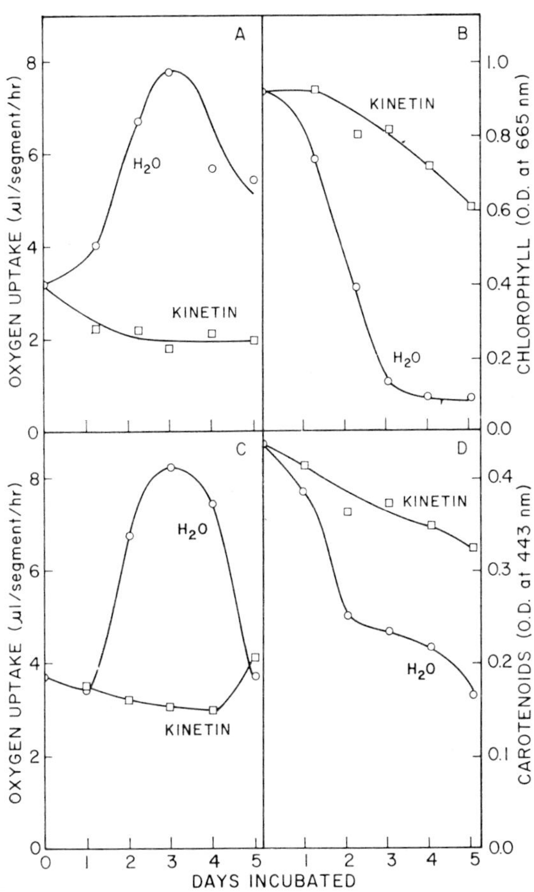

FIGURE 4. A and C, left, the progress of respiration in detached oat leaves (7 days old) in water and kinetin (3 ppm). B and D, right, the disappearance of chlorophyll, B, or carotenoids, D, in the same period. A and B, green leaves; C and D, etiolated leaves. All leaves in darkness at 25°C. From Tetley, R. M. and Thimann, K. V., *Plant Physiol.*, 54, 294—303, 1974, cf. Figure 2. With permission.)

remains low and even gradually falls (Figure 4). The same is true in wheat seedlings sprayed with zeatin.[179] It seems surprising that depression of respiration should go along with the promotion of synthesis, but the explanation offered rests on the previously mentioned idea that partial uncoupling takes place in senescence, and kinetin is postulated to maintain the coupling, which would in consequence keep the respiratory rate low.

Regreening due to decapitation and debudding was referred to above. Regreening due to cytokinin application appears to follow similar lines; in yellowed tobacco leaves DNA, RNA, and protein are all formed along with the chlorophyll; the lamellae of chloroplasts and mitochondria all become "more distinct" and the nucleolus increases in size.[81] Thus the effect of decapitation and disbudding may well be due to the redirection of the stream of cytokinin to the remaining organs (cf. the discussion above). Changes perhaps comparable to those in regreening occur when cytokinins are applied to very young leaves; the thylakoids seem to multiply till the chloroplasts swell up into a mass of fine membranes.[37] A contrast to regreening is shown by etiolated leaves, which show senescent changes very much like those in green leaves (Figure 4).

C. Gibberellic Acids (GA)

In leaves of *Taraxacum* (dandelion), kinetin has been found to have little effect on senescence, but GA acts powerfully.[50] In leaves of *Tropaeolum* (nasturtium) both GA and kinetin are effective.[13,14,176] The latter workers even proposed the response to GA as a bioassay method for GA. Correspondingly, as *Tropaeolum* leaves senesce their GA content steadily decreases;[24] a similar decrease occurs in lettuce leaves under water stress.[4] The striking threefold parallel between the effects of GA and kinetin on chlorophyll, protein, and RNA losses in *Tropaeolum* leaves is shown in Figure 5. *Rumex* (sorrel) presents a similar case; GA and kinetin being about equally effective in delaying senescence, while zeatin is only effective if the solution is frequently renewed.[56,91]

It is curious that three hormones of such different chemical types and characteristic physiological effects as auxin, cytokinin, and GA should all act in the same general direction on senescence. Since chlorophyll, protein, and RNA are involved in all three cases (see also Böttger and Wollgiehn[15] and Osborne[114]) the cause(s) of leaf senescence must lie at a very fundamental level.

D. Abscisic Acid (AbA)*

Effects in the opposite direction from those above, that is, hastening of senescence, are exerted by AbA. The varied effects of AbA on plants were reviewed by Addicott and Lyon in 1969,[2] but at that time there was little appreciation of its action in senescence. Very soon after AbA was identified (1966 to 1967), Wareing's group[44,175] as well as Aspinall et al.[8] and Sankhla and Sankhla[139] reported that it promoted senescence in leaves. AbA is known to be present in leaves, especially when fully grown (see the review of Milborrow[97]). Its level increases in detached tobacco leaves, rapidly in the dark, slowly in the light, then decreases again (Table 2). Kinetin prevents the rise. It is also well known to increase rapidly in leaves that are water-stressed,[183,184] which would correspond to the often-observed tendency of water-stressed leaves to turn yellow. Not only does AbA increase under this condition, but bioassays show that cytokinins decrease at the same time.[67,112,47,4] Thus AbA, if not proved to be a cause, is certainly an endogenous factor in leaf senescence. AbA clearly inhibits protein synthe-

* The more usual abbreviation for Abscisic acid is ABA, but this is illogical, for a capital letter represents the first letter of a word or syllable, as Indole-Acetic Acid. The B in ABA has no such meaning, but is simply the second letter; it should therefore be written in "lower case" and will be so written here throughout.

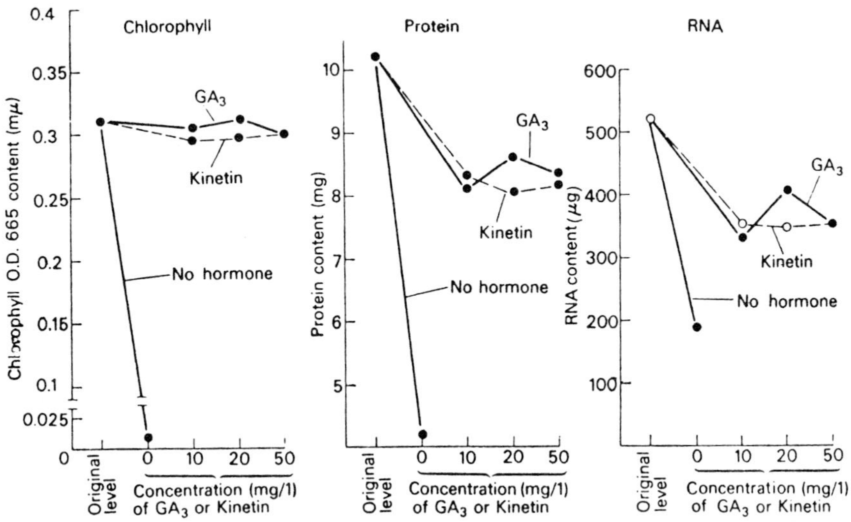

FIGURE 5. Delaying effect of kinetin and GA_3 on senescence of excised leaf discs of *Tropaeolum majus*. Chlorophyll, protein, and RNA measured at start and after 10 days. Plotted from data of Beevers and Guernsey[14] by Phillips.[128a] With permission.

sis in *Pennisetum* leaves, and kinetin reverses this effect.[43] AbA also inhibits the formation of glutamate dehydrogenase, and again kinetin reverses the effect. In leaves that respond both to GA and to kinetin, AbA is more effective against GA than against kinetin; kinetin at 0.01 m*M* could completely annul the action of 0.04 m*M* AbA.[10]

AbA exerts other effects typical of senescence in general, especially effects on nucleic acids. Thus it increases the losses of protein and RNA from isolated leaves and leaf discs.[14,154,36,120,30,91] It prevents part of the increase in RNA formation that occurs in radish leaf discs.[30] In several such studies it was shown to increase the activity of RNase, notably in *Rhoeo* leaves, whose RNase increases fivefold in 24 hr on water and eightfold on AbA.[36] Of the four nucleases present in young oat leaves, AbA increases the activity only of that one that normally increases during senescence.[48] Like kinetin in the opposite direction, the action of AbA is less dramatic in leaves attached to the plant,[44,152] and here it is difficult to invoke the explanation that was offered for kinetin, since AbA is synthesized in leaves and so far is not known to be transported about in the plant.

The concentrations of AbA needed for its effects on senescence are much greater than for cytokinins, though naturally there is wide variation in sensitivity among species and even among cultivars.[149] Exogenous AbA is commonly applied at 1 to 5 m*M*, whereas cytokinins act in the micromolar range and even below; as a result kinetin can usually antagonize the action of AbA at concentrations 1% or less of that of the AbA. Also AbA can act only when the other circumstances are favorable, e.g., it acts on mature cotton leaves, but not on young ones;[152] on *Rhoeo* leaves grown under low light, but not on really healthy ones.[36] In the last case it increases the leakage of anthocyanin out of the vacuole, which was mentioned above as typical of normal senescence in *Rhoeo*. It accelerates the breakdown of membranes, including the chloroplast envelope, plasmalemma, and tonoplast, in wheat leaves.[103]

Thus, with a possible reservation in regard to the rather high concentrations needed, AbA appears to be a true endogenous senescence-accelerating factor. Indeed, its action in this process is stronger than the action on Abscission for which AbA was named.

TABLE 2

Abscisic acid content (in nanograms per gram fresh weight) of detached tobacco leaves held either in dark or in light. Data of Even-Chen and Itai,[47]

Day number	Dark	Light
0	*ca.* 0.1	0.2
1	0.3	0.2
3	8.5	0.3
4	—	2.0
5	8.0	3.3
7	2.2	12.0
9	1.5	—
11	—	3.5

From data of Even-Chen, Z. and Itai, C., *Physiol. Plant.*, 34, 97—100, 1975. With permission.

E. Ethylene

Ethylene has long been known to promote chlorophyll loss in fruits (see Chapters 8 and 9), but with a few exceptions most investigators have found it to have small or negligible effects on leaf senescence. It was used to promote the blanching of celery by Denny as long ago as 1924, and became widely adopted for this purpose. There have been few studies of its action on leaves. Goldney and Van Steveninck[53] found that detachment of the leaves of the water plant *Nymphoides indica* greatly increased its ethylene production, but that applied ethylene (25 ppm) did not modify the slow decline of the chlorophyll and protein contents of the leaves over a 10-day senescence period. In tomato leaf segments the rise in respiration is paralleled by a rise in ethylene production, but whether this has any effect on the senescence was not studied.[96] In our own work with detached oat leaves, neither the application of ethrel nor the placement of leaves in reduced air pressures ("hypobaric storage") has had detectable effects on senescence in darkness. However, illuminated spinach leaves offer one case where ethylene clearly promotes chlorophyll loss ("deverdissage").[128] Ethylene at 50 or 100 ppm has only a small effect on these leaves in the dark, but in bright light (220,000 erg cm^{-2} sec^{-1}) it roughly doubles the rate of chlorophyll loss. Kinetin, and to a lesser degree, BAP, antagonize the ethylene effect but, curiously enough, they both *increase* the ethylene production of the leaves. One really has to deduce that the ethylene production of the leaves is not a factor in their senescence. Ethylene has very recently been shown to be converted to ethylene oxide in *Vicia faba* cotyledons by Jerie and Hall,[69] and it was suggested that this metabolic removal of ethylene from plant tissues may be very general, and, what is more, it may increase with age.

V. INFLUENCES OF OTHER COMPOUNDS

A. Amino Acids

Following proteolysis, the resulting amino acids are transported out of the leaf, as described in Section 3B. But the amino acids themselves are by no means inactive in the senescence syndrome. The stream of amino acids down the petiole promotes *abscission* of the leaf,[136] though this is not a strong effect. There are more specific effects on senescence itself, as was shown subsequently. In looking for evidence of cytokinin production in roots, Shibaoka and Thimann[144] found instead that root extract actually *promoted* senescence; the active factor was soluble in aqueous butanol and was traced to the amino acid L-serine, which in many plants is one of the most plentiful amino acids. Both chlorophyll loss and proteolysis were promoted, and L-serine at 100 m*M* could completely negate the maximum effect of kinetin. Tobacco leaves showed the same promotion,[72] but barley leaves showed little response to serine.[126] Similar (but weaker) effects were exerted by cysteine and glycine, but D-serine, homoserine, and other related compounds were inactive.[94] Since proteases clearly increase in activity in these senescing oat leaves, the action of L-serine was attributed to its promoting the formation of serine proteases (cf. Chapter 6). Cysteine could similarly lead to a cysteine protease, while glycine is on the biosynthetic pathway to both. It is in any event striking that serine and cysteine, the only amino acids that strongly promote senescence, should also be those at the active site of the two main classes of protease. It is suggestive that another hydroxy derivative, N-hydroxy-urea, also promotes senescence, both in cucumber cotyledons and in barley leaves, and the cytokinin BAP partly reverses its effects.[132]

Arginine acts in the opposite direction to serine, reversing the action of serine in oat leaves,[144] and correspondingly it somewhat lowers the level of protease.[94] In these seedling leaves arginine has little effect by itself, but in the fourth leaves of grown oat plants arginine, as well as ornithine and lysine, clearly inhibit senescence.[1] Several other dibasic compounds, especially spermine and spermidine, inhibit senescence, and these have been used by Galston and co-workers[52] to maintain vitality in oat protoplasts, which otherwise deteriorate rapidly. It is hardly surprising that senescence or its opposite, juvenility, should be related to the maintenance of ability to grow and divide in isolated organs and organelles, and this aplication of Galston's may have far-reaching utility.

B. Inhibitors of Protein Synthesis

Reasoning from the senescence-promoting action of L-serine and the increase in protease in senescing oat leaves, Martin and Thimann[94] looked for direct evidence that protease was being synthesized and not merely liberated from storage in vacuoles or membranes, or being converted from a pro-enzyme form. Cycloheximide, which inhibits the activity of cytoplasmic ribosomes, proved to inhibit the onset of senescence in the dark very powerfully, and at concentrations that inhibited over 80% of the incorporation of ^{14}C into the alcohol-insoluble fraction ("protein"). Other inhibitors of protein synthesis, such as chloramphenicol and streptomycin, which act on the ribosomes of chloroplasts, were without effect, but 6-methyl-purine caused a moderate inhibition. Cycloheximide was equally effective on detached barley leaves.[126] These inhibitions indicate that the breakdown enzymes causing senescence originate in the cytoplasm. Indeed, the inability of chloramphenicol to control the loss of chlorophyll stands in complete contrast to its strong inhibition of the *formation* of chlorophyll (e.g.,[92,11]).

In a similar vein Makovetzki and Goldschmidt[89] found cycloheximide to prevent the loss of chlorophyll from leaflets of *Anacharis canadensis* ("Elodea") in the dark. Again streptomycin and chloramphenicol had no similar effect, indeed they even slightly promoted chlorophyll loss. (In isolated oat chloroplasts they slightly promoted the loss of system I of photosynthesis also.[26]) What was remarkable in the *Anacharis* system was that cycloheximide (10 mg/ℓ) arrested the chlorophyll loss even though it was applied after several days, when most of the chlorophyll had already disappeared (Figure 6). For this suggests that the protein or enzyme whose synthesis is necessary for senescence could hardly be a protease, since once a protease had been formed it

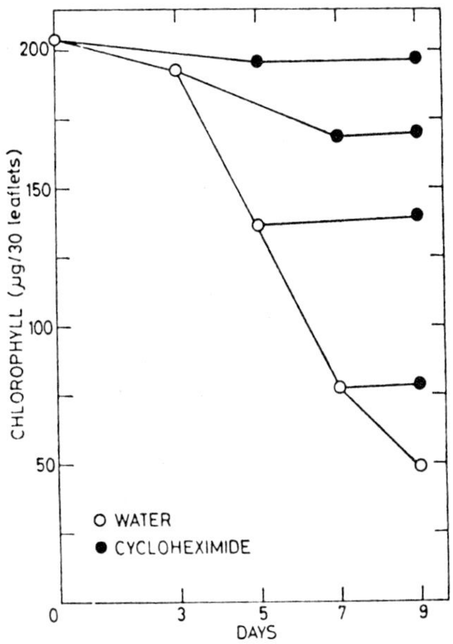

FIGURE 6. The arrest of chlorophyll loss in *Anacharis* leaflets by cycloheximide (10 ppm). Leaflets in water in the dark were transferred to cycloheximide solution at 0, 3, 5, or 7 days. (From Makovetski, S. and Goldschmidt, E. E., *Plant Cell Physiol.*, 17, 859—862, 1976. With permission.)

would continue to hydrolyze protein. Instead, the cycloheximide-sensitive protein would have to be subject to rapid turnover, but yet be in control of senescence at all times. Since the published data[89] so far refer only to chlorophyll and not to total protein, the cycloheximide-sensitive system might conceivably be a peroxidase, for, at least in rice leaves, catalase decreases markedly in the dark and kinetin acts to prevent this decrease.[70] Kinetin increases the peroxidase also, by 70%. Thus the enzymes might perhaps act to remove an accumulation of peroxide, or of oxidizing free radicals, to which chlorophyll is susceptible. Since the chlorophyll of isolated chloroplasts bleaches readily in the light, in a reaction prevented by ascorbate or glutathione,[25] it is evidently susceptible to such systems.

VI. PROBABLE MECHANISMS OF LEAF SENESCENCE

The first theory of senescence, that of Molisch, based on the redirection of nutrients toward flowers and fruits, was applicable only to whole plants (see Chapters 1 and 10). For the senescence of leaves attached to the plant the same theory might well be applied, but in view of the rapid onset of senescence changes in isolated leaves such concepts are beside the point. This is not to say that such redirection (and, in general, transport movements) may not play a part in the senescence of attached leaves, but merely that they are not *causal.* The leaf does, of course, acquire water, salts, and organic compounds from the rest of the plant, and it exports sugars, vitamins, and (as we noted above) amino acids (the export of cytokinins is open to doubt); these exchanges, while varying diurnally, are always going on and must therefore be continually modifying the leaf's vitality. But for *control,* applicable both to attached and detached leaves, we must evidently look deeper.

Many workers have accordingly looked to nucleic acids. As shown in Section 3C, total RNA declines during senescence, roughly in parallel with the decline in protein, and DNA also declines, though more slowly than RNA. At the same time RNA continues to be synthesized (refs. above) and it is worth noting that the synthesis may be both from adenine and from kinetin.[134] The idea that kinetin owes its action to being incorporated into RNA is still controversial. That there is some incorporation is clear, however, for the Wisconsin group has shown that tobacco callus incorporates ^{14}C-kinetin, via its riboside, into both ribosomal and transfer RNA; BAP was incorporated about one quarter as effectively as kinetin.[7,109] But the data of Richmond et al.[134] show that while labeled kinetin or adenine are converted to RNA in detached leaves, labeled BAP was not incorporated into any nucleic acid fraction, and they concluded that the idea that kinetin acts via conversion to t-RNA in leaves "is not supported." Another objection was made by Paranjothy and Wareing[120] when they followed the incorporation of ^{3}H-cytidine into the r-RNA and s-RNA of radish leaf discs. For while kinetin increased this incorporation, and AbA decreased it, both acting in parallel with their effects on senescence, yet 5-fluorouracil, which inhibited the incorporation, did not change their actions on senescence.

The behavior of the enzyme RNAase during senescence is also not as would be expected if the integrity of RNA were critical. It is true that in several instances one or more of the leaf RNAases does increase during senescence, and that cytokinins inhibit this increase;[154,144,185] in the work of Wyen et al.[185] the increase was promoted by AbA too. But these parallels are offset in several ways. For one thing, there are RNAases in chloroplasts,[66,86] yet when chloroplasts are separated from the leaf they senesce only at a very slow rate.[25,26] In soybean cotyledons the chloroplast RNAase acts specifically on only 2 of the 4 tyrosyl t-RNAs, an action which might perhaps inhibit the synthesis of some proteins, but seems hardly adapted to cause the very generalized breakdown that characterizes senescence.[86,87] More strange is the fact that in these soybean cotyledons the RNAase *decreases* along with the total RNA.[77] In any event, the relationship between nucleases and actual nucleic acid content may not be a simple one; for instance, in the pea seedling DNAase actually increases 400% just before the onset of DNA replication.[68] The increased RNA in senescing radish leaf discs was noted above.[30]

All in all the rapid senescence of leaves in the dark, occurring in a tissue where cell division has already nearly stopped, seems most unlikely to be *caused* by changes in nucleic acids, though such changes certainly take part in the general syndrome.

The newest approach is through the comparison between senescence in light and in dark. As described in Section 4B, light delays both chlorophyll and proteolysis in nearly all species in which its action on both processes has been studied. Light virtually always delays chlorophyll loss. At least in detached oat leaves, an important feature of the effect is that on transference to darkness the influence of light does not carry over. Figure 7 shows that after exposure for 1 or 2 days to white light, during which time senescence proceeded at the slow rate characteristic of light, transference to darkness immediately reinstates the faster rate characteristic of the dark. This shows that light does not act by inducing formation of a stable substance such as a cytokinin. Transference from dark to light causes a similar immediate change to the light rate, showing that darkness does not induce formation of a stable senescence-accelerating substance. If part of the leaf is covered, the green color of the exposed part, even after 4 days, does not spread into the etiolated color of the dark part, showing that light does not produce a diffusible product either, i.e., the effect of light is both *localized* and *unstable.*

From this point on, two types of explanations can be offered, the one cellular, the other organismic. Both have considerable support.

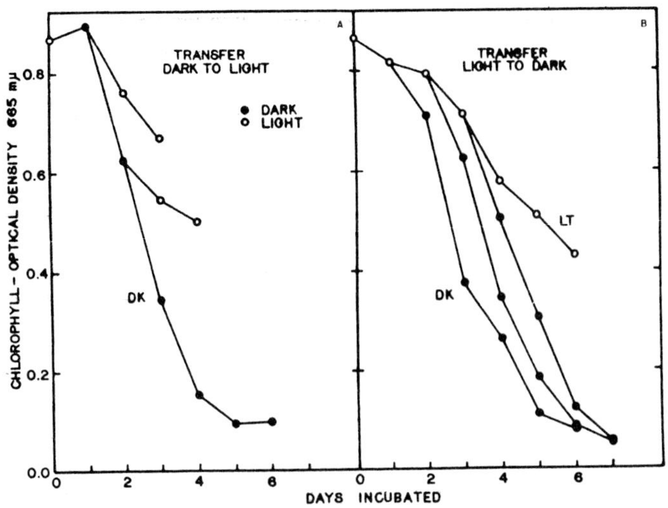

FIGURE 7. Loss of chlorophyll from oat leaves in darkness (filled circles) or in light (open circles) or after transfer from one to the other on day 1, day 2, or (light to dark) day 3. (From Thimann, K. V., Tetley, R. M., and Krivak, B. M., *Plant Physiol.*, 59, 448— 454, 1977. With permission.)

A. The Cellular Explanation

One light product that would satisfy the above criteria is, of course, ATP. It can be produced both by cyclic and by non-cyclic photophosphorylation. Of importance is the fact, shown in different systems by Haber et al.[60] and Thimann et al.[166] that light can exert its effect on senescence in presence of the inhibitor DCMU (3,4-dichlorophenyl-1,1-dimethylurea), at a concentration that completely prevents formation of sugars (Table 3). Thus the action of light cannot be ascribed to the photosynthetic production of sugar, even though the feeding of glucose does somewhat delay senescence, an observation dating back to Vickery,[173] but studied carefully by Marrè et al.[93] and Udvardy et al.[172] Table 3 shows that DCMU prevents both the formation of soluble sugars and the characteristic rise in respiration, yet it permits the almost total maintenance of chlorophyll. Since DCMU is known not to inhibit the cyclic process of photophosphorylation (Duysens and Sweep[40] and later workers), it was concluded that the effect of light is exerted by forming ATP through cyclic photophosphorylation. In line with this interpretation is the fact that the action of light, unlike that of kinetin, is not antagonized by L-serine, which shows that they have different sites of action, kinetin presumably acting on dark respiratory phosphorylation and light on the cyclic photo-process. Also this would explain why senescence is more rapid in the dark than in the light, since photophosphorylation is much more active than mitochondrial phosphorylation. Direct measurements of ATP have shown, however, that its level in detached oat leaves tends to follow the changes in respiration, and thus to go through a maximum in the dark, but remain low in kinetin.[89a] This suggests that the rate of *usage* of ATP may be more determining than the rate of its production. Inhibitors of cyclic photophosphorylation like ammonia and methylamine do decrease the effect of light, but the matter is far from settled. It would be attractive to view the whole syndrome of leaf senescence as a phenomenon of the deprival of chemical energy.

B. The Organismic Explanation

Another light product that satisfies the criteria of localization and instability would

TABLE 3

Effect of DCMU [3-(3,4-dichlorphenyl)-1,1-dimethylurea] on the accumulation of sugars and the retention of the chlorophyll in *Avena* leaves. Concentration of DCMU, 0.1 m*M*. All leaves except dark controls held in continuous white light of 200 ft-candles.

Treatment	Reducing sugar nmoles/segment	"Sucrose" nmoles/segment	Chlorophyll A/665
Initial values	146	86	0.866
3 days on water	760	531	0.832
3 days on DCMU	175	4	0.793
3 days on DCMU in dark	—	—	0.23

From Thimann, K. V., Tetley, R. M., and Krivak, B. M., *Plant Physiol.*, 59, 448—454, 1977. With permission.

be a reversible *structural* change. Several lines of evidence have recently shown a relationship between senescence of leaves and the opening or closing of their stomata.

In 1976 Kasamo[71] noted that peeling the epidermis of tobacco leaf discs decreased or nearly nullified the effect of kinetin on senescence. At about the same time Kuraishi[79] found that in *Paederia chinensis* and 4 other species whose stomata occur only on the lower side, kinetin had no effect on senescence, whether the leaves were floated with the stomata up, i.e., in air, or down, in the solution. (*Fatsia* did show a small kinetin effect, about 10% more chlorophyll with the stomata up, but the values are only just significant.) Kinetin promoted the transpiration of leaf discs of *Brassica* and *Nicotiana*, whose stomata occur on both sides, but not that of *Paederia*. Vaselining the leaves promoted chlorophyll loss in the light in both types. Furthermore, peeling the epidermis of several species of leaves decreased the normal senescence and largely reduced the effect of kinetin on it. These results have recently been confirmed and considerably extended, using agents that close stomata in the light and that open them in the dark.[163,164]

1. *Abscisic acid* is well known to close stomata in the light;[183,184] for instance it reduced the aperture in *Commelina* epidermal strips, in the light, to 10% of the control value.[168a] According to Raschke[131] it sensitizes the stomata to the CO_2 of the air. We find that it largely prevents the delaying action of light on the senescence of oat leaves.
2. *Phenyl mercuric nitrate,* which was shown by Zelitch and Waggoner in 1962 to act powerfully on stomatal closure in the light, is equally effective, accelerating senescence in light to almost its rate in the dark. Although the leaves floated on solution, their stomata remained tightly closed for about 3 days.
3. *Fusicoccin,* which was reported to cause stomatal opening in the dark,[170,171] has now been found to prevent or delay the onset of senescence in the dark. An example of the parallel between senescence and stomatal opening in the dark is shown in Figure 8.
4. *Kinetin,* the "classical" anti-senescence agent, is now found also to cause opening in the dark, as well as *maintenance* of opening in light. Even in leaves floating on solution the stomata gradually close in white light, but kinetin keeps them open, and correspondingly its action in preventing senescence is additive to that of light. Incidentally, this may explain the observation of Kazaryan et al.[74] that the apparent photosynthetic rate in isolated leaves is greatly increased when their petioles dip into collected xylem sap.

Three more clear-cut parallels have come to light. The chelator *α,α-dipyridyl* had

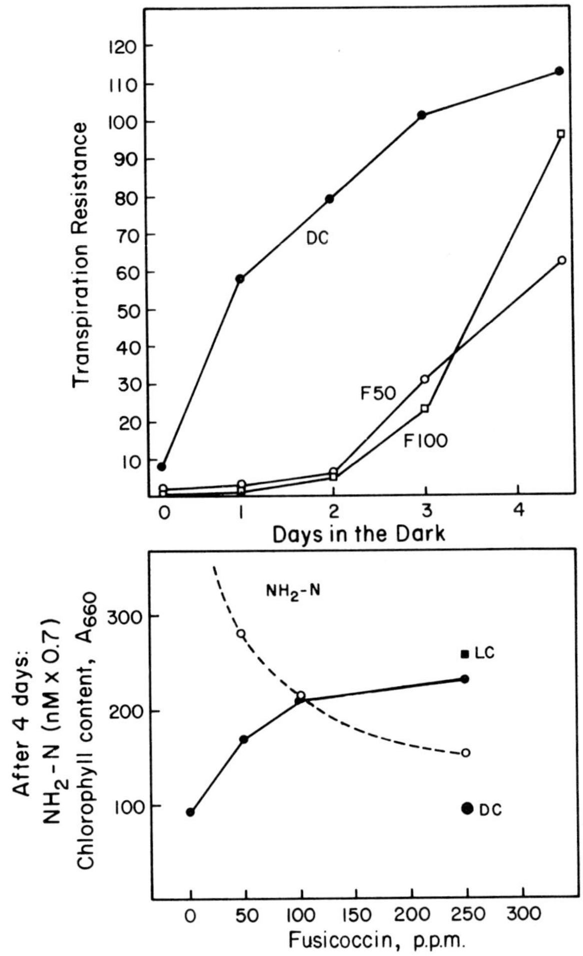

FIGURE 8. Above, time course of diffusion resistance, *r*, of oat leaves for 4 days in the dark. Fusicoccin 50 or 100 ppm maintains opening fully for 2 days and partially for the 3rd. Below, chlorophyll (solid line) and free amino acids (dashed line) after the 4th day in the dark, as function of Fusicoccin concentration. LC and DC, chlorophyll contents in light and dark, respectively. (From Reference 164, with permission.)

been reported to delay senescence in the dark;[160] it is now found highly effective in opening the stomata. Also in leaves floating on water at 2°, which delays senescence almost indefinitely, the stomata are indeed found to be wide open, and to stay open for at least 10 days in the dark. Objections based on possible toxicity of chemicals cannot hold in this case.

Thirdly, simple floating of the leaves on 1 *M* sucrose or mannitol causes stomatal closure, and typical senescence follows. Not only does this confirm the concept of stomatal control, but it opens the way to another step in interpretation. For these osmotic effects are reversed by kinetin. But since kinetin at 3 ppm could not antagonize 1 *M* solution osmotically, it was proposed[163] that stomatal closure might not of itself cause senescence; rather it might lead to an increase of some internal factor such as AbA which is more direct cause. Determinations of AbA in these leaves (current work) tend to support this interpertation.

In this work stomatal opening was measured by the diffusion resistance porometer, since direct observation of stomatal aperture is made difficult by their narrow linear form. Besides, some stomata open only at their extreme ends,[145] which would make direct measurement of aperture almost impossible. The opening is very sensitive to temperature, but under most laboratory conditions significant temperature fluctuations would probably be small.

C. A Possible Combination

The evidence that stomatal aperture controls, or varies in parallel with, senescence is thus rather complete, and the evidence that phosphorylation is also involved, while not so firm, is highly suggestive. It is essential, therefore, to look for relationships between these functions. A number of inhibitor studies on leaves already led Das and Raghavendra[35] to suggest that transpiration, i.e., stomatal aperture, is connected with cyclic photophosphorylation. Raschke[131] doubts this and considers that light acts only by lowering the CO_2 concentration in the intercellular spaces. Yet the arguments in (A) above, together with the obvious consideration that both the opening of stomata — mediated by the uptake of K^+ ions into the guard cells, and the closure — due to the reverse process, require some source of energy, point to the suggestion that the energy for stomatal movement is provided by phosphorylation and that this is at least largely under hormonal control. For opening in the dark the energy must of course come from mitochondrial phosphorylation, and we see how the earlier proposal (Section 4B above) that kinetin or other cytokinin acts to maintain or tighten the coupling between respiration and phosphorylation fits in well with this concept. In regard to the light effect, it is worth noting also that there is increasing evidence that the guard cells do not carry out total photosynthesis, but only some fraction of the process. This would fit well with their causing mainly cyclic photophosphorylation.

As yet the evidence on the role of phosphorylation is by no means firm, for actual determinations of ATP in senescing leaves are complicated by the balance between the production and utilization of the ATP.[89a] But the relationship between stomatal aperture and senescence seems very clear.

The accumulation of labeled amino acids in a leaf area treated with a cytokinin (the "Mothes" effect, described in Section IV.B above) can now find its explanation in stomatal opening. Indeed, Kuraishi and Ishikawa[80] have shown that vaselining part of the kinetin-treated area prevents such accumulation (see Figure 9). This obviously means that the accumulation takes place by passage of the labeled amino acid into the xylem and its subsequent movement in the transpiration stream to that part of the leaf which has open stomata.

Another and broader group of observations may be accommodated here. For it has long been known that water stress[4] and salinity stress[121] in intact plants, commonly cause leaf yellowing and often other symptoms of senescence. These symptoms may thus be simply due to stomatal closure. In this way three major aspects of the physiology of leaves — photophosphorylation, stomatal aperture, and senescence — may well be drawn together into a single syndrome. Which of the three is primary, and what is the order of sequence, remain to be determined.

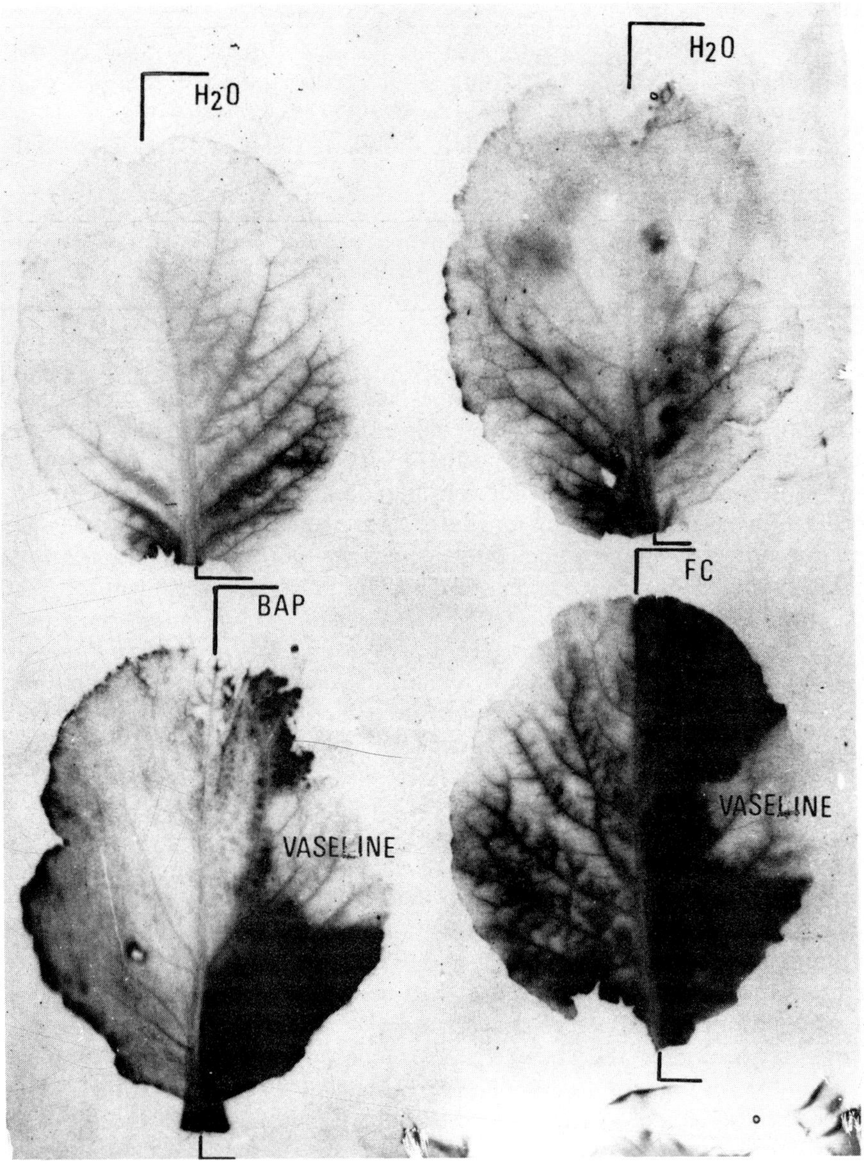

FIGURE 9. Radioautographs showing the effect of Vaseline® on the accumulation of ^{14}C label by cytokinin treatment (cf. Figure 3). ^{14}C-α-aminoisobutyric acid applied to petioles of four *Brassica* leaves; the right half of the leaf treated with water (both above), benzylaminopurine (below, left) or Fusicoccin (below, right). Part of the right side of the right-hand leaf above and of both the leaves below was coated with Vaseline®. Photographed after 3 days (1 day after the ^{14}C amino acid). (From Kuraishi and Ishikawa.[80] Original photo supplied by Dr. Kuraishi. With permission.)

REFERENCES

1. **von Abrams, G. J.**, An effect of ornithine on degradation of chlorophyll and protein in excised leaf tissue, *Z. Pflanzenphysiol.*, 72, 410—421, 1974.
2. **Addicott, F. T. and Lyon, J. L.**, Physiology of AbA and related substances, *Ann. Rev. Plant Physiol.*, 20, 139—164, 1969.
3. **Adedipe, N. O., Hunt, L. A., and Fletcher, R. A.**, Effect of benzyladenine on photosynthesis, growth, and senescence of bean plant, *Physiol. Plant.*, 25, 151—153, 1971.
4. **Aharoni, N., Blumenfeld, A., and Richmond, A. E.**, Hormonal activity in detached lettuce leaves as affected by leaf water content, *Plant Physiol.*, 59, 1169—1173, 1977.
5. **Anderson, J. B.**, Adenylate metabolism of embryonic axes from deteriorated soybean seeds, *Plant Physiol.*, 59, 610—614, 1977.
6. **Anderson, J. W. and Rowan, K. S.**, Activity of peptidase in tobacco leaf tissue in relation to senescence, *Biochem. J.*, 97, 741—746, 1965.
7. **Armstrong, D. J., Murai, N., Taller, B. J., and Skoog, F.**, Incorporation of cytokinin N^6-Benzyladenine into tobacco callus transfer-RNA and ribosomal RNA preparations, *Plant Physiol.*, 57, 15—22, 1976.
8. **Aspinall, D., Paleg L. G., and Addicott, F. T.**, Abscisin II and some hormone-regulated plant responses, *Aust. J. Biol. Sci.*, 20, 869—882, 1967.
9. **Atkin, R. K. and Srivastava, B. I. S.**, Changes in soluble proteins of excised barley leaves during senescence and kinetin treatment, *Physiol. Plant.*, 22, 242—250, 1969; 23, 304—315, 1970.
10. **Back, A. and Richmond, A. E.**, Interrelation between GA, cytokinins, and AbA in retarding leaf senescence, *Physiol. Plant.*, 24, 76—79, 1971.
11. **Banerji, D. and Laloraya, M. M.**, Chlorophyll formation in isolated pumpkin cotyledons in presence of kinetin and chloramphenicol, *Plant Cell Physiol.*, 8, 263—268, 1967.

11a. **Barton, R.**, Fine structure of mesophyll cells in senescing leaves of *Phaseolus, Planta,* 71, 314—325, 1966.

12. **Beever, J. E. and Woolhouse, H. W.**, Increased cytokinin export from the roots of *Perilla frutescens* following disbudding or floral induction, in *Mechanics of Regulation of Plant Growth,* Bielecki, R. L., Ferguson, A. R., and Cresswell, M. M., Eds., Roy. Soc. of N. Z. Bull., 12, 681—686, 1974.
13. **Beevers, L.**, Effect of gibberellic acid on the senescence of leaf discs of Nasturtium *(Tropaeolum majus), Plant Physiol.*, 41, 1074—1076, 1966.
14. **Beevers, L. and Guernsey, F. S.**, Interaction of growth regulators in the senescence of Nasturtium leaf disks, *Nature,* 214, 941—942, 1967.
15. **Bottger, I. and Wollgiehn, R.**, Untersuchangen über den Zusammenhang zwischen Nucleinsäure und Eiweisstoffwechsel in grunen Blattern, *Flora (Jena),* 146, 302—320, 1958.
16. **Brady, C. J., Patterson, B. D., Tung, H. F. and Smillie, R. M.**, Protein and RNA synthesis during aging of chloroplasts in wheat leaves, in *Autonomy and Biogenesis of Mitochondria and Chloroplasts,* North-Holland, Amsterdam, 1971, 453—465.
17. **Brennan, T. and Frenkel, C.**, Involvement of hydrogen peroxide in the regulation of senescence in peas, *Plant Physiol.*, 59, 411—416, 1977.
18. **Bruce, M. I., Zwar, J. A., and Kefford, N. P.**, Chemical structure and plant kinin activity — the activity of urea and thiourea derivatives, *Life Sci.*, 4, 461—466, 1965.
19. **Butler, R. D.**, The fine structure of senescing cotyledons of cucumber, *J. Exp. Bot.*, 18, 535—543, 1967.
20. **Butler, R. D. and Simon, E. W.**, Ultrastructural aspects of senescence in plants, *Adv. Gerontol. Res.*, 3, 73—129, 1971.
21. **Callow, M. E. and Woolhouse, H. W.**, Changes in nucleic acid metabolism in regreening leaves of *Perilla, J. Exp. Bot.*, 24, 285—294, 1973.
22. **Carr, D. J. and Pate, J. S.**, Ageing: an orderly development programme, *Symp. Soc. Exp. Biol.*, 21, 559—588, 1967.
23. **Chibnall, A. C.**, Investigations on the nitrogenous metabolism of higher plants, *Biochem. J.*, 18, 395—404, 1924.
24. **Chin, S. T.-Y. and Beevers, L.**, Changes in endogenous growth regulators in Nasturtium leaves during senescence, *Nature (London),* 92, 178—188, 1970.
25. **Choe, H. T. and Thimann, K. V.**, The senescence of isolated chloroplasts, *Planta,* 121, 201—203, 1974.
26. **Choe, H. T. and Thimann, K. V.**, The metabolism of oat leaves during senescence. III. The senescence of isolated chloroplasts, *Plant Physiol.*, 55, 828—834, 1975.
27. **Cockshull, K. E. and Hughes, A. P.**, Distribution of dry matter to flowers in *Chrysanthemum morifolium, Nature (London),* 215, 780—781, 1967.

28. **Cocucci, S. and Marrè, E.**, Ricerche sull'invecchiamento in foglie recise. II. Sull' effetto di protezioni degli zuccheri, della luce e della cinetine sugli acidi ribonucleici e sulla capacita fotosintetica, *Rend. Accad. Naz. (Lincei), Classe Sci. Fis., Mat., Nat., ser. 8,* 40, 1095—1102, 1966.
29. **Cocucci, M. C. and Marrè, E.**, The effect of cycloheximide on respiration, protein synthesis, and adenosine nucleotide levels in *Rhodotorula gracilis, Plant Sci. Lett.*, 1, 293—301, 1973.
30. **Colquhoun, A. J. and Hillman, J. R.**, The effects of AbA on senescence in leaf discs of radish, *Raphanus sativus* L., *Planta,* 105, 213—224, 1972.
31. **Colquhoun, A. J., Hillman, J. R., Crewe, C., and Bowes, B. G.**, An ultrastructural study of the effects of AbA on senescence of leaves of radish *(Raphanus sativus* L.), *Protoplasma,* 84, 205—221, 1975.
32. **Crowther, F.**, Studies on growth analysis of cotton plants under irrigation in the Sudan, *Ann Bot. (London),* 48, 877—913, 1934.
33. **Dale, J. E. and Felippe, G. M.**, Nitrogen movement into and out of the first leaf of barley, *Z. Pflanzenphysiol.*, 84, 27—53, 1977.
34. **Das, T. M. and Leopold, A. C.**, Physiological changes with leaf senescence, in *Plant Growth & Development,* Leopold, A. C., Ed., McGraw-Hill, New York, 1964.
35. **Das, V. S. R. and Raghavendra, A. S.**, Role of cyclic photophosphorylation in the control of stomatal opening, in *Mechanisms of Regulation of Plant Growth,* Bieleski, R. L., Ferguson, A. R., and Cresswell, M. M., Eds., Roy Soc. N. Z., Bull., 12, 455—460, 1974.
36. **DeLeo, P. and Sacher, J. A.**, Control of RNase and acid phosphatase by auxin and AbA during senescence of *Rhoeo* leaf sections, *Plant Physiol.*, 46, 806—811, 1970.
37. **Dennis, D. T., Stubbs, M., and Coultate, T. P.**, The inhibition of brussels sprout leaf senescence by kinins, *Can. J. Bot.*, 45, 1019—1024, 1967.
38. **Dove, L. D.**, Short term responses and chemical control of ribonuclease activity in tomato leaflets, *New Phytol.*, 70, 397—401, 1971.
39. **Draper, S. R. and Simon, E. W.**, Changes in free fatty acid content and respiratory activity during the senescence of cotyledons of cucumber, *J. Exp. Bot.*, 22, 481—486, 1971.
40. **Duysens, L. M. M. and Sweers, H. E.**, Mechanism of two photochemical reactions in algae as studied by means of fluorescence, in *Studies on Microalgae and Photosynthetic Bacteria,* Japan Society for Plant Physiology, Tokyo University Press, 1963, 353—372.
41. **Dyer, T. A. and Osborne, D. J.**, Leaf nucleic acids. II. Metabolism during senescence and effect of kinetin, *J. Exp. Bot.*, 22, 552—560, 1971.
42. **Ecklund, P. R. and Moore, T. C.**, Quantitative changes in gibberellin and RNA correlated with senescence of shoot apex in the Alaska pea, *Amer. J. Bot.*, 55, 494—503, 1968.
43. **Eder, A. and Huber, W.**, Zur Wirkung von AbA und Kinetin auf biochemische Verӓnderungen in *Pennisetum typhoides* unter Stresseinwirkungen, *Z. Pflanzenphysiol.*, 84, 303—311, 1977.
44. **El-Antably, H. M. M., Wareing, P. F., and Hillman, J.**, Some physiological responses to D,L-Abscisin (Dormin), *Planta,* 73, 74—90, 1967.
45. **Engelbrecht, L.**, Ueber Kinetinwirkung bei intakten Blӓttern von *Nicotiana rustica, Flora (Jena),* 154, 57—69, 1964.
46. **Etherton, B. and Rubinstein, B.**, Evidence of amino acid — proton co-transport in oat coleoptiles, *Plant Physiol.*, 61, 933—937, 1978.
47. **Even-Chen, Z. and Itai, C.**, The role of abscisic acid in the senescence of detached tobacco leaves, *Physiol. Plant.*, 34, 97—100, 1975.
48. **Uvardy, J. and Farkas, G. L.**, ATP stimulates the formation of nucleases in excised *Avena* leaves, *Z. Pflanzenphysiol.*, 69, 394—401, 1973.
49. **Fletcher, R. A.**, Retardation of leaf senescence by benzyladenine in intact bean leaves, *Planta,* 89, 1—8, 1969.
50. **Fletcher, R. A. and Osborne, D. J.**, Gibberellin as a regulator of protein and ribonucleic acid synthesis during senescence in leaf cells of *Taraxacum officinale, Can. J. Bot.*, 44, 739—745, 1966.
51. **Freeland, R. O.**, Effect of age on the rate of photosynthesis in some conifers, *Plant Physiol.*, 27, 685—690, 1952.
52. **Galston, A. W., Altman, A., and Kaur-Sawhney, R.**, Polyamines, ribonuclease, and the improvement of oat leaf protoplasts, *Plant Sci. Lett.*, 11, 69—79, 1978.
53. **Goldney, D. C. and van Steveninck, R. F. M.**, Ethylene production and biochemical changes in detached leaves of Nymphoides indica, in *Plant Growth Substances, 1970,* Carr, D. J., Ed., Springer-Verlag, Berlin, 1970, 604—610.
54. **Goldschmidt, E. E., Goren, R., Even-Chen, Z., and Bittner, S.**, Increase in free and bound abscisic acid during natural and ethylene-induced senescence of citrus fruit peel, *Plant Physiol.*, 51, 879—882, 1973.

55. **Goldsmith, M. H. M. and Thimann, K. V.**, Some characteristics of movement of IAA in coleoptiles of Avena. I. Uptake, destruction, immobilization, and distribution of IAA during basipetal translocation, *Plant Physiol.*, 37, 492—505, 1962.
56. **Goldthwaite, J. J. and Laetsch, W. M.**, Regulation of senescence in bean leaf disks by light and chemical growth regulators, *Plant Physiol.*, 42, 1757—1762, 1967.
57. **Goldthwaite, J. J. and Laetsch, W. M.**, Control of senescence in *Rumex* leaf discs by gibberellic acid, *Plant Physiol.*, 43, 1855—1858, 1968.
58. **Gorsheva, N. P.**, Age-related changes in RNase activity and RNA content in perennial plant *Libanotis intermedia, Byull. Gl. Bot. Sada.*, 83, 92—96, 1972, from *Chem. Abstr.*, 77, 149752, 1972.
59. **Gregory, F. G.**, Mineral nutrition of plants, *Annu. Rev. Biochem.*, 6, 557—578, 1937.
60. **Haber, A. H., Thompson, P. J., Walne, P. L., and Triplett, L. L.**, Non-photosynthetic retardation of chloroplast senescence by light, *Plant Physiol.*, 44, 1619—1628, 1969.
61. **Hamzi, H. Q. and Skoog, F.**, Kinetin-like growth-promoting activity of 1-substituted adenines (1-benzyl-6-aminopurine and 1-6,6-dimethylallyl-6-aminopurine), *Proc. Natl. Acad. Sci. U.S.A.*, 51, 76—83, 1964.
62. **Hardwick, K., Wood, M., and Woolhouse, H. W.**, Photosynthesis and respiration in relation to leaf age in *Perilla frutescens* (L.) Britt., *New Phytol.*, 67, 79—86, 1968.
63. **Hernandez-Gil., R. and Schaedle, M.**, Functional and structural changes in senescing *Populus deltoides* chloroplasts, *Plant Physiol.*, 51, 245—249, 1973.
64. **Hewett, E. W. and Wareing, P. F.**, Cytokinins in *Populus* X *robusta* (Schneid.); Light effects on endogenous levels, *Planta*, 114, 119—129, 1973.
65. **Hopkinson, J. M.**, Studies on the expansion of the leaf surface. VI. Senescence and the usefulness of old leaves, *J. Exp. Bot.*, 17, 762—770, 1966.
66. **Howe, R. C. and Ursino, D. J.**, RNA degrading activity associated with ribosomes from spinach leaf chloroplasts, *Can. J. Bot.*, 50, 691—695, 1972.
67. **Itai, C. and Vaadia, Y.**, Cytokinin activity in water-stressed shoots, *Plant Physiol.*, 47, 87—90, 1971.
67a. **Jacobi, G., Klemme, B., Krapf, G., and Postius Ch.**, Dark starvation and plant metabolism. IV. The alteration of enzyme activities, *Biochem. Physiol. Pflanz.*, 168, 242—256, 1975.
68. **Jenns, S. M. and Bryant, J. A.**, Correlation between DNase activity and DNA replication in the embryonic axes of germinating peas, *Planta*, 138, 99—103, 1978.
69. **Jerie, P. H. and Hall, M. A.**, The identification of ethylene oxide as a major metabolite of ethylene, *Proc. Roy. Soc. London, Ser. B*, 200, 87—94, 1978.
70. **Kar, M. and Mishra, D.**, Catalase, peroxidase, and polyphenoloxidase activities during rice leaf senescence, *Plant Physiol.*, 57, 315—319, 1976.
71. **Kasamo, K.**, The role of the epidermis in kinetin-induced retardation of chlorophyll degradation in tobacco leaf discs during senescence, *Plant Cell Physiol.*, 17, 1297—1307, 1976.
72. **Kawashima, N.**, Protein metabolism in higher plants. Effect of light on incorporation of C^{14}-labeled leucine into chloroplastic and cytoplasm proteins by tobacco leaf discs, *Hatano Tabako Shikenjo Hokoku*, 71, 117—120, from *Chem. Abstr.*, 79, 2872, 1972.
73. **Kawashima, N. and Mitake, T.**, Studies on protein metabolism in higher plants. XI, *Agric. Biol. Chem.*, 33, 537—543, 1969.
74. **Kazaryan, V. O., Davtyan, V. O., and Chilingaryan, A. A.**, Role of the roots in action of the leaves during the stage of generative development of plants, *Sov. Plant Physiol.*, 20, 588—594, 1973.
75. **Kende, H.**, Kinetin-like factors in the root exudate of sunflowers, *Helianthus annuus, Proc. Natl. Acad. Sci. U.S.A.*, 53, 1302—1307, 1965.
76. **Krotkov, G.**, Carbohydrate and respiratory metabolism in the isolated starving leaf of wheat, *Plant Physiol.*, 14, 203—229, 1939.
77. **Krul, W. A.**, Nucleic acid and protein metabolism of senescing and regenerating soybean cotyledons, *Plant Physiol.*, 54, 36—40, 1974.
78. **Kuraishi, S.**, The effect of kinetin on protein level of *Brassica* leaf discs, *Physiol. Plant.*, 21, 78—83, 1968.
79. **Kuraishi, S.**, Ineffectiveness of cytokinin-induced chlorophyll retention in hypostomatous leaf discs, *Plant Cell Physiol.*, 17, 875—885, 1976.
80. **Kuraishi, S. and Ishikawa, F.**, Relationship between transpiration and amino acid accumulation in *Brassica* leaf discs treated with cytokinins and fusicaccin, *Plant Cell Physiol.*, 18, 1273—1279, 1977.
81. **Kursanov, A. L., Kulaeva, O. N., Sveshrikova, I. N., Popova, E. A., Volyakina, Y. P., Klyachko, N. L., and Vorolieva, I. P.**, Restoration of cellular structures and metabolism in yellow leaves due to the action of 6-benzylaminopurine, *Fiziol. Rast.*, 11, 838—847, 1964.
82. **Laurière, C., Skakoun, A., and Daussant, J.**, Adaptation de methodes immunochimiques a l'étude de l'évolution autogènique de protéines: fraction I protéique et malate deshydrogenase des feuilles de Ble en voie de senescence, *Physiol. Vég.*, 13, 467—478, 1975.
83. **Lee, P. K.-W., Kessler, B., and Thimann, K. V.**, The effect of Hadacidin on bud development and its implications for apical dominance, *Physiol. Plant.*, 31, 11—14, 1974.

84. **Leopold, A. C., Niedergang-Kamien, E., and Janick, J.,** Experimental modification of plant senescence, *Plant Physiol.*, 34, 570—573, 1959.
85. **Letham, D. S.,** Regulators of cell division in plant tissues. V. A comparison of the activities of Zeatin and other cytokinins in five bioassays, *Planta*, 74, 228—242, 1967.
86. **Locy, R. D. and Cherry, J. H.,** Evidence for a specific tyrosyl-tRNA degrading activity, *Biochem.*
87. **Locy, R. D. and Cherry, J. H.,** Purification and characterization of two tyrosyl-tRNA synthetase activities from soybean cotyledons, *Phytochemistry*, 17, 19—28, 1978.
88. **MacNicol, P. K., Young, R. E., and Biale, J. B.,** Metabolic regulation in the senescing tobacco leaf. I. Changes in pattern of ^{32}P incorporation into leaf disc metabolites, *Plant Physiol.*, 51, 793—797, 1973.
89. **Makovetski, S. and Goldschmidt, E. E.,** A requirement for cytoplastic protein synthesis during chloroplast senescence in the aquatic plant *Anacharis canadensis*, *Plant Cell Physiol.*, 17, 859—862, 1976.
89a. **Malik, N. and Thimann, K.,** The metabolism of oat leaves during senescence. VI. Changes in ATP levels, *Plant Physiol.*, in press.
90. **Malik, N. S. A. and Berrie, A. M. M.,** Correlative effects of fruits and leaves in senescence of pea plant, *Planta*, 124, 169—175, 1975.
91. **Manos, P. J. and Goldthwaite, J.,** A kinetic analysis of the effects of GA, Zeatin, and AbA on leaf tissue senescence in *Rumex*, *Plant Physiol.*, 55, 192—198, 1975.
92. **Margulies, M. M.,** Effect of chloramphenicol on light-dependent synthesis of proteins and enzymes of leaves and chloroplasts of *Phaseolus vulgaris*, *Plant Physiol.*, 39, 579—585, 1964.
93. **Marrè, E., Schwendiman, M., and Lado, P.,** Ricerche sull'in schiamento in foglie recise. I. Effetto protettivo degli zuccheri e della luce sui sistemi fotosintetico e respiratorio de foglie recise de *Solanum tuberosum*, *Rend. Accad. Naz. (Lincei), Classe Sci. Fis., Mat., Nat., Ser. 8*, 40, 1089—1094, 1966.
94. **Martin, C. and Thimann, K. V.,** The role of protein synthesis in the senescence of leaves. I. The formation of protease. II. The influence of amino acids on senescence, *Plant Physiol.*, 49, 64—71; 52, 294—303, 1972.
95. **Matile, P.,** *The Lytic Compartment of Plant Cells*, Springer-Verlag, Berlin, 1975.
96. **McGlasson, W. B., Poovaiah, B. W., and Dostal, H. C.,** Ethylene production and respiration in aging leaf segments and in disks of fruit tissue of normal and mutant tomatoes, *Plant Physiol.*, 56, 547—549, 1975.
97. **Milborrow, B. V.,** The chemistry and physiology of Abscisic acid, *Annu. Rev. Plant Physiol.*, 25, 259—307, 1974.
98. **Millikan, D. F. and Ghosh, B. N.,** Changes in nucleic acid associated with maturation and senescence in *Hedera helix*, *Physiol. Plant.*, 24, 10—13, 1971.
99. **Mingo-Castel, A. M., Young, R. E., and Smith, O. E.,** Kinetin-induced tuberization of potato in vitro: on the mode of action of kinetin, *Plant Cell Physiol.*, 17, 557—570, 1976.
100. **Mishra, D. and Misra, B.,** Retardation of induced senescence of leaves from crop plants by benzimidazole and cytokinins, *Exp. Gerontol.*, 8, 235—239, 1973.
101. **Mishra, D. and Pradhan, P. K.,** Regulation of senescence in detached rice leaves by light, benzimidazole and kinetin, *Exp. Gerontol.*, 8, 153—155, 1973.
102. **Misra, G. and Biswal, U. C.,** Factors concerned in leaf senescence. I. Effects of age, chemicals, petiole, and photoperiod on senescence in detached leaves of *Hibiscus rosa-sinensis* L., *Bot. Gaz. (Chicago)*, 138, 5—11, 1973.
103. **Mittelheuser, C. J. and van Steveninck, R. F. M.,** Stomatal closure and inhibition of transpiration induced by (RS)-abscisic acid, *Nature (London)*, 221, 281—282, 1969.
104. **Mizrahi, Y., Amir, J., and Richmond, A. E.,** The mode of action of kinetin in maintaining the protein content of detached *Tropaeolum majus* leaves, *New Phytol.*, 69, 355—361, 1970.
105. **Mlodzianowski, F. and Ponitka, A.,** Ultrastructural changes in chloroplasts in detached parsley leaves yellowing in darkness and the influence of kinetin on that process, *Z. Pflanzenphysiol.*, 69, 13—15, 1973.
106. **Molisch, H.,** *Die Lebensdauer der Pflanzen*, Eng. transl., Fulling, H., *The Longevity of Plants*, Science Press, Lancaster, 1938.
107. **Mothes, K. and Engelbrecht, L.,** Kinetin induced directed transport of substances in excised leaves in dark, *Phytochem.*, 1, 58—62, 1961.
108. **Muller, K. and Leopold, A. C.,** Correlative aging and transport of P^{32} in corn leaves under the influence of kinetin, *Planta*, 68, 167—185, 1966.
109. **Murai, N., Taller, B. J., Armstrong, D. J., Skoog, F., Micke, M. A., and Schnoes, H. K.,** Kinetin incorporated into tobacco callus r-RNA and t-RNA preparations, *Plant Physiol.*, 60, 197—202, 1977.
110. **Nishimura, M. and Beevers, H.,** Hydrolases in vacuoles from castor bean endosperm, *Plant Physiol.*, 62, 44—48, 1978.
111. **Oota, Y.,** RNA in developing plant cells, *Annu. Rev. Plant Physiol.*, 15, 17—36, 1964.

112. **Oritani, T. and Yoshida, R.,** Studies on nitrogen metabolism in crop plants. XII. Cytokinins and AbA -like substance levels in riceand soybean leaves during their growth and senescence, *Proc. Crop. Sci. Soc. Japan,* 42, 280—287, 1973.

113. **Osborne, D. J.,** Acceleration of abscission by a factor produced in senescent leaves, *Nature (London),* 176, 1161—1163, 1955.

114. **Osborne, D. J.,** The effect of kinetin on protein and nucleic acid metabolism in *Xanthium* leaves during senescence, *Plant Physiol.,* 37, 595—602, 1962.

115. **Osborne, D. J. and Hallaway, H. M.,** Auxin control of protein levels in detached autumn leaves, *Nature (London),* 188, 240—241, 1960.

116. **Osborne, D. J.,** Hormonal regulation of leaf senescence, *Symp. Soc. Exp. Biol.,* 21, 179—213, 1967.

117. **Osborne, D. J. and Hallaway, M.,** The auxin 2,4D as a regulator of protein synthesis and senescence in detached leaves of *Prunus, New Phytol.,* 63, 334—346, 1964.

118. **Osborne, D. J.,** Hormonal mechanisms regulating senescence and abscission, in *Biochemistry and Physiology of Plant Growth Substances,* Wightman, F. and Setterfield, G., Eds., Runge Press, Ottawa, 1968, 815—840.

119. **Osborne, D. J., Jackson, M. B., and Milborrow, B. V.,** Physiological properties of abscission accelerator from senescent leaves, *Nature (London) New Biol.,* 240, 98—101, 1972.

120. **Paranjothy, K. and Wareing, P. F.,** The effects of AbA, kinetin and 5-fluorouracil on RNA and protein synthesis in senescing radish leaf disks, *Planta,* 99, 112—119, 1971.

121. **Parisco, J. T. and O'Leary, J. W.,** Enhancement of intact bean leaf senescence by NaCl salinity, *Physiol. Plant.,* 27, 95—100, 1972.

122. **Parish, R. W.,** Studies on senescing tobacco leaf disks with special reference to peroxidase. I. The effects of cutting and of inhibition of nucleic acid and protein synthesis. II. The effects of interactions of proline, hydroxyproline and kinetin, *Planta,* 82, 1—13; 14—21, 1968.

123. **Parish, R. W.,** The lysosome concept in plants. II. Location of acid hydrolases in maize root tips, *Planta,* 123, 15—31, 1975.

124. **Peoples, M. B. and Dalling, M. J.,** Degradation of ribulose bisphosphate carboxylase by proteolytic enzymes from crude extracts of wheat leaves, *Planta,* 138, 153—160, 1978.

125. **Person, C., Samboradi, D. J., and Forsyth, F. R.,** Effect of benzimidazole on detached wheat leaves, *Nature (London),* 180, 1294—1295, 1975.

126. **Peterson, L. W. and Huffaker, R. C.,** Loss of RDP carboxylase and increase of proteolytic activity in senescence of detached primary barley leaves, *Plant Physiol.,* 55, 1009—1015, 1975.

127. **Peterson, L. W., Kleinkopf, G. E., and Huffaker, R. C.,** Evidence for lack of turnover of RuDP carboxylase in barley leaves, *Plant Physiol.,* 51, 1042—1045, 1973.

128. **Phan, C. T. and Hsu, H.,** L'ethylêne et le déverdissage des tissus végêtaux, *Physiol. Vég.,* 13, 427—434, 1975.

128a. **Phillips, I. D. J.,** *The Chemistry and Physiology of Plant Growth Hormones,* McGraw-Hill, New York, 1971.

129. **Raafat, A., Hofner, W., and Linser, H.,** $^{14}CO_2$ Assimilation during photosynthesis of ageing bean seedlings, *Z. Pflanzenphysiol.,* 64, 22—33, 1971.

130. **Racusen, D. W. and Aronoff, S.,** Metabolism of soya bean leaves. VI. Exploratory studies in protein metabolism, *Arch. Biochem. Biophys.,* 51, 68—78, 1954.

131. **Raschke, K.,** Stomatal action, *Annu. Rev. Plant Physiol.,* 26, 309—340, 1975.

132. **Rennert, A. and Knypl, J. S.,** Inhibition of chlorophyll and RNA synthesis by N-hydroxyurea, *Acta Soc. Bot. Pol.,* 40, 669—679, 1971.

133. **Richardson, S. D.,** The effect of leaf age on the rate of photosynthesis in detached leaves of tree seedlings, *Acta Bot. Neerl.,* 6, 445—457, 1957.

134. **Richmond, A. E., Back, A., and Sachs, B.,** A study of the hypothetical role of cytokinins in completion of tRNA, *Planta,* 90, 57—65, 1970.

135. **Richmond, A. E. and Lang, A.,** Effect of kinetin on protein content and survival of detached *Xanthium* leaves, *Science,* 125, 650—651, 1957.

136. **Rubinstein, B. and Leopold, A. C.,** The nature of leaf abscission, *Q. Rev. Biol.,* 39, 356—372, 1964.

137. **Sacher, J. A.,** Relationship between auxin and membrane integrity in tissue senescence and abscission, *Science,* 125, 1199—1200, 1957.

138. **Sacher, J. A.,** Studies of auxin-membrane permeability relations in fruit and leaf tissues, *Plant Physiol.,* 13, 365—372, 1959.

139. **Sankhla, N. and Sankhla, D.,** Abscisin II — Kinetin interaction of leaf senescence, *Experientia,* 24, 294—295, 1968.

140. **Schreyer, S. L. and Noodén, L. D.,** Screening nutrients and growth promoters or inhibitors for ability to mimic or antagonize the death signal in soy beans, *Plant Physiol.,* 56 (Suppl.), 83, 1975.

141. **Sestak, Z.,** Changes in the chlorophyll content as related to photosynthetic activity and age of leaves, *Photochem. Photobiol.,* 2, 101—110, 1963.

142. **Shaw, M., Bhattacharya, P. K., and Quick, W. A.,** Chlorophyll, protein, and nucleic acid levels in detached senescing wheat leaves, *Can. J. Bot.*, 43, 739—746, 1965.
143. **Shaw, M. and Manocha, M. S.,** Fine structure in detached senescing wheat leaves, *Can. J. Bot.*, 43, 747—755, 1965.
144. **Shibaoka, H. and Thimann, K. V.,** Antagonisms between kinetin and amino acids: experiments on the mode of action of cytokinins, *Plant Physiol.*, 46, 212—220, 1970.
145. **Shiraishi, M., Hashimoti, Y., and Kuraishi, S.,** Cyclic variations of stomatal aperture observed under the scanning electron microscope, *Plant Cell Physiol.*, 19, 637—645, 1978.
146. **Simon, E. W.,** Types of leaf senescence, *Symp. Soc. Exp. Biol.*, 21, 215—230, 1967.
147. **Singh, B. N. and Lal, K. N.,** Investigation of the effect of age on assimilation of leaves, *Ann. Bot. (London)*, 49, 291—307, 1935.
148. **Singh, N. and Mishra, D.,** The effect of benzimidazole and red light on the senescence of detached leaves of *Oryza sativa* cv. Ratna., *Physiol. Plant.*, 34, 67—74, 1965.
149. **Sloger, C. and Caldwell, B. E.,** Response of cultivars of soybean to synthetic AbA, *Plant Physiol.*, 46, 634—635, 1970.
150. **Smillie, R. M. and Krotkov, G.,** Changes in the dry weight, protein, nucleic acid, and chlorophyll contents of growing pea leaves, *Can. J. Bot.*, 39, 891—900, 1961.
151. **Smillie, R. M.,** Photosynthetic and respiratory activities of growing pea leaves, *Plant Physiol.*, 37, 716—721, 1962.
152. **Smith, O. E., Lyon, J. L., Addicott, F. T., and Johnson, R. E.,** Abscission physiology of Abscisic acid, in *Biochemistry and Physiology of Plant Growth Substances*, Wightman, F. and Setterfield, G., Eds., Runge Press, Ottawa, 1968, 1547—1560.
153. **Sorokin, E. and Krauss, R. L. S.,** Relative efficiency of photosynthesis in the course of cell development, *Biochem. Biophys. Acta*, 48, 314—319, 1961.
154. **Srivastava, B. I. S.,** Increase in chromatin-associated nuclease activity of excised barley leaves during senescence and its suppression by kinetin, *Biochem. Biophys. Res. Commun.*, 32, 533—538, 1968.
155. **Srivastava, B. I. S. and Ware, G.,** The effect of kinetin on nucleic acids and nucleases of excised barley leaves, *Plant Physiol.*, 40, 62—64, 1965.
156. **Storey, R. and Beevers, L.,** Proteolytic activity in relation to senescence and cotyledonary development in *Pisum sativum* L., *Planta*, 124, 77—87, 1977.
157. **Stogey, G.hand Bzevzrs, L.,** Enzymology of glutamine metabolism related to senescence and seed development in the pea *(Pisum sativum* L.), *Plant Physiol.*, 61, 494—500, 1978.
158. **Takegami, T.,** A study on senescence in tobacco leaves. II. Chloroplast and cytoplasmic t-RNA's, *Plant Cell Physiol.*, 16, 417—425, 1975.
159. **Tetley, R. M. and Thimann, K. V.,** The metabolism of oat leaves during senescence. I. Respiration, carbohydrate metabolism, and the action of cytokinins, *Plant Physiol.*, 54, 294—303, 1974.
160. **Tetley, R. M. and Thimann, K. V.,** The metabolism of oat leaves during senescence. IV. The effects of α,α-dipyridyl and other metal chelators on senescence, *Plant Physiol.*, 56, 140—142, 1975.
161. **Theologis, A. and Laties, G. G.,** Membrane lipid integrity as a prerequisite element of cyanide-resistant respiration in potato slices, *Plant Physiol.*, 57 (Suppl.), 483, 1976.
161a. **Thimann, K. V.,** *Hormone Action in the Whole Life of Plants*, University of Massachusetts Press, Amherst, 1977, 448 pp.
162. **Thimann, K. V.,** Senescence, *Bot. Mag.*, Tokyo Special Issue 1, 19—43, 1978.
163. **Thimann, K. V. and Satler, S.,** Interrelation between leaf senescence and stomatal aperture, *Proc. Nat. Acad. Sci.*, 76, 2295—2298; 2770—2773, 1979.
164. **Thimann, K. V., Malik, N., and Satler, S.,** Stomatal Aperture and the Senescence of Leaves, in Proc. NATO Adv. Study Institute, Plant Regulation and World Agriculture, Scott, T. K., Ed., in press, 1979.
165. **Thimann, K. V., Tetley, R. M., and Tran Thanh Van, M.,** The metabolism of oat leaves during senescence. II. Senescence in leaves attached to the plant, *Plant Physiol.*, 54, 858—862, 1974.
166. **Thimann, K. V., Tetley, R. M., and Krivak, B. M.,** The metabolism of oat leaves during senescence. V. Senescence in light, *Plant Physiol.*, 59, 448—454, 1977.
167. **Thomas, H. and Stoddart, J. L.,** Separation of chlorophyll degradation from other senescence processes in leaves of a mutant genotype of meadow fescue *(Festuca pratensis* L.), *Plant Physiol.*, 56, 438—443, 1975.
168. **Tran Thanh Van, M. and Driva, A.,** Definition of a simple experimental system of directed organogenesis *de nuevo*, in, Coll. Int. C.N.R.S. No. 193 (Les Cultures de Tissus de Plantes), C.N.R.S., Paris, 1971, 169—176.
168a. **Tucker, D. J. and Mansfield, T. A.,** A simple bio-assay for detecting "Antitranspirant" activity of naturally occurring compounds such as abscisic acid, *Planta*, 98, 157—163, 1971.

169. **Tulecke, W. and Rutner, A.**, Changes in amino acid composition of medium and cells of a plant tissue culture during growth in a liquid medium containing arginine, in *Proceedings, International Conference on Plant Tissue Culture,* White, P. and Grove, A. R., Eds., McCutcheon Publishing Corp., Berkeley, Calif., 1965, 103—116.

169a. **Tung, H. F. and Brady, C. J.**, Kinetin treatment and protein synthesis in detached wheat leaves, in *Plant Growth Substances 1970,* Carr, D. J., Ed., Springer, Berlin, 1970, 589—597.

170. **Turner, N. C. and Graniti, A.**, Fusicoccin: a fungal toxin that opens stomata, *Nature (London),* 223, 1070—1071, 1969.

171. **Turner, N. C.**, Potassium ion uptake of guard cells stimulated by fusicoccin, *Nature (London),* 235, 341—342, 1972.

172. **Udvardy, J., Farkas, G. L., Marré, E., and Fortes, G.**, The effects of sucrose and light on the level of soluble and particle-bound ribonuclease activities in excised *Avena* leaves, *Plant Physiol.,* 20, 781—785, 1967.

172a. **Uvardy, J. and Farkas, G. L.**, ATP stimulates the formation of nucleases in excised *Avena* leaves, *Z. Pflanzenphysiol.,* 69, 394—401, 1973.

173. **Vickery, H. B., Pucher, G. W., Wakeman, A. J., and Leavenworth, C. S.**, Chemical investigations of the tobacco plant. VI. Chemical changes that occur in leaves during culture in light and in darkness, *Conn. Agr. Exp. Sta. New Haven Bull.,* 399, 757—832, 1937.

174. **Vickery, H. B., Pucher, G. W., Wakeman, A. J., and Leavenworth, C. S.**, Chemical investigations of the rhubarb plant, *Conn. Agr. Exp. Sta. New Haven Bull.,* 424, 1—157, 1939.

175. **Wareing, P. F., El-Antably, H. M. M., Good, J., and Manuel, J.**, The possible role and mode of action of Abscisin II (Dormin) in the regulation of plant growth and development, Proc. Int. Conf. Growth Regulators in Plants, Rostock, 1967.

176. **Whyte, P. and Luckwill, L. C.**, A sensitive bioassay for gibberellins based on retardation of leaf senescence in *Rumex obtusifolius, Nature (London),* 210, 1360, 1966.

177. **Williams, R. F.**, Redistribution of mineral elements during development, *Annu. Rev. Pl. Physiol.,* 6, 25—42, 1955.

178. **Wilson, C. M.**, Purification of corn ribonuclease, *J. Biol. Chem.,* 242, 2260—2263, 1967.

179. **Wittenbach, V. A.**, Induced senescence of intact wheat seedlings and its reversibility, *Plant Physiol.,* 59, 1039—1042, 1977.

180. **Wollgiehn, R.**, Nucleic acid and protein metabolism of excised leaves, *Symp. Soc. Exp. Biol.,* 21, 231—266, 1967.

181. **Wood, J. G. and Cruikshank, D. H.**, The metabolism of starving leaves, *Aust. J. Exp. Biol. Med. Sci.,* 22, 111—123, 1944.

182. **Woolhouse, J. W.**, The nature of senescence in plants, *Symp. Soc. Exp. Biol.,* 21, 179—213, 1967.

183. **Wright, S. T. C. and Hiron, R. W. P.**, (+)-Abscisic acid, the growth inhibitor induced in detached wheat leaves following a period of wilting, *Nature (London),* 224, 719—720, 1969.

184. **Wright, S. T. C. and Hiron, R. W. P.**, The accumulation of abscisic acid in plants during wilting and under other stress conditions, in *Plant Growth Substances 1970,* Carr, D. J., Ed., Springer, Berlin, 1972, 291—298.

185. **Wyen, N. V., Erdei, S., Udvardy, J., Bagi, G., and Farkas, G. L.**, Hormonal control of nuclease level in excised *Avena* leaf tissues, *J. Exp. Bot.,* 23, 37—44, 1972.

186. **Yemm, E. W.**, Respiration of barley leaves. II. Carbohydrate concentration and CO_2 production in starving leaves, *Proc. R. Soc. London Ser. B,* 117, 504—525, 1935.

187. **Yemm, E. W.**, Respiration of barley leaves. III. Protein catabolism in starving leaves, *Proc. R. Soc. London Ser. B,* 123, 243—273, 1937.

188. **Yemm, E. W.**, The metabolism of senescent leaves, in *Ciba Found. Colloq. on Ageing,* 2, 202—210, 1956.

189. **Yoshida, Y.**, The role of the cell nucleus in relation to the senescence of chloroplasts in detached *Elodea* leaves, *Symp. Cell Biol.,* 25, 131—140, 1974.

190. **Yoshida, Y.**, Nuclear control of chloroplast activity in *Elodea* leaf cells, *Protoplasma,* 54, 476—492, 1961.

Chapter 6

THE ROLE OF PEPTIDE HYDROLASES IN LEAF SENESCENCE

Graeme J. T. Frith and Michael J. Dalling

TABLE OF CONTENTS

I. INTRODUCTION

A. Historical

The earliest studies of peptide hydrolases in green leaves were made at the beginning of this century,[1,2] but the presence of protein degrading enzymes in plants was recorded at least 150 years earlier (cited by Vines[3]), and no doubt their existence was known before then. While the validity of the experimental procedure of several early investigations may be questioned today, the approach in many cases was sound and showed considerable forethought. Consequently, some of the early studies and the deductions made at the time are still valid now.

In 1919, Fisher[4] recorded that extracts from the leaves of plants in the post-anthesis stage degraded protein such as legumin and peptone more rapidly than leaf extracts from vegetative plants. Further, Mothes[5] showed that in water-stressed plants, the older leaves, which wilted first, had higher proteolytic activity than turgid leaves, and he claimed that protein degradation products from the older leaves were translocated to the younger leaves. The ubiquitous nature of leaf peptide hydrolases was demonstrated by Schaede[6] in 1934, but with few exceptions[7-10] these enzymes were hardly studied until a great surge of interest in the past 10 to 15 years.

B. Scope of Review

This review will be restricted to the discussion of leaf enzymes (including those from petals) that degrade protein, and in particular their role in senescence. However, we draw the reader's attention to a considerable body of work since Ryan's review in 1973,[11] which includes studies of peptide hydrolases in almost every part of the plant including fruits[12-16] and roots.[17] In particular, several investigations have been made of peptide hydrolases in cotyledons, e.g., References 18 to 21. While there is some inclination to consider cotyledons in a review of leaves, degradation of cotyledonary protein is more akin to that of endosperm than to that of leaves. The storage proteins of cotyledons and endosperm differ from leaf proteins in both character and location, and whereas the onset of leaf senescence is associated with high protein level, the opposite is the case in cotyledons.

C. Nomenclature

The nomenclature of peptide hydrolases has always been confusing and prompted comment by Fisher[4] as early as 1919. Of particular note is the confusion between *protease* and *proteinase*. *Protease* originally was the general term for enzymes hydrolyzing peptide bonds, but now, in many instances, is used as a synonym for proteinase. It is not recommended by the Nomenclature Committee of the International Union of Biochemistry and is not indexed by several of the larger abstracting journals.

Further confusion exists where authors have assumed that enzymes degrading proteins such as hemoglobin and casein are *ipso facto*, proteinases (i.e., endopeptidases: EC 3.4.21-24) despite the fact that there are several plant and animal peptidases (i.e., exopeptidases: EC 3.4.11-17), which also degrade such proteins.

In this review we will conform to the recommendations of the 1978 Committee; using *peptide hydrolase* for enzymes hydrolyzing the peptide bond, *proteinase* if the hydrolysis releases a peptide of several amino acid residues and *peptidase (amino-* and *carboxy-*) if only terminal residues are released.

D. Source Plants

Investigations of leaf peptide hydrolases have been made almost exclusively in plants of economic significance, usually those important for food production. The cereals have received the greatest attention, wheat[22-37] being the most extensively studied, but also studied have been oats,[38-42] maize,[43-45] rice,[46-49] and barley.[50] Apart from the cereals, tobacco[51-58] has received considerable attention, as have some legumes.[23,59-64] Less attention has been paid to perennial fruit trees[65-69] and *Agave*[70-72] and there have been single reports on several other species.[73-77] Plants for the study of petal peptide hydrolases have been chosen for their scientific suitability rather than their agricultural importance.[78-81]

II. ENZYME CHARACTERISTICS

A. Wheat

Peptide hydrolases capable of degrading protein substrates at acid and neutral pH have been extracted from wheat leaves. Almost all reports indicate the presence of enzymes degrading hemoglobin, casein, RuBPCase, (Ribulose 1,5-bisphosphate carboxylase), or other protein at acid pH, usually pH 5 or below.[22,24,25,29-31,34-37] It appears that up to six peptide hydrolases with acid pH optima may occur in wheat leaves[25] (Figure 1), and that at least three of these are carboxyl proteinases that degrade hemoglobin at about pH 4.5. On the basis of its inhibition by diazoacetyl-DL-norleucine methyl ester and 1,2-epoxy-3-(p-nitrophenoxy)-propane, one of these carboxyl proteinases appeared to have the characteristics of the "pepsin-like" group,[27] although a serine residue at the active site was also implicated.[28]

Degradation of RuBPCase by carboxyl proteinases from wheat leaves occurred at about pH 5, a slightly higher value than that for the degradation of hemoglobin.[26,36,37] Acid peptide hydrolase(s) degrading RuBPCase were inactivated by sulfhydryl inhibitors and were stimulated by cysteine and dithiothreitol, suggesting the presence of a SH-proteinase.[37]

Neutral peptide hydrolases from wheat leaves have also been identified in several instances, usually by their degradation of casein at pH 7.[23,29,31,34] It is of interest that several of the reports of enzymes with acid pH optima do not record the presence of any activity at neutral pH.

B. Oats

When hemoglobin was used as substrate, oat leaves exhibited two distinct regions of peptide hydrolase activity, one at acid pH and the other at neutral pH,[38-41] although the degradation of phytochrome by green leaf extracts occurred at the latter pH.[42] Degradation of hemoglobin at acid pH seemed to be predominantly due to the presence of a proteinase with a serine residue at the active site, but the neutral enzyme, which was stimulated by sulfhydryl reagents, exhibited peptidase activity. Both enzymes have molecular weights estimated at 76,000.[38]

C. Maize

Both aminopeptidase and carboxypeptidase activities have been detected in maize leaves,[43] as well as enzymes with acid and neutral pH optima for casein degradation.[43-45] The optimal degradation of casein by maize leaf extracts at acid pH occurred at a higher pH than casein degradation by wheat leaves[24,30] or hemoglobin degradation by wheat[24] or oat leaves.[38]

D. Rice and Barley

To date, only peptide hydrolases active at neutral pH have been reported in rice leaves[47-49] and barley.[50] Activity at this pH was stimulated by sulfhydryl reagents such as cysteine and dithiothreitol.[49]

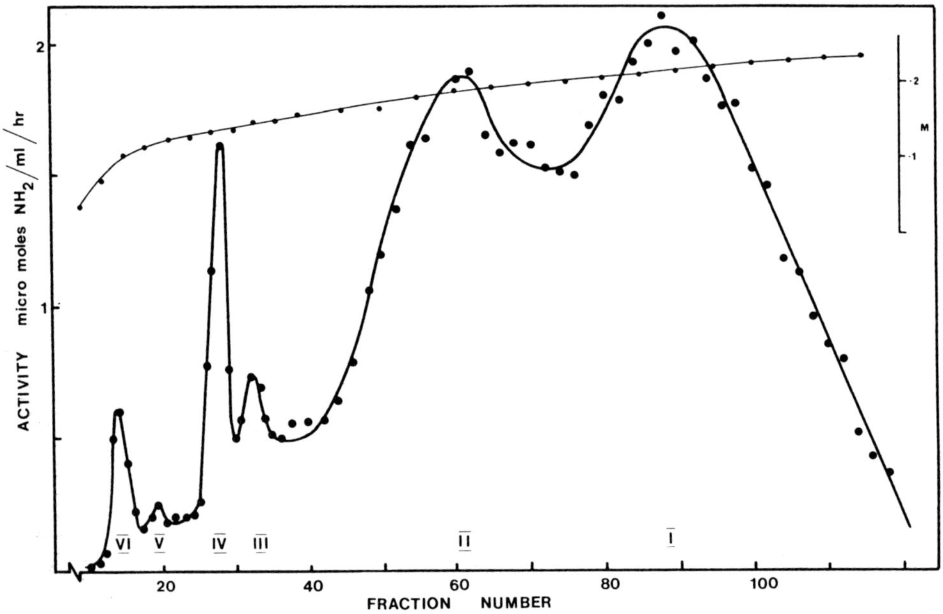

FIGURE 1. Separation of wheat-leaf peptide hydrolases on DEAE cellulose. Crude extract, partially purified on Sephadex® G200, was adsorbed on a 1.3 × 42.5 cm column of DEAE cellulose equilibrated with 0.025 *M* Tris buffer pH 8.1. After washing the column with 700 mℓ of the same buffer, the chromatogram was developed with a convex gradient of 600 mℓ 0.1 *M* Tris start buffer and 0.3 *M* Tris buffer, both pH 8.1. Elution was accomplished at a pumping speed of approximately 45 mℓ/hr and 5.0 mℓ fractions collected. The protein elution profile did not provide useful information and is omitted for clarity. (From Frith, G. J. T., Gordon, K. H. J., and Dalling, M. J., *Plant Cell Physiol.*, 19, 491—500, 1978. Reproduced with permission.)

E. Tobacco

In all studies, peptide hydrolases have been detected, which degrade a variety of substrates at acid pH; gelatin,[51] hemoglobin,[52] myoglobin,[54] and casein,[55-57] but activity against casein has also been detected at pH 7[56] and pH 8.5.[56] Activity at acid pH was stimulated by sulfhydryl reagents.[51,55,57]

F. Legumes

There have been two reports of serine carboxypeptidase activity in legume leaves[61,63] and one report of an aminopeptidase.[59] Degradation of plant protein by pea leaf extracts was optimal at acid pH,[62] but casein and azocoll were degraded by bean and soybean peptide hydrolases at neutral[61,64] or alkaline pH.[60,61]

G. Other Species

Citrus leaves have been shown to contain a serine carboxypeptidase[67,69] and an aminopeptidase has been purified from *Agave* leaves.[70-72] At neutral pH, peptide hydrolases degrading casein have been detected in betel leaves[74] and enzymes degrading artificial substrates in tomato[76] and *Perilla* leaves,[73] while acid peptide hydrolases degrading hemoglobin at pH 3.7 and casein at pH 5.0 have been found in Scots pine[75] and tea leaves,[77] respectively.

Activity of peptide hydrolases from petals has been found to be optimal over the range from acid to alkaline. Using various artificial substrates maximum activity was found to vary through the range pH 5.2-8.6 in extracts from tulip petals,[80] while a serine carboxypeptidase and an enzyme degrading azocoll at pH 7.0 were found in

Hippeastrum petals.[78] The peptide hydrolases studied in *Ipomoea* petals had an acid pH optimum of 4.2 with hemoglobin substrate.[81]

III. ROLE IN SENESCENCE

A. Vegetative Plants

Activity against hemoglobin and RuBPCase at acid pH rose to a maximum in naturally senescing (yellowing) wheat leaves and then fell.[24,32] Similar changes occurred when casein and RuBPCase were used as substrates, at acid pH, in dark-induced senescing leaves.[37] In dark-induced senescence it was noted that the process was reversible for the first two days in the dark, and during that period there was no increase in proteolytic activity.[37]

In dark-induced senescence in oat leaves, the pattern of peptide hydrolase activity against hemoglobin at acid pH was the same as in wheat. There was an initial rapid rise for the first four days, followed by a decline to the sixth day.[38,40] However, activity against hemoglobin at neutral pH did not rise as steeply as acid pH activity, and did not decline at the same rate as the acid activity.[40] Peptide hydrolase activity against hemoglobin at pH 7 did not increase significantly in detached rice leaves,[47] but activity of barley enzyme did increase during senescence of detached leaves, and this increase was negatively correlated with the loss of RuBPCase protein.[50]

In tobacco, an early study indicated that carboxyl proteinase activity did not change significantly with physiological age,[51] but later reports indicated that acid peptide hydrolase activity did increase with the onset of senescence,[52,54-56] although activity against casein at neutral and alkaline pH did not alter.[56] Detached betel leaves, however, did show an increased peptide hydrolase activity at pH 7.[74] In addition to the changes in senescing tissues, peptide hydrolase activity has been shown to change with the stage of leaf development, activity being higher in mature or fully expanded leaves[24,30,60,69] of vegetative plants, although the level has been reported not to change in *Perilla*.[73] Further, in wheat leaves, degradation of RuBPCase at pH 5 showed a distinct diurnal fluctuation.[32]

B. Flowering Plants

In three separate studies of wheat, the pattern of peptide hydrolase activity measured by the degradation at acid pH of hemoglobin,[22] bovine serum albumin,[31] and RuBPCase[36] was found to follow essentially the same pattern. After anthesis, peptide hydrolase activity rose for a few days, then remained constant or declined slightly, and then rose to a maximum at between 20 and 30 days, before eventually declining (Figure 2). The initial rise in activity coincided with rapid loss of nitrogen from the leaf,[22,31] although it preceded detectable loss of RuBPCase,[36] but in all cases the maximum activity corresponded to low levels of nitrogen in the leaf. Activity measured against casein[31] or azocasein[23] at neutral pH followed a not dissimilar pattern to that of the acid enzyme(s), but it appears that the wheat leaf peptidases decreased before or during senescence,[23] as did those from bean leaves.[23]

Peptidase activity similarly declined during grain development of maize, but acid peptide hydrolase activity against casein increased during grain filling until two thirds of the leaf protein was lost; thereafter the activity declined.[43]

In rice leaves neutral peptide hydrolase activity reached a maximum towards anthesis,[49] but in bean leaves the neutral enzyme peaked during senescence.[23] However, in soybean leaves activity against casein at pH = 6.0 actually declined before yellowing of leaves,[61] and in peas acid peptide hydrolase activity did not increase during senescence.[62]

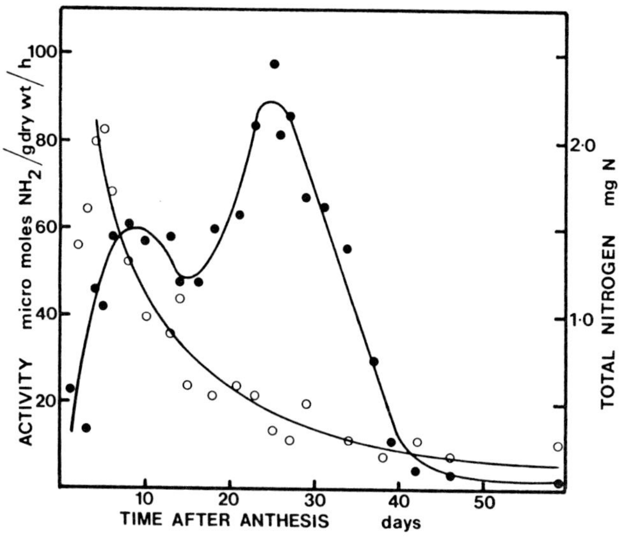

FIGURE 2. Changes in acid peptide hydrolase activity (•) and nitrogen content (o) of the flag leaf of wheat during grain development. (From Dalling, M. J., Boland, G., and Wilson, J. H., *Aust. J. Plant Physiol.*, 3, 721—730, 1976. Reproduced with permission.)

Activity against hemoglobin at pH 4.2 changed little in senescing corollas of *Ipomoea*,[79,81] but in the tulip[80] it was shown that maximum activity of peptide hydrolases occurred before the bulk of the protein was hydrolyzed.

C. Effect of Kinetin

Exogenous kinetin, which delayed senescence in detached oat leaves in the dark, markedly decreased the activity of peptide hydrolase toward hemoglobin at pH 3 and pH 7.5.[40] However, serine reduced the effect of kinetin by about 50%.

In barley and rice the effect of kinetin is similar to that in oats. The increases in peptide hydrolase activity against casein at pH 6 in detached barley leaves,[50] and against hemoglobin at pH 7 in detached rice leaves,[47] were prevented by application of kinetin.

D. Agricultural Significance

Leaf senescence and the accompanying breakdown of protein is an important physiological determinant of yield in many agriculturally important plants. As such, an understanding of the nature of the enzymes involved and the genetic and physiological control of their activity will have important implications in those programs aimed at increasing plant and crop productivity.

1. Nitrogen Redistribution

In almost all annual plants the protein content of the developing seed is accumulated as a direct consequence of senescence of the vegetative tissues. Such a reliance is absolute in those situations where there is no available soil nitrogen during seed development, but may be minimal in those instances where supplemental fertilizer nitrogen is applied. Grain legumes probably represent an intermediate example in that nitrogen fixation during seed growth augments the nitrogen derived from the soil and senescence of the vegetative tissues.

In the case of wheat,[82-85] oats,[86] corn,[45] and rice[87,88] there is evidence that the efficiency of nitrogen redistribution can vary with genotype, with high grain-N cultivars typically having a high efficiency for nitrogen redistribution. Evidence that such differences may be related to activity levels of the protein-degrading enzymes of the senescing tissues was first reported by Rao and Croy,[33,34] who observed a tentative relation between hemoglobin-degrading activity at pH 6.2 and grain-N percent in a small number of wheat varieties. Although other studies with wheat[22,24,25,29-31,35-37] suggest that nitrogen mobilization is mainly carried out by enzymes with a lower pH optimum, Rao and Croy's findings are important in providing preliminary support for the idea that efficiency of nitrogen redistribution may be determined by a relatively simple procedure.

With rice it has been shown that genotypes with high grain protein have a high rate of leaf protein degradation,[87,88] and this appears to be related to the level of leaf proteinase.[48]

More recently Hageman's group at the University of Illinois have reported a significant relation between maximum casein degradation (pH 5.4 or 7.5) in the ear leaf of a wide range of corn cultivars and grain-N percentage.[45] Those varieties with the highest percent grain-N had the highest proteinase activity.

2. Photosynthesis

Current photosynthesis is the major source of dry matter with which to sustain seed growth.[89] Although photosynthesis is a complex process involving several integrated systems, one component, namely the enzyme RuBPCase, assumes particular importance in this discussion. Although this enzyme is responsible for CO_2-fixation and hence dry matter accumulation, it also plays a fundamental role in the nitrogen economy of plants. This is because of its very high concentration in plant tissues, usually in the range of 6 to 13 mg/g fresh weight,[37,50,90-92] which can represent up to 50% of the leaf soluble protein. Therefore, in those situations where soil nitrogen is not available during grain development the RuBPCase of the leaves represents the single most important source of nitrogen to the grain.[32]

Clearly there is a continuing conflict of interest in the plant: to maintain photosynthesis and dry matter accumulation, or to initiate protein breakdown in order to satisfy the nitrogen requirement of the developing grain. This dilemma expresses itself in determinate plants such as soybean, where those proteins essential to photosynthesis are degraded in order to support the developing seed.[93] One indirect consequence of this phenomenon is a reduction in symbiotic nitrogen fixation.[94] In order to overcome this conflict Abu-Shakra et al.[94] have undertaken studies of soybean selections that exhibit delayed leaf senescence. They find that nitrogen fixation and RuBPCase activity are maintained throughout most of the pod filling period, and suggest that incorporation of this trait into an agronomically acceptable genetic background may help to increase the yield of seed and seed protein.

Delayed leaf senescence, especially when seasonal conditions are suitable or irrigation is available, has also been discussed by Thorne[89] as a means of increasing grain yield in temperate cereals. This may be achieved agronomically through the application of nitrogen fertilizer at flowering. Such a treatment satisfies the nitrogen demand of the developing seed and presumably alleviates the need for senescence by the rest of the plant.

In each of these examples, delayed or reduced senescence may be associated with reduced levels of the leaf peptide hydrolase enzymes or with a postponement of their rate of appearance. Alternatively, because the mechanism by which the plant gauges the relative balance between *sink* and *source* tissues is not understood, then any impediment to senescence could just as easily be invoked at this point rather than as a direct effect on the enzymes concerned.

Whatever the cause, there is sufficient agronomic evidence to conclude that elucidation of the mechanism and nature of the control of senescence has great agricultural importance.

IV. DISCUSSION

In plant leaves, the group of acid peptide hydrolases exceeds that of any others, both in number and depth of study. This, combined with the information that some of them have been identified as proteinases,[25] their occurrence in nearly all species examined, their ability to degrade plant protein,[26,32,36,37,62] and their increase in activity early in senescence, makes it tempting to suggest that the carboxyl proteinases are the initiators of protein degradation during senescence. The suggestion that the carboxyl proteinases have this role is supported by the finding of a highly significant relation between the actual rate of nitrogen removal from several senescing tissues of two wheat cultivars during grain development and the rate estimated from the carboxyl proteinase activity against hemoglobin (Figure 3).[22] In fact, if these enzymes are associated with the initiation of terminal protein hydrolysis, then not only would generation of a localized zone of low pH within the cell be necessary for their activity, but this requirement could be a means of controlling them. One example of how this might be achieved is suggested in the discussion of RuBPCase degradation by carboxyl proteinases.[32]

Nevertheless, the role of neutral proteinases as initiators of terminal protein hydrolysis is not discounted, and it has been suggested by Nair et al.[31] that carboxyl proteinases are involved in normal protein turnover, while the neutral proteinases are responsible for mobilization of protein during senescence. However, the scant information on neutral peptide hydrolases makes speculation on their role difficult. One enzyme has almost certainly been identified as a peptidase.[39] The fact that this neutral enzyme is a peptidase, combined with the observation that its activity continues to increase while acid pH activity is declining,[40] fits the general opinion that degradation is initiated by proteinases and completed by peptidases. Disturbingly, the activity of other peptidases that have been identified appears to decrease before anthesis or at the onset of senescence,[43] and we find it impossible to ascribe a role to them that involves senescence.

In almost all cases, the onset of senescence is correlated with an increase in peptide hydrolase activity, probably proteinase activity. Senescence, however, may be the consequence of anthesis in annual species or associated with the onset of autumn in deciduous perennials. Also there are several environmentally induced states of senescence, such as those caused by water, light, or nutrient stress — and there is insufficient evidence yet to say that the ontogeny of senescence is the same in each case.

Instances where peptide hydrolase activity does not increase at or during senescence may reflect a real difference in the process of senescence, but may also be due to other factors. As the triggering mechanism for a rise in peptide hydrolase activity is not known, it is possible that in some species the rise may precede anthesis rather than follow it. Also the choice of substrate, pH, or other experimental conditions may well decide whether a rise is observed or not.

It has been suggested[95] that differences between detached and attached leaves with

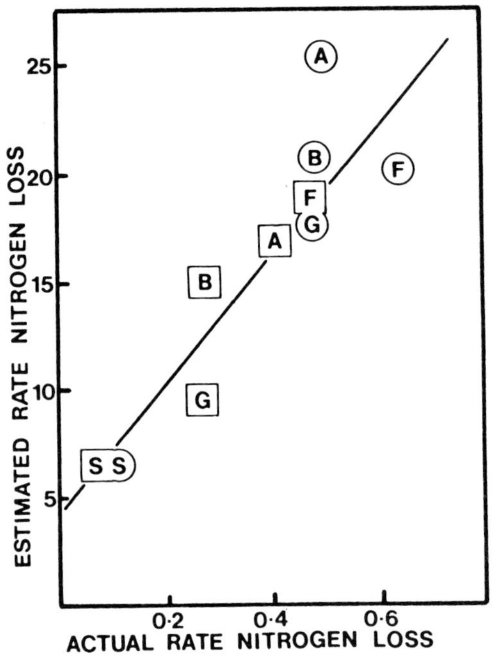

FIGURE 3. Relation between the actual rate of nitrogen loss from senescing tissues and the rate estimated from acid peptide hydrolase activity for tissues of Argentine IX (o) and Insignia (□). F, Flag leaf; A, Leaf A; B, Leaf B; G, Glumes; S, Stem. The units for estimated and actual rates of nitrogen loss are mg N/g dry weight per day. (From Dalling, M. J., Boland, G., and Wilson, J. H., *Aust. J. Plant Physiol.*, 3, 721—730, 1976. Reproduced with permission.)

regard to changes in peptide hydrolase levels during senescence may be related to changes in amino acid content.[40,41] This is an attractive theory, and we feel that it could be extended as a possible explanation in several other instances where there does not appear to be a relationship between peptide hydrolase levels and senescence.

Nonetheless, the study of plant peptide hydrolases generally lacks sufficient biochemical precision to speculate with certainty. Indications are that there may be several proteinases and peptidases in leaves, and it is conceivable that each may have specific and possibly interdependent roles in both protein turnover and terminal degradation. Thus, the simple degradation of an unspecific substrate by a crude extract is unlikely to give more than a suggestion of what actually occurs within the plant, and we feel that this type of study has now reached the limit of its usefulness. In the future, the positive distinction between proteinases and peptidases, at least, will be essential. Preferably, the number and type of each of these enzymes should be determined and their likely common occurrence in different species decided. Only then will meaningful comparison and speculation become possible.

It is still not possible to say with any certainty whether terminal protein degradation and the mobilization of the products of this degradation are the consequences of increased degradation above normal turnover rates, decreased synthesis, or a combination of both. Further, if the level of degradation increases, as the evidence to date indicates, this could be attributed to several factors such as synthesis of more peptide hydrolase enzymes, activation of existing enzyme, removal of inhibitors, or access to the substrate.

Endogenous inhibition of yeast peptide hydrolases has been demonstrated, and a model of possible control by them constructed.[96] However, while proteinase inhibitors have been isolated from barley leaves that inhibit chymotrypsin and an alkaline proteinase of *Aspergillus*,[97] endogenous inhibition of plant proteinases or peptidases has yet to be shown. Similarly, in yeast, compartmentation of hydrolytic enzymes in vacuoles has been shown,[98] but biochemical support for the notion of vacuolar location of peptide hydrolases in plants is still not clear. Vacuoles isolated from petals of *Hippeastrum* were found to lack proteinase and some glycosidases but did contain several phosphohydrolases.[78] Vacuoles from wheat leaves also appear to be devoid of proteinase activity, but in this tissue some carboxyl proteinase activity appears to be associated with the chloroplast.[99] However, a major portion, if not all, the proteinase of suspension-cultured tobacco cells,[100] pineapple leaves,[100] and pea leaves[101] appears to be located in the vacuole.

Separation of enzyme and substrate is a relatively simple mechanism on which to base control of protein degradation. However, it is difficult to envisage how this would allow for selectivity of protein hydrolysis to occur during senescence. Similar difficulties apply equally to the other cellular organelles as ultrastructural evidence indicates that the internal structure of the chloroplast and mitochondrion can be reduced before their external membranes are breached, or before the rupture of the vacuole tonoplast.[102-104]

In view of these difficulties we consider that emphasis on the exclusive role of the vacuole is too restrictive, and that other sites of degradation within the cell should also be considered.

REFERENCES

1. **Loew, O.**, in U.S. Dep. Agric. Rep. No. 65, 1900.
2. **Vines, S. H.**, Proteolytic enzymes in plants, *Ann. Bot. (London)*, 17, 237—264, 1903.
3. **Vines, S. H.**, The proteases of plants. (VI), *Ann. Bot. (London)*, 23, 1—18, 1909.
4. **Fisher, E. A.**, Contributions to the study of the vegetable proteases. I. Introductory, *Biochem. J.*, 13, 124—134, 1919.
5. **Mothes, K.**, Die Wirkung des Wassermangels auf den Eiwessumsatz in höheren Pflanzen, *Ber. Dtsch. Bot. Ges.*, 46, 59—67, 1928.
6. **Schaede, R.**, Proteolytische Enzyme in Pflanzen, *Ber. Dtsch. Bot. Ges.*, 52, 378—383, 1934.
7. **Bamann, E. and Schimke, O.**, "d-Peptidase" in wachsenden Teilen älterer Pflanzen. IV. Mitteilung: Zur Kenntnis der Peptidasen, *Biochem. Z.*, 310, 302—310, 1942.
8. **Tracey, M. V.**, Leaf protease of tobacco and other plants, *Biochem. J.*, 42, 281—287, 1948.
9. **Brady, C. J.**, The leaf protease of *Trifolium repens*, *Biochem. J.*, 78, 631—640, 1961.
10. **Singh, N.**, Proteolytic activity of leaf extracts, *J. Sci. Food Agric.*, 13, 325—332, 1962.
11. **Ryan, C. A.**, Proteolytic enzymes and their inhibitors in plants, *Annu. Rev. Plant Physiol.*, 24, 173—196, 1973.
12. **Dal Belin Peruffo, A. and Pallavicini, C.**, Enzymatic changes associated with ripening of grape berries, *J. Sci. Food Agric.*, 26, 559—566, 1975.
13. **Kaneda, M. and Tominaga, N.**, Isolation and characterization of a proteinase from the sarcocarp of melon fruit, *J. Biochem. (Tokyo)*, 78, 1287—1296, 1975.
14. **Kubota, Y., Funakoshi, T., Shimooki, O., and Shoji, S.**, Carboxypeptidases from the exocarp of mandarin orange *(Citrus unshiu* Mare.). I. Purification and properties of two carboxypeptidases, *Seikagaku*, 47, 1115—1124, 1975.
15. **Matoba, T. and Doi, E.**, Carboxypeptidase activity of tomato fruit during the ripening process and some enzymatic properties, *Agric. Biol. Chem.*, 38, 1901—1905, 1974.

16. **Zuber, H.**, Reinigung und Eigenschaften der Carboxypeptidase aus Citrus fruchten, *Hoppe-Seyler's Z. Physiol. Chem.*, 349, 1337—1352, 1968.
17. **Robinson, E.**, Proteolytic enzymes in growing root cells, *J. Exp. Bot.*, 7, 296—305, 1956.
18. **Baumgartner, B. and Chrispeels, M. J.**, Purification and characterization of vicilin peptidohydrolase, the major endopeptidase in the cotyledons of mung bean seedlings, *Eur. J. Biochem.*, 77, 223—233, 1977.
19. **Ihle, J. N. and Dure, L. S.**, The developmental biochemistry of cottonseed embryogenesis and germination. I. Purification and properties of a carboxypeptidase from germinating cotyledons, *J. Biol. Chem.*, 247, 5034—5040, 1972.
20. **Yomo, H. and Varner, J. E.**, Control of the formation of amylases and proteases in the cotyledons of germinating peas, *Plant Physiol.*, 51, 708—713, 1973.
21. **Yomo, H. and Srinivasan, K.**, Protein breakdown and formation of protease in attached and detached cotyledons of *Phaseolus vulgaris* L., *Plant Physiol.*, 52, 671—673, 1973.
22. **Dalling, M. J., Boland, G., and Wilson, J. H.**, Relation between acid proteinase activity and redistribution of nitrogen during grain development in wheat, *Aust. J. Plant Physiol.*, 3, 721—730, 1976.
23. **Feller, U.**, Proteolytic enzymes in leaves of field grown beans and wheat, *Plant Physiol.*, (Suppl.) 61, S-145, 1978.
24. **Frith, G. J. T., Bruce, D. G., and Dalling, M. J.**, Distribution of acid proteinase activity in wheat seedlings, *Plant Cell Physiol.*, 16, 1085—1091, 1975.
25. **Frith, G. J. T., Gordon, K. H. J., and Dalling, M. J.**, Proteolytic enzymes in green wheat-leaves. I. Isolation on DEAE-cellulose of several proteinases with acid pH optima, *Plant Cell Physiol.*, 19, 491—500, 1978.
26. **Frith, G. J. T., Peoples, M. B., and Dalling, M. J.**, Peptide hydrolases isolated from wheat leaves, *Plant Physiol.*, (Suppl.) 61, S-135, 1978.
27. **Frith, G. J. T., Peoples, M. B., and Dalling, M. J.**, Proteolytic enzymes in green wheat-leaves. III. Inactivation of acid proteinase II by diazoacetyl-DL-norleucine methyl ester and 1,2-epoxy-3-(p-nitrophenoxy)-propane, *Plant Cell Physiol.*, 19, 819—824, 1978.
28. **Frith, G. J. T., Swinden, L. B., and Dalling, M. J.**, Proteolytic enzymes in green wheat-leaves. II. Purification by affinity chromatography, and some properties of proteinases with acid pH optima, *Plant Cell Physiol.*, 19, 1029—1041, 1978.
29. **Morris, W. G.**, Some aspects of nitrogen metabolism of wheat with special regard to nitrate reductase and protease effects on grain protein, *Diss. Abstr.*, 35B, 4740, 1975.
30. **Myczkowski, J.**, Investigations on protease activity in leaves during growth and development of wheat, *Acta Agrobot.*, 23, 159—170, 1970.
31. **Nair, T. V. R., Grover, H. L., and Abrol, Y. P.**, Nitrogen metabolism of the upper three leaf blades of wheat at different soil nitrogen levels. II. Protease activity and mobilization of reduced nitrogen to the developing grains, *Physiol. Plant.*, 42, 293—300, 1978.
32. **Peoples, M. B. and Dalling, M. J.**, Degradation of ribulose-1,5-bisphosphate carboxylase by proteolytic enzymes from crude extracts of wheat leaves, *Planta*, 138, 153—160, 1978.
33. **Rao, S. C. and Croy, L. I.**, Protease levels in "high" vs. "low" grain protein wheats and their association with the production of amino acids, tryptophan, and IAA during early growth, *Crop Sci.*, 11, 790—791, 1971.
34. **Rao, S. C. and Croy, L. I.**, Protease and nitrate reductase seasonal patterns and their relation to grain protein production of "high" vs. "low" protein wheat varieties, *J. Agric. Food Chem.*, 20, 1138—1141, 1972.
35. **Sopanen, T. and Laurière, C.**, Activities of various peptidases in the first leaf of wheat, *Physiol. Plant.*, 36, 251—254, 1976.
36. **Wittenbach, V. A.**, Changes in proteolytic activity and the level of RuBPCase in the flag leaf of wheat during grain development and senescence, *Plant Physiol.*, (Suppl.) 61, S-136, 1978.
37. **Wittenbach, V. A.**, Breakdown of ribulose bisphosphate carboxylase and change in proteolytic activity during dark-induced senescence of wheat seedlings, *Plant Physiol.*, in press.
38. **Drivdahl, R. H. and Thimann, K. V.**, Proteases of senescing oat leaves. I. Purification and general properties, *Plant Physiol.*, 59, 1059—1063, 1977.
39. **Drivdahl, R. H. and Thimann, K. V.**, Proteases of senescing oat leaves. II. Reaction to substrates and inhibitors, *Plant Physiol.*, 61, 501—505, 1978.
40. **Martin, C. and Thimann, K. V.**, The role of protein synthesis in the senescence of leaves. I. The formation of protease, *Plant Physiol.*, 49, 64—71, 1972.
41. **Martin, C. and Thimann, K. V.**, The role of protein synthesis in the senescence of leaves. II. The influence of amino acids on senescence, *Plant Physiol.*, 50, 432—437, 1972.
42. **Pike, C. S. and Briggs, W. R.**, Partial purification and characterization of a phytochrome-degrading neutral protease from etiolated oat leaves, *Plant Physiol.*, 49, 521—530, 1972.

43. **Feller, U. K., Soong, T-S. T., and Hageman, R. H.**, Leaf proteolytic activities and senescence during grain development of field grown corn *(Zea mays* L.), *Plant Physiol.*, 59, 290—294, 1977.
44. **Klein, M. and Harpaz, I.**, Proteolytic activity in maize leaves, *Phytochemistry*, 4, 327—332, 1965.
45. **Soong, T-S. T., Feller, U. K., Lambert, R. J., and Hageman, R. H.**, Relation of proteolytic activities in leaves to grain nitrogen of corn *(Zea mays* L.), *Plant Physiol.*, (Suppl.) 61, S-134, 1978.
46. **Gupta, V. K. and Pawar, V. S.**, Leucine aminopeptidase activity in tall and dwarf cultivars of rice at successive stages of development, *Ann. Bot. (London)*, 38, 205—218, 1974.
47. **Kar, M. and Mishra, D.**, Protease activity during rice leaf senescence, *Biol. Plant.*, 19, 365—369, 1977.
48. **Perez, C. M., Cagampang, G. B., Esmama, B. V., Monserrate, R. U., and Juliano, B. O.**, Protein metabolism in leaves and developing grains of rices differing in grain protein content, *Plant Physiol.*, 51, 537—542, 1973.
49. **Villareal, R. M. and Juliano, B. O.**, Some properties of rice leaf protease, *Kimika*, 1, 20—25, 1976.
50. **Peterson, L. W. and Huffaker, R. C.**, Loss of ribulose-1,5-diphosphate carboxylase and increase in proteolytic activity during senescence of detached primary barley leaves, *Plant Physiol.*, 55, 1009—1015, 1975.
51. **Anderson, J. W. and Rowan, K. S.**, Activity of peptidase in tobacco-leaf tissue in relation to senescence, *Biochem. J.*, 97, 741—746, 1965.
52. **Balz, H. P.**, Intrazellular Lokalisation und Funktion von hydrolytischen Enzymen bei Tabak, *Planta*, 70, 207—236, 1966.
53. **De Jong, D. W.**, Detergent extraction of enzymes from tobacco leaves varying in maturity, *Plant Physiol.*, 50, 733—737, 1972.
54. **Hochkeppel, H-K.**, Isolierung einer Endopeptidase aus alterden Tabakblättern und ihre Beziehung zum Vergilben, *Z. Pflanzenphysiol.*, 69, 329—343, 1973.
55. **Kawashima, N.**, Protein metabolism in higher plants. VII. Degradation mechanisms of fraction I by a proteinase increased in tobacco leaves during curing, *Hatano Tabako Shikenjo Hokoku*, 71, 111—116, 1972.
56. **Kawashima, N., Fukushima, H., and Tamaki, E.**, Studies on protein metabolism in higher plant leaves. II. Variation in amino acid composition of protein in autolysed tobacco leaves, *Phytochemistry*, 6, 339—345, 1967.
57. **Kawashima, N., Fukushima, H., Imai, A., and Tamaki, E.**, Studies on protein metabolism in higher plants. V. Some properties of a tobacco leaf protease increased during curing, *Agric. Biol. Chem.*, 32, 1141—1145, 1968.
58. **Vasyunina, A. D.**, Water-insoluble nitrogen compounds in autolysed tobacco, *Tabak (Moscow)*, 23, 54—56, 1962.
59. **Buchner, J. S.**, The purification and properties of two amino peptidases from *Pisum sativum*, *Diss. Abstr.*, 32B, 4390, 1972.
60. **Racusen, D. and Foote, M.**, An endopeptidase of bean leaves, *Can. J. Bot.*, 48, 1017—1021, 1970.
61. **Ragster, L., Chrispeels, M. J., Andersen, J., and Hageman, R. H.**, Temporal changes in proteolytic activities in the leaves of field-grown soybean plants, *Plant Physiol.*, (Suppl.) 61, S-148, 1978.
62. **Storey, R. and Beevers, L.**, Proteolytic activity in relationship to senescence and cotyledonary development in *Pisum sativum* L., *Planta*, 137, 37—44, 1977.
63. **Wells, J. R. E.**, Purification and properties of a proteolytic enzyme from french beans, *Biochem. J.*, 97, 228—235, 1965.
64. **Wells, J. R. E.**, Characterization of three proteolytic enzymes from french beans, *Biochim. Biophys. Acta*, 167, 388—398, 1968.
65. **Kenis, J. D. and Edelman, M. O.**, Algunos aspectos bioquimicos del reposo y su ruptura en yemas florales de duraznero *(Prunus persicae* var Dixie Red). I. Cambios en la actividad de RNAases y proteinasas, actividad y composicion de polifenol oxidasas y peroxidasas y modificaciones en la concentracion de ARA proteinas solubles y nitrozeno total, *Phyton (Buenos Aires)*, 34, 133—142, 1976.
66. **Spencer, P. W. and Titus, J. S.**, Biochemical and enzymatic changes in apple leaf tissue during autumnal senescence, *Plant Physiol.*, 49, 746—750, 1972.
67. **Sprössler, B., Heilmann, H-D., Grampp, E., and Uhlig, H.**, Eigenschaften der Carboxypeptidase C aus Orangeblättern, *Hoppe-Seyler's Z. Physiol. Chem.*, 352, 1524—1530, 1971.
68. **Zuber, H.**, Purification and properties of a new carboxypeptidase from citrus fruit, *Nature (London)*, 201, 613, 1964.
69. **Zuber, H. and Matile, Ph.**, Acid carboxypeptidases: their occurrence in plants, intracellular distribution, and possible function, *Z. Naturforsch.*, 23B, 663—665, 1968.
70. **Du Toit, P. J.**, Isolation and partial characterization of a protease from *Agave americana variegata*, *Biochim. Biophys. Acta*, 429, 895—911, 1976.

71. **Schabort, J. C. and Du Toit, P. J.**, An aminopeptidase from *Agave americana variegata*. II. Aspects of enzyme kinetics, specificity, inhibition, and mechanism of action, *Int. J. Biochem.*, 9, 171—177, 1978.
72. **Schabort, J. C., Du Toit, P. J., and Kempff, P. G.**, An aminopeptidase from *Agave americana variegata*. I. Isolation, purification, and physical and chemical characterization, *Int. J. Biochem.*, 9, 161—170, 1978.
73. **Kannangara, C. G. and Woolhouse, H. W.**, Changes in the enzyme activity of soluble protein fractions in the course of foliar senescence in *Perilla frutescens* (L.) Britt., *New Phytol.*, 67, 533—542, 1968.
74. **Mishra, S. D. and Gaur, B. K.**, Role of petiole in protein metabolism of senescing betel *(Piper betle* L.) leaves, *Plant Physiol.*, 59, 961—964, 1977.
75. **Salmia, M. A., Nyman, S. A., and Mikola, J. J.**, Characterization of the proteinases present in germinating seeds of Scots pine, *Physiol. Plant.*, 42, 252—256, 1978.
76. **Stern, M. S.**, The purification and characterization of a protease-esterase complex from tomato leaves, *Diss. Abstr.*, 30B, 4498, 1970.
77. **Sanderson, G. W. and Roberts, G. R.**, Peptidase activity in shoot tips of the tea plant *(Camellia sinensis* L.), *Biochem. J.*, 93, 419—423, 1964.
78. **Butcher, H. C., Wagner, G. J., and Siegelman, H. W.**, Localization of acid hydrolases in protoplasts: examination of the proposed lysosomal function of the mature vacuole, *Plant Physiol.*, 59, 1098—1103, 1977.
79. **Matile, Ph. and Winkenbach, F.**, Function of lysosomes and lysosomal enzymes in the senescing corolla of the morning glory *Ipomoea purpurea*, *J. Exp. Bot.*, 22, 759—771, 1971.
80. **Sopanen, T. and Carfantan, N.**, Activities of various peptidases in the senescing petals of tulip, *Physiol. Plant.*, 36, 247—250, 1976.
81. **Winkenbach, F.**, Zum Stoffwechsel der aufbluhenden und welkenden Korolle der Prunkwinde *Ipomea purpurea*, *Ber. Schweiz. Bot. Ges.*, 80, 391—406, 1970.
82. **Dalling, M. J., Halloran, G. M., and Wilson, J. H.**, The relation between nitrate reductase activity and grain nitrogen productivity in wheat, *Aust. J. Agric. Res.*, 26, 1—10, 1975.
83. **Eilrich, G. L. and Hageman, R. H.**, Nitrate reductase activity and its relationship to accumulation of vegetative and grain nitrogen in wheat *(Triticum aestivum* L.), *Crop Sci.*, 13, 59—66, 1973.
84. **Johnson, V. A., Mattern, P. J., and Schmidt, J. W.**, Nitrogen relations during spring growth in varieties of *Triticum aestivum* L. differing in grain protein content, *Crop. Sci.*, 7, 664—667, 1967.
85. **Beech, D. F. and Norman, M. T. J.**, A preliminary assessment of the adaptation of semi-dwarf wheat varieties to the Ord River Valley, *Aust. J. Exp. Agric. Anim. Husb.*, 8, 349—357, 1968.
86. **Peterson, D. M., Schrader, L. E., Cataldo, D. A., Youngs, V. L., and Smith, D.**, Assimilation and remobilization of nitrogen and carbohydrates in oats, especially as related to groat protein concentration, *Can. J. Plant Sci.*, 55, 19—28, 1975.
87. **Anon.**, *Int. Rice Res. Inst. (Los Baños), Ann. Rep.*, 13, 1971.
88. **Anon.**, *Int. Rice Rest. Inst. (Los Baños), Ann. Rep.*, 11, 1972.
89. **Thorne, G. N.**, Physiology of grain yield of wheat and barley, *Rothamsted Exp. Stn. Rep.*, Part 2, 5—25, 1973.
90. **Brady, C. J. and Steele Scott, N.**, Chloroplast polyribosomes and synthesis of fraction I protein in the developing wheat leaf, *Aust. J. Plant Physiol.*, 4, 327—335, 1977.
91. **Gordon, K. H. J., Peoples, M. B., and Murray, D. R.**, Ageing-linked changes in photosynthetic capacity and in fraction I protein content of the first leaf of pea, *Pisum sativum* L., *New Phytol.*, 81, 35—42, 1978.
92. **O'Toole, J. E. and Dalling, M. J.**, unpublished data, 1978.
93. **Sinclair, T. R. and de Wit, C. T.**, Photosynthate and nitrogen requirements for seed production by various crops, *Science*, 189, 565—567, 1975.
94. **Abu-Shakra, S. S., Phillips, D. A., and Huffaker, R. C.**, Nitrogen fixation and delayed leaf senescence in soybeans, *Science*, 199, 973—975, 1978.
95. **Beevers, L.**, *Nitrogen Metabolism in Plants*, Edward Arnold, London, 1976, 282.
96. **Holzer, H. and Saheki, T.**, Mechanism and regulation of proteolytic processes in yeast, *Tokai J. Exp. Clin. Med.*, 1, 115—125, 1976.
97. **Kirsi, M. and Mikola, J.**, Occurrence and heterogeneity of chymotrypsin inhibitors in vegetative tissues of barley, *Physiol. Plant.*, 39, 110—114, 1977.
98. **Matile, Ph.**, Biochemistry and function of vacuoles, *Annu. Rev. Plant Physiol.*, 29, 193—213, 1978.
99. **Peoples, M. B. and Dalling, M. J.**, Intracellular localization of acid peptide hydrolases in wheat leaves, *Plant Physiol.*, (Suppl.) 63, S-882, 1979.
100. **Bollen, T. and Kende, H.**, Hydrolytic enzymes in the central vacuole of plant cells, *Plant Physiol.*, 63, 1123—1132, 1979.

101. **Noble, E. R. and Dalling, M. J.,** unpublished data, 1979.
102. **Barton, R.,** Fine structure of mesophyll cells in senescing leaves of *Phaseolus, Planta,* 71, 314—325, 1966.
103. **Butler, R. D. and Simon, E. W.,** Ultrastructural aspects of senescence in plants, *Adv. Gerontol. Res.,* 3, 73—129, 1971.
104. **Shaw, M. and Manocha, M. S.,** Fine structure in detached, senescing wheat leaves, *Can. J. Bot.,* 43, 747—755, 1965.

Chapter 7

FLOWER SENESCENCE*

S. Mayak and A. H. Halevy

TABLE OF CONTENTS

* Literature search terminated on January 1, 1978.

I. INTRODUCTION

Flowering, often associated with senescence and death, is a dramatic phenomenon in monocarpic plants, where the whole plant dies after flowering or fruiting. In polycarpic plants, death is restricted to parts of the flower itself — those which senesce and often abscise soon after flowering. The flower is in most cases the organ with the shortest period of longevity. The reasons underlying this fact are not known and have been little studied.

Relatively little has been published up to now on the physiology of flower and petal senescence in comparison with that of other plant organs, such as leaves. This article seems to be the first devoted exclusively to flower senescence. It is difficult to understand why this is so, since petals seem to be an excellent model system for the study of fundamental senescence processes. The whole natural period from maturity to senescence and death is much shorter in petals than in leaves. Hence, the processes occurring during natural senescence may be studied without recourse to artificial "aging" treatments so common in studies of leaf senescence. The metabolic processes occurring in "aged" excised leaves may not be analogous to those occurring in normal leaves.[82,127a] Since chloroplast senescence does not necessarily follow the same path as that of the rest of the cell, the petal tissue may serve as an excellent model for the study of senescence processes not related to and not influenced by the presence of chloroplasts. For example, chloroplast degradation may be reversed up to a certain stage,[150] whereas petal senescence is an irreversible process, thus better fitting Sacher's (1973) definition of senescence as the last phase in the life in which "a series of normally irreversible events is initiated that leads to cellular breakdown and death." The effects of CO_2 on senescence, other than those related to chloroplast fixation processes, may be better studied in a tissue in which the interference by chloroplasts has been minimized.

The flower is a more complex organ than leaves, fruits, roots, or seeds on which most studies on plant senescence have been conducted. It includes in many cases a green calyx, achlorophyllous corolla, androecium, gynoecium, and a peduncle. Each of these components is complex in its own right, and they all differ from each other both in structure and physiology. The interrelationship between the various flower parts may determine their rate of senescence. For example, in many flowers, accelerated senescence of the corolla occurring after pollination is manifested by very rapid fading, as in orchids[7] and *Petunia*,[51] or by abscision as in sweet peas and snapdragon. The gynoecium simultaneously starts to grow and to draw metabolites from other flowering parts.

Not all flower parts follow the same path of senescence. In some plants, the green sepals persist after the petals have wilted or abscised, and may even continue photosynthesis.

The ovule of the gynoecium usually develops to form the fruits, when all other flower parts senesce. In some plants, perianth segments — and even the style — may persist to form structures encapsulating or adjoining the fruit. Because senescence of the whole flower is so complex, we will concentrate mainly on changes occurring during the senescence of the corolla.

II. ULTRASTRUCTURAL, BIOCHEMICAL, AND BIOPHYSICAL CHANGES DURING PETAL SENESCENCE

The most comprehensive and detailed studies on the ultrastructural and biochemical processes during petal senescence were carried out by Matile and associates in the ephemeral corolla of the Japanese morning glory, *Ipomoea tricolor*. This flower follows a strictly timed development and senescence sequence (Figure 1). Buds open at 5 to 6 in the morning and remain open until early afternoon (1 to 3 p.m.) of the same day. At this time, the corolla starts to curl up and change its color from blue to purple. Curling and wilting continue in the afternoon, and the corolla abscises 1 to 2 days later.

The first observed sign of aging was invagination of the tonoplast, indicating autophagic activity of the vacuole, which is considered by Matile[87] to represent the lysosomal cell compartment. Abolition of the compartmentation of the vacuole and the release of hydrolytic enzymes results in cell death. This concept is supported by the presence of cytoplasmic material such as disintegrated mitochondria and various types of membranes in the aging vacuole (Figure 2). Later, the tonoplast is broken down, followed by autolysis of the cell. These changes are schematically illustrated in Figure 3.

Recently, techniques have been described for isolation of intact vacuoles of petals[83,142] which enable direct biochemical analysis of the vacuole. In the examination of mature and senescent vacuoles and protoplasts of *Hippeastrum* flower petals, it was found[35] that acid hydrolases commonly associated with animal lysosomes (esterase, protease, carboxypeptidase and β-glycosidase) were part of the soluble cytoplasmic fraction, and not of the vacuole lysate. Acid phosphatase, RNase, and DNase were present in both fractions. These findings do not support the concept of the vacuole functioning as a generalized lysosome of mature and senescent cells, and further work is needed.

The corolla of morning glory may be representative of changes occurring in the majority of the flowers in which the petals contain no plastids. However, petals of some flowers (mainly yellow) do contain plastids. Changes in the fine structure of tissue containing chromoplasts have been described for cucumber petals.[130] The most noticeable changes during development and senescence occur in the plastids. The change from green chloroplasts (in the young petals) to yellow chromoplasts is characterized by the gradual disappearance of the thylakoids and the appearance of bundles of tubules in their place in the stroma. It was suggested that the tubules are largely composed of material derived from the thylakoids, the chromoplast being five times larger than the chloroplast from which it was derived. The senescent chromoplasts show invaginations in the plastid envelope. During maturation and senescence, disappearance of the ribosomes was observed; first of the free, single ribosomes, then of those aggregated into clusters, and finally of those attached to the endoplasmic reticulum which vesiculates. This pattern of changes in ribosomes is very similar to that found in other plant tissues.[34]

Two major metabolic events occur in senescing petals; increase in respiration and hydrolysis of cell components. The enzymic changes found during petal senescence are associated mainly with these two processes. Increase in peroxidases was found in senescing petals of several plants.[30,37,138] The increased activity of peroxidases is apparently related to an increase in peroxides and free radicals, which react with cellular constituents[48] and are involved in promotion of senescence[31,101] and possibly also in ethylene production.[21] Indeed, treatment of flowers with free radical scavengers (such as sodium benzoate) delayed senescence of carnation flowers.[15]

In *Ipomoea* a sharp increase was found in the activity of some hydrolytic enzymes (Figure 4) — RNase, DNase, and hydrolases of cell wall polysaccharides.[146] An increase in RNase was found also in roses.[56,95]

During the course of petal aging there is a drop in the level of macromolecular components (Figure 5): starch,[68] cell wall polysaccharides,[146] proteins,[26,112,114-115] and

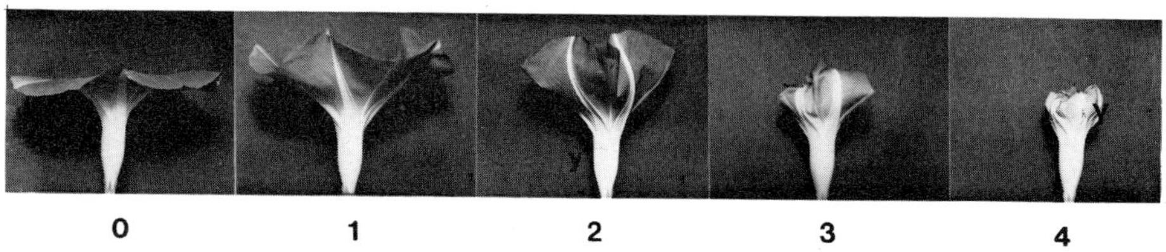

FIGURE 1. Japanese Morning Glory, *Ipomoea tricolor:* Scale of flower fading. Stage 0 represents the fully open corolla, stages 1 to 4 progressive phases of fading. (From Kende, H. and Baumgartner, B., *Planta*, 116, 279—289, 1974. With permission.)

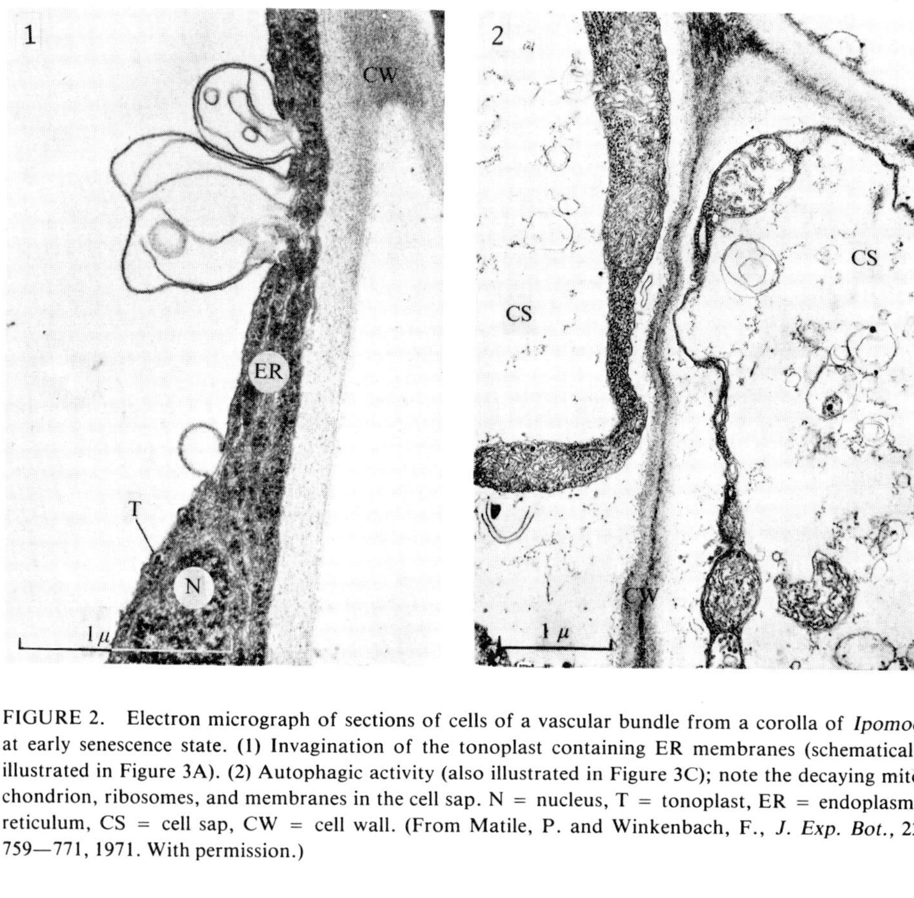

FIGURE 2. Electron micrograph of sections of cells of a vascular bundle from a corolla of *Ipomoea* at early senescence state. (1) Invagination of the tonoplast containing ER membranes (schematically illustrated in Figure 3A). (2) Autophagic activity (also illustrated in Figure 3C); note the decaying mitochondrion, ribosomes, and membranes in the cell sap. N = nucleus, T = tonoplast, ER = endoplasmic reticulum, CS = cell sap, CW = cell wall. (From Matile, P. and Winkenbach, F., *J. Exp. Bot.*, 22, 759—771, 1971. With permission.)

nucleic acids (Figure 5). In several flowers an increase in the pH of the vacuole was observed in aging petals.[26,114,136] This was attributed to proteolysis and an increase in asparagine as the major amino compound in old petals, followed by accumulation of free ammonia.[114,144] These changes are generally similar to those in senescing leaves (see Chapter 5).

The sequence of events demonstrated that a correlation was not always found between hydrolase activity and the level of the corresponding macromolecule. RNA content starts decreasing in *Ipomoea* even before anthesis, whereas the sharp increase in RNase is evident only after the beginning of fading (compare Figures 4 and 5), indicating that the first stage in decline of RNA is caused by reduced synthesis. This seems to be the case also with the proteins. In contrast to RNA, degradation of DNA starts

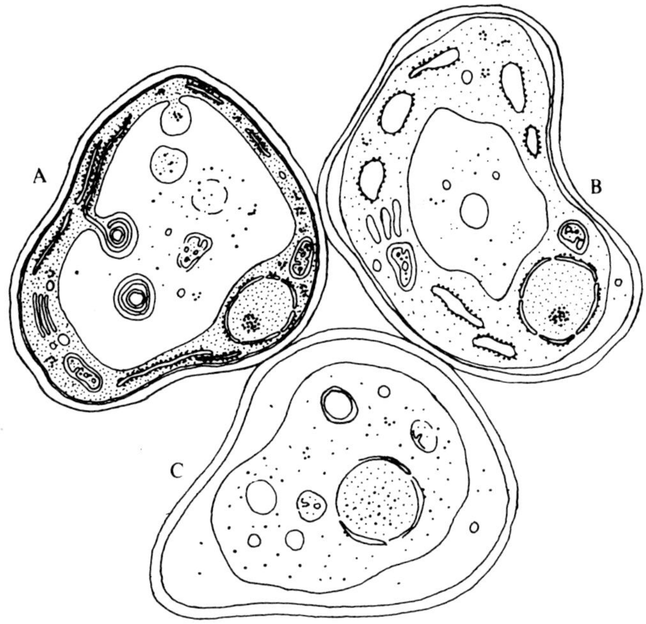

FIGURE 3. Schematic illustration of morphological changes in mesophyll cells of the senescing *Ipomoea* corolla. (A) Autophagic activity of the vacuole; invaginations of the tonoplast result in the sequestration of cytoplasmic material into the lysosomal compartment. (B) Shrinkage of the vacuole, dilution of the cytoplasm and inflation of cytoplasmic membrane systems. (C) Autolysis is initiated by the breakdown of the tonoplast. (From Matile, P. and Winkenbach, F., *J. Exp. Bot.*, 22, 759—771, 1971. With permission.)

only after the onset of wilting (Figure 5), thus indicating that there is autolysis — which signifies the death of the cell.

It was shown that some hydrolytic enzymes: RNase, β-glycosidase, and β-galactosidase were synthesized *de novo,* while the total protein content of the corolla was declining,[19,142] indicating the requirement of protein synthesis for the onset of senescence. Indeed, the application of protein synthesis inhibitors delayed petal aging of several flowers,[9,40] as it does of some leaves (Chapter 5).

The breakdown products of the cell's macromolecules are transported out of the petals to other parts of the plant[148] or to the developing gynoecium.[105,116]

Senescence-associated phenomena were accelerated in *Tradescantia*[69] and morning glory flowers by low pH, which has a very similar effect to that of ethylene.[61] Since the pH of normal cytoplasm is neutral and that of the vacuole is acidic, the disruption of the tonoplast may cause leakage from the vacuole, thus reducing the cytoplasm pH. The low pH is optimal for hydrolases such as RNase[20] or phospholipase.[28] Indeed, in *Ipomoea*[14] and in several other flowers[36,151] a decrease in pH was found in old petals, which was caused by increase in the level of organic acids.

One of the most obvious symptoms of the final stage of senescence in petals is the loss of fresh weight (Figure 5), drying, and shriveling (unless they abscise before this stage). This loss of water from the aging petals occurs also when the cut flowers are held in water. In some flowers an infiltration of the intracellular spaces of the petals is apparent,[69] which suggests that there may have been a loss of membrane integrity,

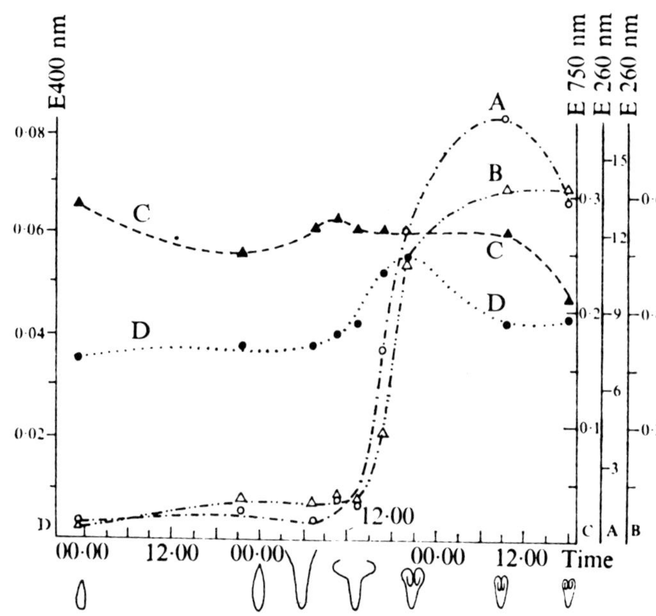

FIGURE 4. Relative activities of acid hydrolases per corolla; (A ○—○) RNase, (B △—△) DNase, (C ▲—▲) protease, and (D •—•) α-glucosidase. The sketches below time axis indicate the shape of corolla. (From Matile, P. and Winkenbach, F., *J. Exp. Bot.*, 22, 759—771, 1971. With permission.)

causing an increase in permeability and leakage. An increase in apparent free space has been recorded in aging spadix of *Arum*[41] and in rose petals,[113] and an increase in membrane permeability was demonstrated during the senescence of several flowers.[61,99,102,122] These changes in the biophysical properties of the membranes were determined by indirect methods — such as measuring the leakage of electrolytes or of metabolites or by compartment analysis, since no direct access to plant cell plasmalemma and tonoplast was feasible until recently. Direct measurements of biophysical properties of plant cell plasmalemma during development and aging have been recently conducted on protoplasts isolated from rose petals.[25,28] A sharp increase in microviscosity of the plasmalemma was observed during aging (Figure 6). This change occurred in protoplasts isolated from intact flowers, as well as from cut flowers or isolated petals. The increase in microviscosity corresponded to an increase in the mole ratio of free sterol to phospholipid. This was attributed to an increase in the activity of phospholipase, which brought about a decrease in the level of phospholipids. A decline in phospholipids was observed also in the *Ipomoea* corolla,[23] even before visible signs of senescence were evident. The decrease in phospholipid level should enhance the permeability of the plasma membrane and make the cell leaky.[128]

III. CHANGES IN PIGMENTATION

Color fading (and discoloration) is a common phenomenon in many flowers during aging. Two major types of pigments contribute to the color of the flowers: carotenoids and anthocyanins. The changes in carotenoid composition were followed in *Strelitzia reginae* flowers during the various developmental stages of the plastids, from the small colorless plastids through the green chloroplasts, and to increasingly large spindle-

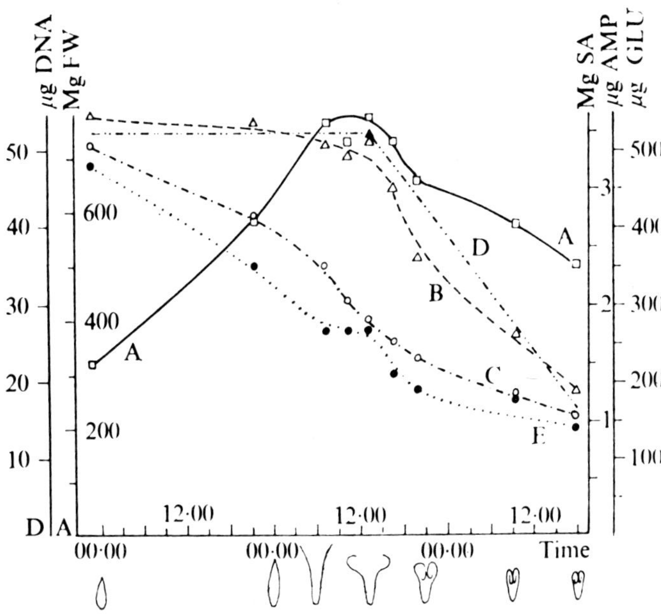

FIGURE 5. Content in (A □—□) freshweight, (B Δ—Δ) protein (mg serum albumin), (C o—o) RNA (µg AMP), and (E •—•) starch (µg glucose) per corolla. Content in (D) DNA (µg DNA) per corolla from another experiment. The sketches below time axis indicate the shape of the corolla. (From Matile, P. and Winkenbach, F., *J. Exp. Bot.*, 22, 759–771, 1971. With permission.)

shaped chromoplasts.[129] An increase in the concentration of oxygenated carotenoids with age was found in *Strelitzia* as well as in rose flowers.[140] This change is considered[52] to be a sign of degenerative and uncontrolled oxidative processes. However, an increase in oxygenated carotenoids was observed, even in maturing sepals that showed no sign of degeneration in their ultrasturcture.[129]

Much more is known about the change in pigmentation due to anthocyanins. There is no uniform trend in the change in anthocyanin content in aging petals. The pigment level stays stable in some flowers,[111,135] declines drastically in others,[137] whereas in some flowers a dramatic synthesis of anthocyanins is evident.[9] Some flowers fade and even turn white upon aging. The ephemeral chicory flower opens at daybreak with bright blue color, which fades in the early afternoon. Proctor and Greasy[119] attributed this to the decolorization by an enzyme system with the properties of catecholase. It was suggested that the increase in β-glucosidase in senescing morning glory corolla is responsible for the hydrolysis of anthocyanins that occurs during wilting.[88]

An opposite trend in color during the course of one day from white (early in the morning) to deep red (in the evening) just before wilting has been known for a long time in *Hibiscus mutabilis.*[79] In the "Masquerade" rose, the petals are orange-yellow when freshly opened and turn deep red upon senescing. More than a tenfold increase in anthocyanin level was measured during that period.[126]

The most important factor determining the color change in senescing petals seems to be a change in the pH of the vacuole.[136] However, only in a very few cases is the color caused by a very low (<3.0) or a very high (>7.0) pH affecting the anthocyanin per se.[11] In most flowers, the decisive factor determining the intensity of the color and its bluing is the copigmentation with other flavonoids and related compounds. The

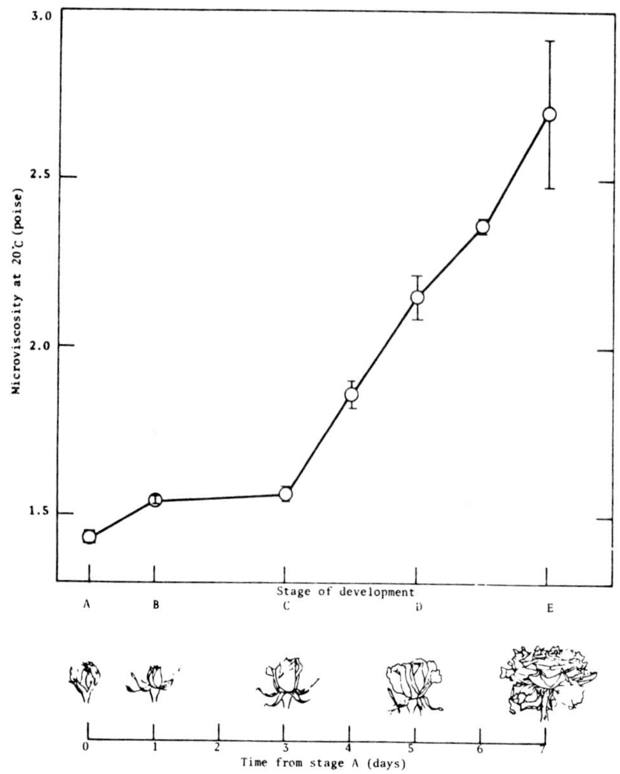

FIGURE 6. Microviscosity of protoplast plasmalemmas obtained from rose petals at different stages of development. Bars represent the range of values obtained with protoplasts from 3 to 5 different flowers. (From Borochov, A. et al., *Nature (London)*, 263, 158—159, 1976a. With permission.)

degree of copigmentation is influenced to a great extent by even slight changes in pH.[13] This may explain the infinite variations in the color of flowers that exist in the pH range of 4 to 6 — which is most prevalent in petals, where anthocyanins per se are virtually colorless. The bluing of red flowers with aging is a well-known phenomenon. A concomitant increase in pH has been demonstrated in roses,[12,26] *Lathyrus*,[111] larkspur,[13] geranium, and other flowers.[136] The increase in pH was attributed to the breakdown of proteins and the release of free ammonia.[114,144] Indeed, treatment of cut flowers with solutions containing sugars, which delayed proteolysis[114] delayed also the increase of the pH and bluing.[12,26,144]

In some other flowers an opposite trend was found. The blue, violet, or purple petals of cornflower,[10] morning glory,[14] and fuchsia[136,151] became more red as they aged. A concurrent decrease in pH has been demonstrated in these cases, which was attributed to an increase in the content of organic acids such as aspartic, malic, and tartaric acids.[151]

It is of interest to note that the color and the pH changes associated with aging may proceed at very different rates in adjacent cells.[136] This may indicate that contiguous cells differ from each other in the rates of their proteolysis and aging.

IV. CARBOHYDRATE AND NITROGEN METABOLISM AND THE ROLE OF APPLIED SUGARS

The final stages of flower development are characterized by a decline in the carbohydrate content and dry weight of petals.[1,95,103,144] The flower is a heterogeneous organ, composed of floral parts, each of which may be at a different physiological developmental stage. Generally, the senescence and wilting of the petals determines the longevity of the flower; we shall, therefore, describe mainly changes in this tissue.

Reducing sugars rather than sucrose were noted as the main constituents of the sugar pool of mature petals.[103,144] The changes in sugars are accompanied by starch hydrolysis.[68]

Respiratory metabolism of cut flowers was reviewed recently by Coorts.[39] The rate of respiration in many flowers rises to a maximum as flowers start to open, followed by a gradual decline as flowers mature and senesce. Later there is a second dramatic increase over a relatively short period, followed by a final decline.

The second peak in the respiration drift is considered to signify the final senescence stage. It was assumed to be analogous to the climacteric rise in respiration of many fruits. Presumably, as in fruits, it is cyanide-insensitive, thus indicating a shift in respiratory metabolic pathways. In fact, many of the first studies on cyanide-insensitive respiration were conducted with the flowers of the Araceae in which there is a dramatic burst of respiration before the final wilting.[134] These processes may lead to the formation of free radicals with high oxidation potential. Free radicals have been found to promote senescence in fruits and other tissues and to be associated with increased sensitivity to ethylene.[31,48]

If the second peak in respiration reflects internal metabolic changes closely associated with senescence, then it can serve to indicate the effectiveness of treatments aimed at postponing senescence. Indeed, it was found that chemicals that delay the occurrence of the second peak also extend longevity of flowers.[17,98] A unique pattern of respiration was recently demonstrated in tropical orchid flowers;[65] a circadian rhythm in CO_2 production started as soon as flowers opened. The period between amplitudes was about 24 hr, and was not affected by continuous darkness. The rhythm was observed also in cut flowers, but the amplitude of the rhythm was dampened by detachment, and partially boosted by an external supply of sucrose.

The gradual decline in respiration of senescing flowers may be caused by short supply of readily respirable substrates. The size of respirable substrate pool, which is composed mainly of sugars, is affected by: (1) the rate of hydrolysis of starch and other polysaccharides,[68,105] (2) translocation to the petals[108] on the one hand, and (3) respiration and translocation out of the flowers to other plant parts on the other hand.[116,142] Translocation within the flower from the petals to the ovary was also demonstrated in senescing flowers.[70,105,107,108] This transport is promoted by pollination and ethylene. Supplying cut flowers with exogenous sugar maintains the pool of dry matter and respirable substrates, especially in petals, thus promoting respiration and extending longevity.[39,121] Several metabolic sugars are active in this respect, but nonmetabolic sugars are not.[1,55] The application of ^{14}C-sucrose to cut roses revealed that sucrose was incorporated into protein,[115] ethanol-soluble carbohydrate, starch and ethanol-insoluble material;[68,123] although the incorporation into polysaccharides declined with flower maturation. At the same time an absolute reduction in starch content was measured, indicating a dynamic balance in which active anabolic processes take place simultaneously with catabolic processes.[106]

In addition to carbohydrates, other macromolecules are also hydrolyzed. The reduction in protein content[37,95,112] involved degradation to a mixture of smaller polypeptides and amino acids.[112] A several-fold increase in ammonia levels was observed in senescing roses.[114,144] In addition, amides such as asparagine and glutamine were accumulated. The formation of amides was proposed as a mechanism of detoxification in which the excess ammonia was neutralized.

Since the increase in free ammonia coincided with a decline in respiratory activity, a sequential order of these events was suggested.[144] According to this model, the onset of hydrolysis of structural cell components (including proteins) is initiated in response to the depletion of the free sugars used as respirable metabolites in order to supply alternative substrates for respiration, such as the carbon skeleton of amino acids. This supposition is supported by the fact that an exogenous supply of sugars delays the onset of excessive protein degradation,[39] and also serves as a substrate for protein synthesis.[115]

Studies with leaves have provided an alternative explanation. It has been suggested that there may be a decreased ATP production resulting from the progressive uncoupling of electron transport and oxidative phosphorylation.[22] This may occur before the total depletion of sugars. The need to fulfill the partial energy shortage created may activate alternative or complementary pathways with the formation of oxidative products that trigger senescence. The cyanide-resistant pathway of senescing tissues mentioned above is in accordance with this suggestion. This concept is supported by the finding of Nichols,[103] who demonstrated that either naturally or ethylene-triggered senescing and wilting carnation flowers contain reducing sugars, some of which were derived from hydrolysis of starch[68] and other structural polysaccharides.[105] They further demonstrated that substantial amounts of sugars were present in senescing carnation.[75,104,123] Hence the carbohydrate substrate limitation can hardly be the cause of the onset of senescence.

Flower petal tissue generally contains high activity of invertase[63] and indeed most of the sugar pool of mature petals is composed of inverted sugars.[103] However, it was recently reported that an increase in the ratio of sucrose to reducing sugars is found in senescing and in ethylene-triggered prematurely wilting carnations.[105] This may be explained by an assumed partial or total inactivation of the invertase. In *Ipomoea* flowers, evidence has been found for the synthesis of an invertase inhibitor at the onset of wilting,[149] thus cutting down the rate of inversion of sucrose.

Kaltaler and Steponkus[75] suggested that the gradual decline in respiration and decrease in respiration efficiency of rose petals is the result of the progressive inability of the mitochondria to utilize the substrate. They measured a respiratory control ratio is isolated mitochondria and found that the change in this parameter probably reflects a change in mitochondrial integrity, thereby affecting its functional capacity. In mitochondria isolated from cut flowers pretreated with sucrose, respiratory control values were maintained over longer periods of time. They concluded that the main effect of applied sugar in extending longevity is to maintain mitochondrial structure and functions. However, the effect of sugar on mitochondria may not be a specific effect and may stem from its general protective effect on membrane integrity.[125] It is interesting to note that mitochondria are among organelles last to disintegrate in aging leaves; they remain functioniong until the terminal phases of senescence.[22]

If senescence is viewed as a programmed process regulated by internal hormonal balance,[22] then sucrose interaction with plant hormones can modulate the regulatory process. It was shown that sucrose enhanced the effect of cytokinins in delaying senescence of flowers and reduced the effect of ethylene in promoting it.[91] This is possibly accomplished either by altering the sensitivity of the tissue to ethylene[97] or by delaying the natural rise in ethylene production.[38] It was also demonstrated that sucrose antagonized the effect of abscisic acid in promoting the senescence of roses[26] and carnations.[91]

Sucrose is known to improve water balance in cut flowers.[25,29] This was attributed to the effect of sugars on the closure of stomata and reduction of water loss.[86] However, this cannot account for the great increase in water uptake in sucrose-treated flowers.[29,53] The water deficit of sucrose-treated rose flowers was initially higher than that of control flowers held in water[27] since less water was absorbed, presumably due to the lower water potential of the sugar solution. At a later stage, most probably after the supplied sugar had reached the flower head, an improvement in water balance was observed. The improvement of water balance was also associated with a reduced endogenous level of abscisic acid,[27] which is a typical response to reduction of water stress.

Shortly after application of sucrose, increased amounts of reducing sugars appeared in the stem,[75,123] thus indicating that the stem possessed high sucrose inversion capacity.[123] Both the xylem and phloem are involved in the translocation of sugars.[74,108] Radial movement of sugars in the stem was demonstrated,[67] thus providing ready access between xylem and phloem along the stem. The glucose that enters the xylem may move radially to the phloem, be combined there to form sucrose, and be transported to the flower bud.

The translocated sugars accumulate in the flowers, so that their osmotic concentration is increased and their ability to absorb water and maintain their turgidity is thereby improved.[53] Maintenance of improved water status seems to be the most important aspect in extension of longevity.[121] Hence, one of the main effects of applied sugars on flower longevity seems to result from their contribution to the osmotic adjustment of the flowers. Observations that not only sugars, but also mineral salts such as NH_4NO_3, KNO_3, and KCl delay the otherwise gradual decline in cell osmotic concentration naturally occurring in cut flowers, support this view.[98] In addition, the treatment with mineral salts promotes water uptake, reduces sensitivity to ethylene, and extends longevity.[53,98] These mineral salts can only partially substitute for metabolic sugars in extending longevity,[53] which indicates some specific metabolic roles for the applied sugar.

V. PLANT HORMONES

Members of each of the five groups of plant hormones have been implicated in the regulation of senescence. However, ethylene seems to have been dealt with most extensively. Most of the early reports are concerned with the damage induced in flowers by ethylene,[152,66,139,18] i.e., the premature senescence and wilting of the corolla. A wide range of flowers are, to various degrees, affected by ethylene, with few responding with typical symptoms.

Some striking examples: (1) the in-rolling of carnation petals, also termed "sleepiness",[102] (2) fading and in-rolling of the corolla in *Ipomoea* flowers,[77] (3) fading and wilting of sepal tips in orchid flowers[6] and induction of anthocyanin formation in both gynostemia and labella.[8] These symptoms are well defined and can be used to assay the effect of ethylene. In fact, the in-rolling of the mid-rib of the corolla of the *Ipomoea* flower is used to measure quantitatively the response to ethylene.

Application of ethylene also catalyzes the synthesis of ethylene by the plant tissue. This phenomenon is termed autocatalysis and is common to both flower senescence[33,77,102] and ripening of climacteric fruits.[124] The tightness of coupling between visual symptoms induced by ethylene and induction of ethylene synthesis varies with the plant model studied.

Exposure to ethylene of young and mature carnation flowers results in a dual response, i.e., in-rolling of petals and increased synthesis of ethylene. Similarly, exposure of mature *Ipomoea* flowers (on day of opening) will also result in the same dual response. However, younger buds (a day before opening) respond only with the in-rolling of the corolla without a concomitant increase in ethylene synthesis.[78]

Carbon dioxide at 4%, a competitive inhibitor of ethylene, completely prevents the development of in-rolling of carnation petal tissue associated with naturally occurring changes in ethylene production.[91,92,102] By contrast, 4% CO_2 delayed but did not prevent the in-rolling of the corolla in *Ipomoea* flowers.[78] This variance in the responses indicates that the responses regulated by ethylene are complex. In *Ipomoea* flowers, the seemingly well-correlated events, ethylene-induced ethylene synthesis, and the in-rolling of the corolla, are not conditionally coupled. Ethylene has its effect by accelerating the in-rolling process. A somewhat different situation pertains with carnations, in which ethylene seems to play the role of a trigger for the in-rolling process.

A wide assortment of flowers producing ethylene[45,57] also seem to possess the capacity to metabolize various substrates into ethylene.[117] In many studies, ethylene emanation by vegetative or flower tissue was thought to reflect changes in ethylene production. Recently, it has been pointed out that this may not necessarily be so, because the emanation might also be influenced by changes either in the capacity of sites of ethylene attachment[73] or ethylene metabolism rates. Although this hypothesis helps to explain the possible involvement of ethylene in the regulation of senescence processes, it should at this point be regarded as tentative and subject to further experimental verification.

The time course of ethylene production follows a typical profile composed of 3 distinct phases: first, a low steady rate of ethylene production, second, an accelerated rise of ethylene production to a peak, followed by a last phase, in which ethylene production is declining. Various events associated with senescence can be studied with reference to these phases. Associated with the second phase, a rise in CO_2 evolution resembling the climacteric respiration of fruits occurs (Figure 7).

Usually, the visual symptoms occur only in the second phase, and are pronounced toward the end of this phase. Thus the onset of this phase is of importance, because it signals the terminal step of senescence. The transition to the second phase is well correlated with longevity of rose flowers. It was demonstrated to occur earlier in short-lived varieties compared with long-lived ones (Figure 8).

Application of CO_2 to carnations during the first phase of ethylene production will delay the terminal phase of senescence. However, application of CO_2 at the start of the onset of the second phase is sometimes ineffective (Figure 9) in delaying the transition into the autostimulation phase of ethylene production. The onset of the second phase may be induced by various means, such as exposure to ethylene,[102] pollination,[57,33,106] and abscisic acid.[91,92] By contrast, it can be delayed by CO_2,[91,92,132,133] low concentration of oxygen,[102] and hypobaric pressure environment.[91,92]

Senescence includes processes leading to cell disorganization. A possibility that can be considered is that change in the degree of organization could occur either spontaneously during senescence or be initiated by a plant hormone, such as ethylene.[100] Cell organization is generally maintained by spatial separation of reactants by partially permeable protoplasmic membranes. Matile and colleagues[88,149] demonstrated cell membrane disintegration, especially in the tonoplast. These events lead to uncontrolled mixing of vacuole contents with the cytoplasm.

Kende and Baumgartner[77] have proposed a model based on changes in compartmentation to account for ethylene-induced ethylene synthesis: by increasing the permeability of the tonoplast, ethylene could enhance the flow of substrates from the vacuole to the cytoplasm, where an ethylene-generating system is located. This model is supported by experiments in which flower tissue was preloaded with radioactive tracers, representing either the vacuole or cytoplasm content, then exposed to ethylene and subjected to compartmental analysis.[84] The results indicated (Figure 10) that one of the first changes to take place in response to ethylene was an enhanced efflux across the tonoplast.[62,99]

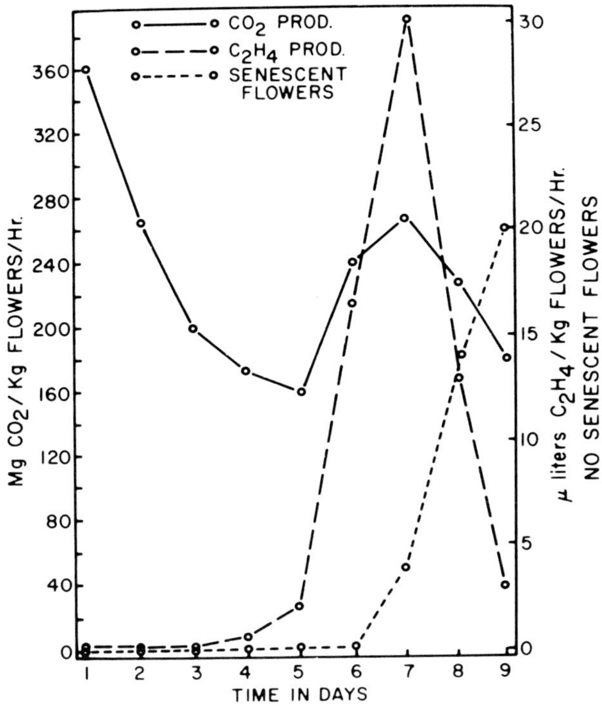

FIGURE 7. Rates of respiration, ethylene production, and senescence of flowers of "Improved White Sim" carnation at 20°C. (From Maxie, E. C. et al., *J. Am. Soc. Hortic. Sci.*, 98, 568—572, 1973. With permission.)

The presence of separate storage and metabolic amino acid pools has been suggested in higher plants.[109] Although clearly demonstrated in yeast,[145] the vacuoles of higher plants probably serve as a storage place for amino acids including methionine.

Numerous studies on the biosynthesis of ethylene have been conducted, in which several ethylene-forming systems were evaluated. Methionine[2] is the only generally accepted precursor of ethylene in the tissue of higher plants.[141] The existence of separate amino acid pools suggests that the tonoplast membrane may act as rate-limiting control in the synthesis of ethylene, thus indirectly lending support to the above proposed model.[77]

An important question is why the model of ethylene-induced ethylene synthesis is not operative during most of the young and early maturation phases of flower petal tissue, although there is a constant low emanation of ethylene. Yet transition into the second phase does not occur. When young flower tissues of *Ipomoea*[78] and carnation[131] are cut or bruised, the ethylene production begins to rise immediately. Later, after relatively short periods of time it will decline without the development of typical senescent symptoms.

Exposure of mature carnation flower tissue to ethylene will generally hasten senescence with the development of typical symptoms. These symptoms will last even when the ethylene is ventilated away, are irreversible,[97] and are presumably associated with a rise in ethylene synthesis. Although the young flower will develop the same symptoms upon exposure to ethylene, if the ethylene is ventilated away, even though it had been present to the point when symptoms become visible, the flowers will recover and resume their natural course of development. Hence the tissues vary in their response to

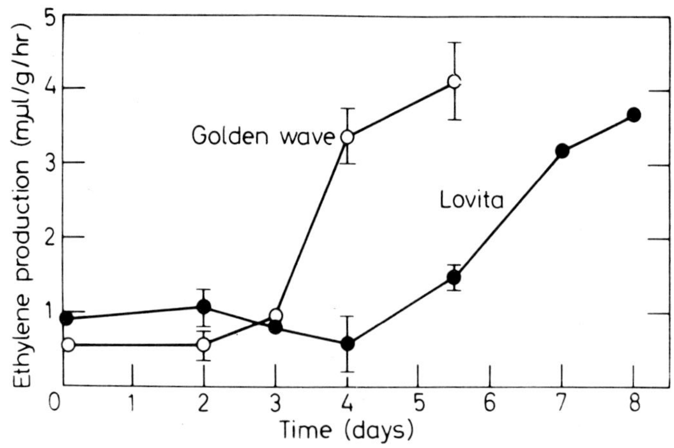

FIGURE 8. Ethylene production by rose flowers (long-lived cv. Lovita, and short-lived cv. Golden Wave). Flowers were tagged at the young stage (a), and were sampled on each of the following days. All values are means of six replications ± SE. (From Mayak, S. et al., *Physiol. Plant.*, 27, 1—4, 1972. With permission.)

the presence of ethylene. Young flowers are not responsive, whereas mature flowers respond — and transition into the second phase occurs. Perhaps while ethylene content in flower tissue is not markedly augmented until the senescence phase, enough gas is present during the presenescence phase. The autostimulation of ethylene production occurring in senescence will be initiated once the sensitivity to ethylene has changed to respond to the existing ethylene. The sensitivity to ethylene can be the result of an intricate, complex interaction between internal factors such as plant hormones[124] like abscisic acid,[92] carbohydrate reserve,[90] and osmotic concentration of the petal tissue.[98] The flower is composed of various organs: petals, sepal, style, ovary, etc., each at a different phase of its development. The compounded result of processes and internal interactions, occurring at each phase of development, may effect the response otherwise referred to as sensitivity to ethylene. The responding systems in each organ gradually become operative at different rates and times as the flower organs mature.

A plausible possibility, though not yet demonstrated in flowers, is that the compounded result leads to a higher oxidation state of the tissue, which may be in the form of accumulation of peroxides[31] or in lipoxygenase activity, resulting in lipid hydroperoxides. A substantial rise in lipoxygenase activity was demonstrated in tomatoes prior to the upsurge in hydroperoxide, which in turn precedes ethylene biosynthesis.[47] Ethylene synthesis could be initiated from methional in the presence of linolenate and lipoxygenase.[85] These results suggest that lipid hydroperoxides may be required for the methionine-dependent synthesis of ethylene.[21,44] However, further research also employing flowers is required to establish the role of these enzymes in the production of ethylene. It is interesting that the oxidation of ethylene, demonstrated also in carnations, was suggested as an important aspect of ethylene action.[36]

The entire process of ethylene-induced senescence may be under programmed regulation or control. However, the responding system may be affected by environmental factors. Temperature,[89,133] water stress to cut flowers imposed by excessive drying, and cold storage used in the practice of handling flowers[97] all result in higher sensitivity to ethylene. Recently we have studied the effect of water stress, imposed by low irrigation of carnation plants, on sensitivity of cut flowers to ethylene. In this situation, flowers were less sensitive. We assume that during growth and development of the flower,

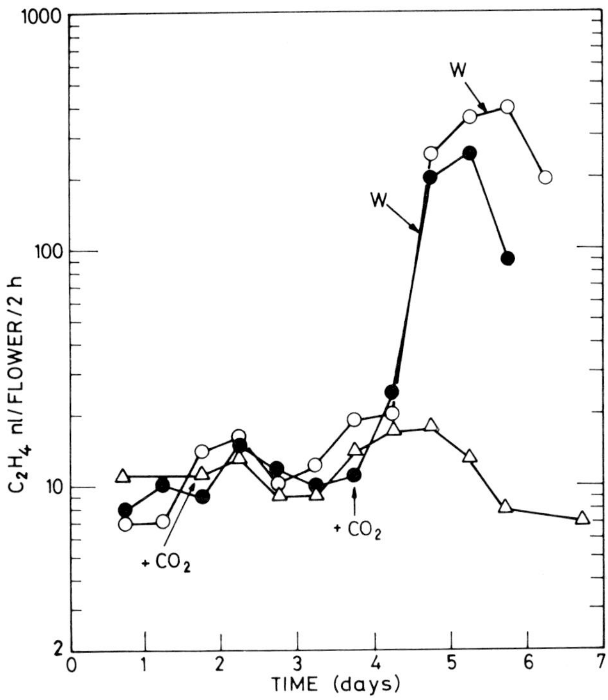

FIGURE 9. Effect of 4% CO_2 on the course of C_2H_4 emanation and senescence of cut carnation flowers. Arrows at times the CO_2 was applied. W: Time of initial wilting as judged by in-rolling of petals. (From Tendler, J. and Mayak, S., unpublished).

osmotic adapation probably occurred, which rendered the flower less sensitive. These results are similar to those of experiments in which the osmotic concentration was artificially maintained.[98] This assumption is supported by experiments with cut flowers, in which treatment with sugars and some mineral ions induces osmotic adjustment and extends the longevity of cut carnation and chrysanthemum flowers. The treatments also decrease their sensitivity to ethylene.[53,98]

Generally, the responding system gradually becomes more sensitive as the flower organs mature. The petals may respond to ethylene generated in the style, which was demonstrated to have a high capacity for ethylene synthesis.[33,57,80] Burg and Dijkman,[33] using *Vanda* orchids in their studies, proposed that following pollination, a transfer of auxin from the pollen to the stigma takes place and is spread through the column. The growth hormone triggers a chain reaction of ethylene synthesis in the cells so that ethylene becomes present throughout the flower, with the net result of petal fading. A similar proposition was put forward to explain the wilting of carnations occurring after pollination.[106] Here the style, through its endogenous ethylene production, is thought to trigger petal wilting. Gillisen,[51] using petunia flowers, reported that the penetrating pollen tubes activate the stylar tissue, which then leads to the wilting of the flower. The rate of wilting depends on the number of viable pollen grains penetrating the stigma.

In conjunction with the petal wilting, swelling of the ovary occurs.[102] Nichols[105] also found that ethylene promotes an accumulation of sugars and inorganic materials in the ovary, accompanied by loss of fresh and dry weight of the petals. The swelling of the ovary appears to result from the enlargement of existing cells. Ethylene assumes the role of a synchronizer; by induction of total wilting it facilitates substantial mobilization of carbohydrates from the petal, providing for the development of the ovary.

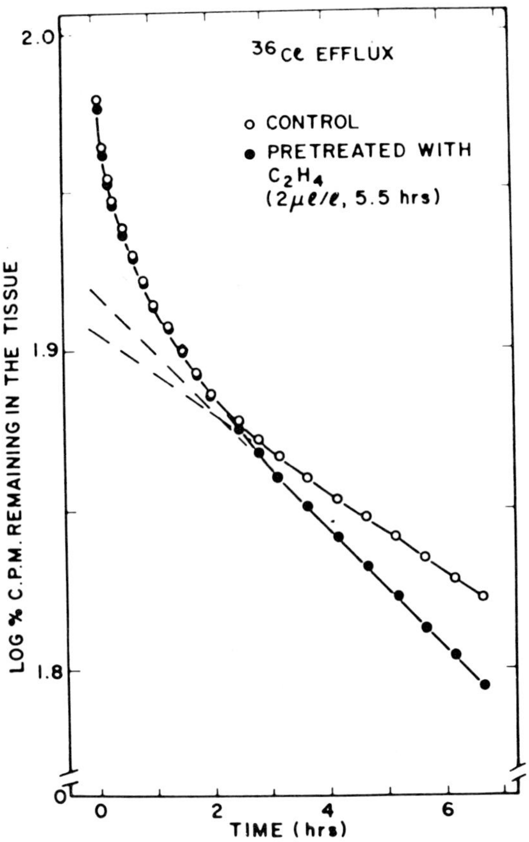

FIGURE 10. Effect of ethylene on ^{36}Cl efflux from carnation petals. Dashed lines are interpolation to zero time. Cut flowers were preloaded with ^{36}Cl and treated with ethylene. Results are means of four replicates. (From Mayak, S. et al., *Plant. Physiol.*, 59, 591—593, 1977. With permission.)

The hormonal regulation of a physiological process may be accompanied by a proportionality mechanism.[141] Thus a process is enhanced as concentration increases, and diminishes as the concentration drops. Accordingly the changes in plant hormone concentrations may be interpreted as being signals controlling the processes. Indeed, the tissue response to ethylene is proportional to the log of the concentration of ethylene present. The level of ethylene in the tissue, as it changes with time, as well as being part of a controlling signal, is determined by the rate of biosynthesis and its diffusion along a concentration gradient into the surrounding tissue and into the atmosphere.

Other plant hormones are also involved in the regulation of petal senescence and they do seem to conform to the same principle. The endogenous level of cytokinins in rose petals decreases as the flower ages, and the level is lower in a short-lived cultivar than in a long-lived one.[93] Results with external application of cytokinins, which delay senescence of various flowers,[64,71,91,93] support the possibility that the reduction of the internal level of this growth regulator is associated with senescence processes in plants. Further support for this supposition may be drawn from experiments[95] in which applied kinetin slowed down the reduction in dry weight that took place as the flower matured. In addition, water uptake was enhanced, and it was postulated that this was achieved by maintenance of cell integrity. However, the effects of kinetin and other cytokinins on the senescence of leaves are more complex (Chapter 5).

The presence of gibberellins and auxins has been reported in carnations.[72] Gibberellic acid increases flower size and extends the longevity of carnations.[76,49] Gilbart and Sink[50] assigned to auxin a central role in the control of senescence in poinsettia.

The auxin level decreased in two poinsettia cultivars, but faster in the short-lived cultivar. Also, the activity of IAA-oxidase and the level of hydrogen peroxide increased with aging. These researchers hypothesized that auxin delays senescence by inducing the synthesis of peroxidase, which then prevents the accumulation of free peroxide associated with aging. Auxin may be linked to factors determining sensitivity to ethylene discussed earlier. However, further studies are needed with various flowers to substantiate the involvement of these plant hormones with senescence.

The role attributed to abscisic acid in petals is that of regulating the process of senescence. The endogenous level of abscisic acid rises as petals senesce. It rises to a higher level in a short-lived variety than in a long-lived one.[94] Also exogenous application of AbA accelerated senescence of carnations,[92] as well as that of roses. The marked development of senescence was reflected in a variety of changes such as red petal bluing, decrease in protein content, and increase in ribonuclease activity.[27,93,56] *De novo* synthesis of this enzyme probably is involved in the rise in its activity during the final stages of petal senescence of *Ipomoea.*[19] If, however, leaves are attached to the flowering stem of a rose, these effects may be reversed. By causing stomatal closure on the leaves, abscisic acid reduced water loss from the flowering shoot, thereby delaying senescence in the flower.[56]

Although not always studied in the same flower, each plant hormone was demonstrated to have a potential role in the hormonal controlling system. It is generally accepted that control is exerted through a balance between plant hormones interacting with each other and with other internal factors. Combined application of benzyladenine and auxin to daffodils was more effective in delaying senescence and climacteric-like respiration[16,17] than the additive effect of the two growth regulators.

Kende and Hanson[78] reported that benzyladenine delayed ethylene production by isolated rib segments of *Ipomoea* flowers. Similar results were reported with carnation flowers, and their responsiveness was also affected.[42] A plant hormone may thus affect the level of another plant hormone, thereby having more intensive effect on the hormonal balance. Evidence for this kind of interaction comes from studies with carnations. Abscisic acid brings about an earlier increase in ethylene production (Figure 11). We hypothesized that abscisic acid accelerated developmental processes associated with maturation. It increases the sensitivity to ethylene, and the tissue may respond to the existing ethylene concentration without having to increase its sensitivity in order to have an effect. Transition into the phase of ethylene autostimulation may occur at an earlier stage. The permeability of the tonoplast[61,99] increases, leading to cell disorganization accompanied by a reduction in water uptake (Figure 12) and plausible development of water or ion stress. The well-established stress response, i.e., increase in the level of abscisic acid,[81] soon follows. In this respect the rise in the level of ABA, in response to ethylene as observed with cut rose flowers,[94] can be interpreted as a secondary result *via* an effect on membrane permeability and water stress imposed by ethylene.

Membrane modification was suggested to be controlled by the cytokinin/ABA balance.[71] The researchers did not rule out the possible participation of ethylene.

Events reflecting the effect of ethylene on the level of abscisic acid are mediated by membrane modifications. It will be of interest to study whether or not the reciprocal effects of cytokinin and ABA on each other's activity, suggested to be a widely occurring phenomenon in plants by Itai and Benzioni,[71] is also mediated by membrane modifications. The observation that kinetin delays wilting of cut rose flowers by protecting

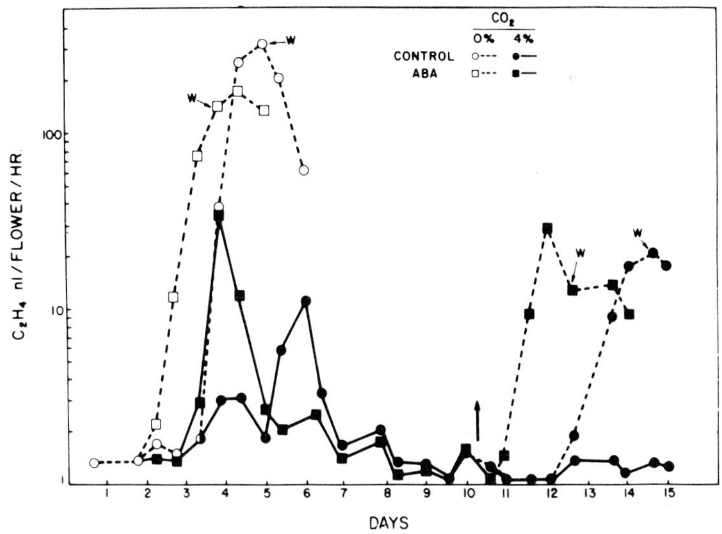

FIGURE 11. Effect of 50 μM ABA and 4% CO_2 on the course of C_2H_4 production and senescence of cut carnation flowers. W: Time of initial wilting as judged by in-rolling of petals; arrow at 10 days; transfer of flowers from CO_2 to air with (—) or without (- - -) 4% CO_2. (From Mayak, S. and Dilley, D. R., *Plant Physiol.*, 58, 663—665, 1976b. With permission.)

cell integrity[95] supports this possibility. It is conceivable that if cell integrity is maintained, the development of a situation of stress is delayed and abscisic acid will be kept at a low level.

Along with the increase in membrane permeability, a reduction in phospholipid content is affected.[23] A reduction in phospholipid was found in carnations[120] also in the presence of rhizobitoxine, a specific inhibitor of the conversion of methionine to ethylene.[110] Hence, the reduction in phospholipid in senescent carnation petals is not necessarily dependent on methionine-dependent ethylene synthesis.

Some tentative generalizations concerning the processes and interactions with plant hormones are summarized in a hypothetical scheme (Figure 13). Changes in sensitivity to ethylene are assumed to play a central role. As the center circle revolves clockwise, the revolution representing progression of developmental processes, a gradual increase in sensitivity to ethylene occurs. The methionine-dependent ethylene generating system is constantly exchanging sensing signals with the center circle. When sensitivity to ethylene reaches a threshold level, an operating signal will be generated thereby inducing certain processes depicted in the scheme: (1) operating signal, possibly in the form of free radicals, was suggested to accelerate the conversion of methionine to ethylene,[44] (2) another operating signal conceivably could modify membrane permeability leading to cell disorganization, consequently causing a reduction in water uptake and of petal wilting,[99] (3) membrane disorganization may cause ion redistribution so that water or ion stress soon develops, (4) in response to stress, the level of abscisic acid increases. The latter phenomenon is well established as a rapid response to water stress in plants.[81] The abscisic acid also is shown to stimulate flower development,[27] as well as to increase sensitivity to ethylene, hence advancing the operating signal to the ethylene synthesis system,[91] (5) a stimulating effect is also induced by the gradual depletion of the osmotic pool. Sugar and certain inorganic salts enrich the osmotic pool, slowing down the development of sensitivity to ethylene.[98]

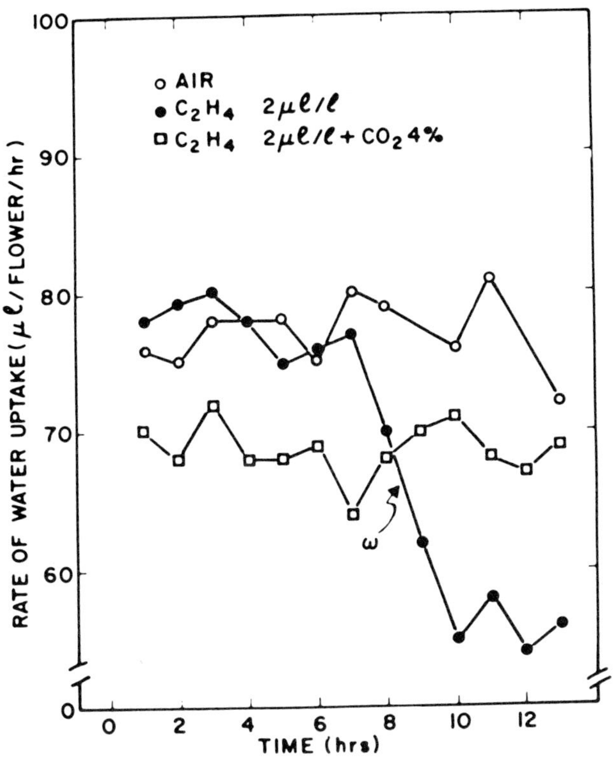

FIGURE 12. Effect of ethylen (2 μℓ/ℓ) with or without 4% CO_2 on the rate of water uptake by carnation flowers. The flowers, excluding the stem, were ventilated with the gas mixtures at 20°C in light at 200 ft-c. W indicates time of initial wilting. (From Mayak, S. et al., *Plant Physiol.*, 59, 591—593, 1977. With permission.)

Cytokinins delay processes associated with senescence, thus maintaining cell integrity.[95] Also, they decrease sensitivity to ethylene.[42]

Although points in this scheme are supported by studies mentioned above, further information is required to substantiate some of the suggested interactions. Some crucial questions in need of answers are (1) what is the nature of sensitivity to ethylene? (2) what is the nature of the interaction between plant hormones and processes they induce, especially the rapid responses? and (3) what are the events leading to the transition into the autostimulation phase of ethylene synthesis?

REFERENCES SUPPORTING THE EFFECTS DESCRIBED IN THE SCHEME

1. Free radicals stimulate conversion to ethylene. (Eskin and Grossman[44]).
2. a. Gradual depletion of osmotic pool.
   b. Enrichment of osmotic pool.
   (Mayak et al.,[98] Halevy[53]).
3. ABA stimulates sensitivity to ethylene. (Mayak and Dilley[92]).
4. Stress induces rise in ABA. (Livne and Vaadia,[81] Borochov et al.[27]).
5. Cytokinin (CK) reduces sensitivity to ethylene. (Eisinger[42]).
6. Reduction in water uptake induced by ethylene. (Mayak and Dilley[91]).

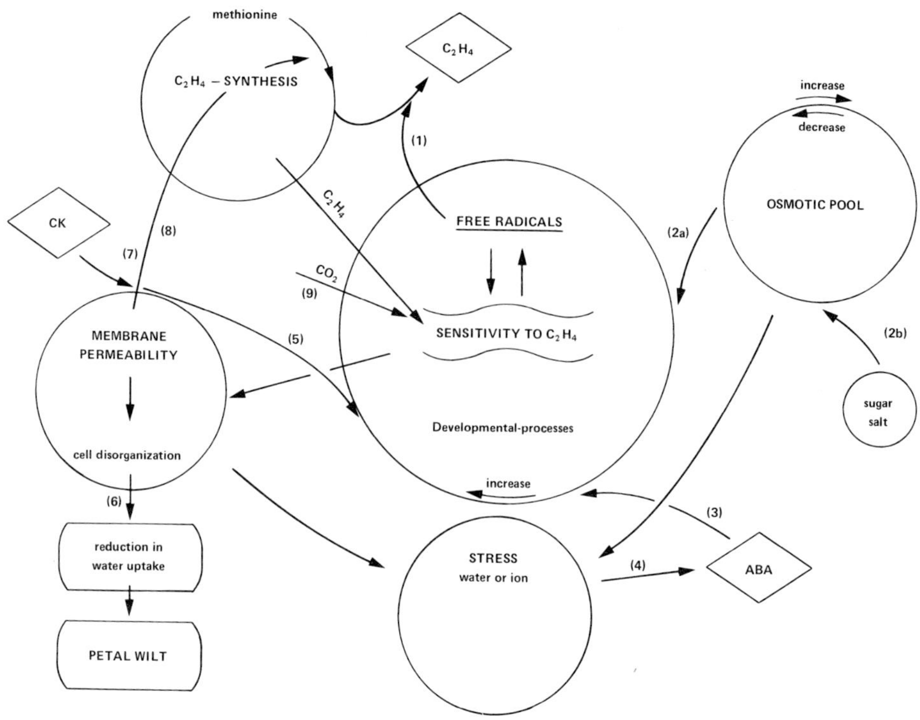

FIGURE 13. An hypothetical scheme describing relations of plant hormones to flower senescence. (Numbers in the scheme refer to supporting pieces of evidence listed below).

7. Cytokinin (CK) maintains membrane integrity (Mayak and Halevy[95]), and extends longevity (Mayak and Halevy[95]).
8. Increased availability of methionine induced by ethylene (Kende and Baumgartner[77]).
9. Carbon dioxide competitively inhibits the ethylene (Burg and Burg[32]), thus the sensing signal is blocked, and the development of response is prevented or delayed (Mayak and Dilley[92]).

VI. FLOWER ABSCISSION

The final stage of flower senescence is accompanied by abscission. Shedding may involve abscission of whole inflorescences, flowers, or parts of flowers. Almost all studies on abscission were carried out with leaves and fruits.[3,5] Some work has also been done with flower bud abscission.[127] The physiology of petal shedding has been, however, almost ignored by researchers. A comprehensive work was published by Fitting in 1911, but the next major work on petal abscission seems to be only that of Addicott in 1977. A question is thus raised as to what extent petal abscission is similar to or different from than leaf abscission. Some facts seem to indicate that the process in petals may not be identical to that of leaves.

When whole flowers or flower buds are shed, an abscission layer is formed as with leaves.[59,127] However cell division usually does not precede petal shedding, no clear abscission layer is apparent, and shedding of petals is caused by softening of the middle lamella.[43] There are indications, however, that petal shedding involves a rising activity of cell wall hydrolytic enzymes, as in leaves.[4]

Fitting (1911) had already observed that external factors such as shaking, wounding, high temperatures, and some gases induced very rapid petal abscission (in sensitive species). This is a much quicker response than is known for leaves. That ethylene promotes abscission of flower buds and petals of many flowers has been known since the work of Zimmerman[152] on roses. However, CO_2, which is known to antagonize ethylene and retard leaf abscission[5] promoted petal abscission.[46]

In many species, pollination — and especially fertilization — promotes the abscission of flower parts, especially the petals.[46] This is apparently associated with a rise in ethylene evolution at this stage.[6,33]

Indeed, treatment with the analogs of rhizobitoxine, an inhibitor of ethylene synthesis, reduced ethylene production and flower abscission of snapdragon.[143] Abscission of flower buds[58] and petals[4] was promoted by ABA, as with leaves.[3] Endogenous ABA was also correlated with abscission in flowers of *Lupinus luteus.*[118] Auxins retarded the abscission of flower buds[128] and petals[4] and endogenous IAA level in begonia flower buds coincided with seasonal fluctuation in flower bud abscission. In roses, cytokinin rather than auxin was a better inhibitor of flower abscission.[54]

The above summary of our knowledge of flower and petal abscission reveal both similarities and dissimilarities with analogous processes in leaves and fruits. Much more work is needed — especially on petal shedding — to clarify the entire situation.

REFERENCES

1. **Aarts, J. F. Th.,** Over de houdbaarheid van snijbloemen, *Meded Landbouwhogesch. Wageningen,* 57, 1—62, 1957.
2. **Adams, D. O. and Yang, S. F.,** Methionine metabolism in apple tissue, *Plant Physiol.,* 60, 892—896, 1977.
3. **Addicott, F. T.,** Plant hormones in the control of abscission, *Biol. Rev.,* 45, 485—524, 1970.
4. **Addicott, F. T.,** Flower behavior in *Linum lewissii:* some ecological and physiological factors in opening and abscission of petals, *Am. Midl. Nat.,* 97, 321—332, 1977.
5. **Addicott, F. T. and Lyon, J. L.,** Physiological ecology of abscission, in *Shedding of Plant Parts,* Kozlowsky, T. T., Ed., Academic Press, New York, 1973, 85—124.
6. **Akamine, E. K.,** Ethylene production in fading vanda orchid blossoms, *Science,* 140, 1217—1218, 1963.
7. **Arditti, J. and Flick, H.,** Post-pollination phenomena in orchid flowers, VI. Excised floral segments of *Cymbidium, Am. J. Bot.,* 63, 201—211, 1976.
8. **Arditti, J., Hogan, N. M., and Chadwick, A. V.,** Post-pollination phenomena in orchid flowers, IV. Effect of ethylene, *Am. J. Bot.,* 60, 883—888, 1973.
9. **Arditti, J. and Knauft, R. L.,** The effect of auxin, actinomycin D, ethionine, and puromycin on post-pollination behavior in *Cymbidium* (Orchidaceae) flowers, *Am. J. Bot.,* 56, 620—629, 1969.
10. **Asen, S.,** Anthocyanins and other flavonoids in red cornflower, *Proc. Am. Soc. Hortic. Sci.,* 91, 653—659, 1967.
11. **Asen, S.,** Known factors responsible for infinite flower color variations, *Acta Hortic.,* 63, 217—223, 1976.
12. **Asen, S., Norris, K. H., and Stewart, R. N.,** Effect of pH and concentration of the anthocyanin-flavanol copigment complex on the color of 'Better Times' roses, *J. Am. Soc. Hortic. Sci.,* 96, 770—773, 1971.
13. **Asen, S., Stewart, R. N., and Norris, K. H.,** Anthocyanin flavanol copigments and pH responsible for larkspur flower color, *Phytochemistry,* 14, 2677—2682, 1975.
14. **Asen, S., Stewart, R. N., and Norris, K. H.,** Anthocyanin and pH involved in the color of 'Heavenly Blue' morning glory, *Phytochemistry,* 16, 1118—1119, 1977.
15. **Baker, J. E., Wang, C. Y., Lieberman, M., and Hardenburg, R.,** Delay of senescence in carnations by a rhizobitoxine analog and sodium benzoate, *HortScience,* 12, 38—39, 1977.
16. **Ballantyne, D. J.,** Senescence of daffodil *(Narcissus pseudonarcissus* L.) cut flowers treated with benzyladenine and auxin, *Nature (London),* 205, 819, 1965.

17. **Ballantyne, D. J.**, Respiration of floral tissue of the daffodil (*Narcissus pseudonarcissus* L.) treated with benzyladenine and auxin, *Can. J. Bot.*, 44, 117—119, 1966.
18. **Barden, L. E. and Hanan, J. J.**, Effect of ethylene on carnation keeping life, *J. Am. Soc. Hortic. Sci.*, 97, 785—788, 1972.
19. **Baumgartner, B., Kende, H., and Matile, P.**, Ribonuclease in senescing morning glory, *Plant Physiol.*, 55, 734—737, 1975.
20. **Baumgartner, B. and Matile, P.**, Isoenzymes of RNase in senescing morning glory petals, *Z. Pflanzenphysiol.*, 82, 371—374, 1977.
21. **Beauchamp, C. and Fridovitch, I.**, A mechanism for the production of ethylene from methional. The generation of the hydroxyl radical by xanthine oxidase, *J. Biol. Chem.*, 245, 2641—4646, 1970.
22. **Beevers, L.**, Senescence, in *Plant Biochemistry*, Bonner, J. and Varner, J. E., Eds., Academic Press, New York, 1976, 771—794.
23. **Beutelmann, P. and Kende, H.**, Membrane lipids in senescing flower tissue of *Ipomoea tricolor*, *Plant Physiol.*, 59, 888—893, 1977.
24. **Blumenfeld, A.**, Ethylene and the annona flower, *Plant Physiol.*, 55, 265—269, 1975.
25. **Borochov, A., Halevy, A. H., and Shinitzky, M.**, Increase in microviscosity with aging in protoplast plasmalemma of rose petals, *Nature (London)*, 263, 158—159, 1976.
26. **Borochov, A., Mayak, S., and Halevy, A. H.**, Combined effects of acid and sucrose on growth and senescence of rose flowers, *Physiol. Plant.* 36, 221—224, 1976.
27. **Borochov, A., Tirosh, T., and Halevy, A. H.**, Abscisic acid content of senescing petals on cut rose flowers as affected by sucrose and water stress, *Plant Physiol.*, 58, 175—178, 1976.
28. **Borochov, A., Halevy, A. H., Borochov, H., and Shinitzky, M.**, Microviscosity of rose petal's plasmalemma as affected by age and environmental factors, *Plant Physiol.*, 61, 812—815, 1978.
29. **Bravdo, B., Mayak, S., and Gavrieli, Y.**, Sucrose and water uptake from concentrated sucrose solutions by gladiolus shoots and the effect of these treatments on floret life, *Can. J. Bot.*, 52, 1271—1281, 1974.
30. **Brendmeijer, G. M. M.**, Peroxidase activities and peroxidases isoenzymes patterns during growth and senescence of the unpollinated style and corolla of tobacco plants, *Acta Bot. Neerl.*, 22, 40—48, 1973.
31. **Brennan, T. and Frenkel, C.**, Involvement of hydrogen peroxide in the regulation of senescence in pear, *Plant Physiol.*, 59, 411—416, 1977.
32. **Burg, S. P. and Burg, E. A.**, Ethylene action and ripening of fruits, *Science*, 148, 1190—1196, 1965.
33. **Burg, S. P. and Dijkman, M. J.**, Ethylene and auxin participation in pollen induced fading of vanda orchid blossoms, *Plant Physiol.*, 42, 1648—1650, 1967.
34. **Butler, R. D. and Simon, E. W.**, Ultrastructural aspects of senescence in plants, *Adv. Gerontol. Res.*, (Strehler, B. L., Ed.,) 3, 73—129, 1971.
35. **Butcher, H. C., Wagner, G. J., and Siegelman, H. W.**, Localization of acid hydrolases in protoplasts, *Plant Physiol.*, 59, 1098—1103, 1977.
36. **Beyer, E. M., Jr.**, $^{14}C_2H_4$: its incorporation and oxidation to $^{14}CO_2$ by cut carnations, *Plant Physiol.*, 60, 203—206, 1977.
37. **Carfantan, N. and Daussant, J.**, Preliminary study of tulip protein during senescence, *Acta Hortic.*, 41, 31—43, 1975.
38. **Carpenter, W. J. and Dilley, D. R.**, Investigations to extend cut flower longevity, *M.S.U. Res. Rpt.*, 263, 1—10, 1975.
39. **Coorts, G. D.**, Internal metabolic changes in cut flowers, *Hort. Sci.*, 8, 195—198, 1975.
40. **Dilley, D. R. and Carpenter, W. J.**, The role of chemical adjuvants and ethylene synthesis on cut flower longevity, *Acta Hortic.*, 41, 117—132, 1975.
41. **Eilam, Y.**, Permeability changes in senescing tissue, *J. Exp. Bot.*, 16, 614—627, 1965.
42. **Eisinger, W.**, Role of cytokinins in carnation flower senescence, *Plant Physiol.*, 59, 707—709, 1977.
43. **Esau, K.**, *Plant Anatomy*, John Wiley, New York, 1965.
44. **Eskin, M. N. A. and Grossman, S.**, Biochemistry of lipoxygenase in relation to food quality, *Crit. Rev. Food Sci. Nutr.*, 9, 1—40, 1977.
45. **Fischer, C. W.**, Ethylene gas a problem in cut flower storage, *N.Y. State Flower Grow. Bull.*, 61, 1, 1950.
46. **Fitting, H.**, Untersuchungen ueber die varzeitige Entbatterung von Blüten, *Jahrb. Wiss Bot.*, 49, 187—263, 1911.
47. **Frenkel, C., Chee-Kok, C., and Eskin, N. A. M.**, Relation of ethylene metabolism to peroxide formation in fruits and potato tubers, 73rd Ann. Meet. Am. Soc. Hortic. Sci., Louisiana, 1976.
48. **Fridovich, I.**, Superoxide dismutases, *Annu. Rev. Biochem.*, 44, 147—159, 1975.
49. **Garrod, J. F. and Harris, G. P.**, Effect of gibberellic acid on senescence of isolated petals of carnation, *Ann. Appl. Biol.*, 88, 309—311, 1978.

50. **Gilbart, D. A. and Sink, K. C.**, Regulation of endogenous indolacetic acid and keeping quality of poinsettia, *J. Am. Soc. Hortic. Sci.*, 96 3—7, 1971.
51. **Gilissen, L. J. W.**, Style controlled wilting of the flower, *Planta*, 133, 275—280, 1977.
52. **Goodwin, T. W.**, The carotenoids, in *Comparative Phytochemistry*, Swain, T. S., Ed., Academic Press, London, 1966, 121—137.
53. **Halevy, A. H.**, Treatments to improve water balance of cut flowers, *Acta Hortic.*, 64, 223—230, 1976.
54. **Halevy, A. H. and Kofranek, A. M.**, The prevention of flower bud and leaf abscission in pot roses during simulated transport, *J. Am. Soc. Hortic. Sci.*, 101, 658—660, 1976.
55. **Halevy, A. H. and Mayak, S.**, Improvement of cut flower quality opening and longevity by pre-shipment treatments, *Acta Hortic.*, 43, 335—347, 1974.
56. **Halevy, A. H., Mayak, S., Tirosh, T., Spiegelstein, H., and Kofranek, A. M.**, Opposing effects of abscisic acid on senescence of rose flowers, *Plant Cell Physiol.*, 15, 813—821, 1974.
57. **Hall, I. V. and Forsyth, F. R.**, Production of ethylene by flowers following pollination and treatments with water and auxin, *Can. J. Bot.*, 45, 1163—1166, 1967.
58. **Hanisch ten Cate, Ch. H. and Bruinsma, J.**, Abscission of flower bud pedicels in Begonia. 1. Effects of plant growth regulating substances on the abscission with intact plants and with explants, *Acta Bot. Neerl.*, 22, 666—674, 1973.
59. **Hänisch ten Cate, Ch. H., Van der Ploeg-Voogd, J. J. L., and Bruinsma, J.**, Abscission of flower bud pedicels in Begonia. III. Anatomical pattern of abscission, *Acta Bot. Neerl.*, 22, 681—685, 1973.
60. **Hänisch ten Cate, Ch. H., Berghof, J., Van der Hoorn, A. M. H., and Bruinsma, J.**, Hormonal regulation of pedicel abscission in Begonia flower buds, *Physiol. Plant.*, 33, 280—284, 1975.
61. **Hanson, A. D. and Kende, H.**, Ethylene enhanced ion and sucrose efflux in morning glory flower tissue, *Plant Physiol.*, 55, 663—669, 1975.
62. **Hanson, A. D. and Kende, H.**, Methionine metabolism and ethylene biosynthesis in senescent flower tissue of morning glory, *Plant Physiol.* 57, 528—537, 1976.
63. **Hawker, J. S., Walker, R. R., and Ruffner, H. P.**, Invertase and sucrose synthesis in flowers, *Phytochemistry*, 15, 1441—1443, 1976.
64. **Heide, O. M. and Oydvin, J.**, Effects of 6-benzylamino-purine on the keeping quality and respiration of glasshouse carnations, *Hortic. Res.*, 9, 26—36, 1969.
65. **Hew, C. S., Thio, Y. C., Wong, S. Y., and Chin, T. Y.**, Rhythmic production of CO_2 by tropical orchid flowers, *Physiol. Plant.*, 42, 226—230, 1978.
66. **Hitchock, A. E., Crocker, W., and Zimmerman, P. W.**, Effect of illuminating gas on the lily narcissus, tulip, and hyacinth, *Contrib. Boyce Thompson Inst.*, 4, 155—176, 1932.
67. **Ho, L. C. and Nichols, R.**, The role of phloem transport in the translocation of sucrose along the stem of carnation cut flowers, *Ann. Bot. (London)*, 39, 439—446, 1975.
68. **Ho, L. C. and Nichols, R.**, Translocation of ^{14}C-sucrose in relation to changes in carbohydrate content in rose corollas cut at different stages of development, *Ann. Bot. (London)*, 41, 227—242, 1977.
69. **Horie, K.**, The behavior of the petals in the fading of the flowers of *Tradescantia reflexa*, *Protoplasma*, 53, 377—386, 1961.
70. **Hsiang, T. H. T.**, Physiological and biochemical changes accompanying pollination in orchid flowers. II. Respiration, catalase activity, and chemical constituents, *Plant Physiol.*, 26, 708—721, 1951.
71. **Itai, C. and Benzioni, A.**, Water stress and hormonal response, in *Ecological Studies, Analysis, and Synthesis*, Vol. 19, Lange, O. L., Kappen, L., and Scholze, E. D., Eds., Springer-Verlag, Berlin, 1976, 225—242.
71a. **Jeffcoat, B.**, Influence of the cytokinin, 6-benzylamino-9-(tetrahydropyran-2-yl)-9H-purine on the growth and development of some ornamental crops, *J. Hort. Sci.*, 52, 143—153, 1977.
72. **Jeffcoat, B. and Harris, G. P.**, Hormonal regulation of the distribution of ^{14}C-labeled assimilates in the flowering shoot of carnation, *Ann. Bot. (London)*, 36, 356—361, 1972.
73. **Jerie, P. H., Hall, M. A., and Zeroni, M.**, Aspects of the role of ethylene in fruit ripening, *Acta Hortic.*, 325—332, 1979.
74. **Kaltaler, R. E. L. ad Steponkus, P. L.**, Uptake and metabolism of sucrose in cut roses, *J. Am. Soc. Hortic. Sci.*, 99, 490—493, 1974.
75. **Kaltaler, R. E. L., and Steponkus, P. L.**, Factors affecting respiration in cut roses, *J. Am. Soc. Hortic. Sci.*, 101, 352—354, 1976.
76. **Kelley, J. D. and Schlamp, A. L.**, Keeping quality, flower size, and flowering response of three varieties of easter lilies to gibberellic acid, *Proc. Am. Soc. Hortic. Sci.*, 85, 631—634, 1964.
77. **Kende, H. and Baumgartner, B.** Regulation of aging in flowers of *Ipomoea tricolor* by ethylene, *Planta*, 116, 279—289, 1974.
78. **Kende, H. and Hanson, A. D.**, Relationship between ethylene evolution and senescence in morning glory flower tissue, *Plant Physiol.*, 57, 523—527, 1976.

79. **Kuijper, J.,** Über den Farbwechsel und the Ephemerie bei den Bluten von *Hibiscus mutabilis* L., *Rec. Trav. Bot., Neerl.,* 28 1—19, 1931.
80. **Lipe, J. A. and Morgan, P. W.,** Location of ethylene production in cotton flowers and dehiscing fruits, *Planta,* 115, 93—96, 1973.
81. **Livne, A. and Vaadia, Y.,** Water deficits and hormone relations, in *Water Deficits and Plant Growth,* Vol. 3, Kozlowski, T. T., Ed., Academic Press, New York, 1972, 255—275.
82. **Lonati, I. C., van Loon, L. C., and Bruinsma, J.,** Effects of auxin on the activity of RNA-hydrolysing enzymes from senescing and aging barley leaves, *Z. Pflanzenphysiol.,* 67, 146—154, 1972.
83. **Lörz, H., Harms, C. T., and Potrykus, I.,** Isolation of "vacuoloplasts" from protoplasts of higher plants, *Biochem. Physiol. Pflanz.,* 169, 617—620, 1976.
84. **Macrobbie, E. A.,** Fluxes and compartmentation in plant cells, *Annu. Rev. Plant Physiol.,* 221, 75—96, 1971.
85. **Mapson, L. W. and Wardale, D. A.,** Enzymes involved in the synthesis of ethylene from methionine, or its derivatives, in tomatoes, *Phytochemistry,* 10, 29, 1968.
86. **Marousky, F. J.,** Inhibition of vascular blockage and increased moisture retention in cut roses induced by pH, 8-hydroxyquinoline citrate, and sucrose, *J. Am. Soc. Hortic. Sci.,* 96, 38—41, 1971.
87. **Matile, P.,** *The Lytic Compartment of Plant Cells,* Springer-Verlag, New York, 1975.
88. **Matile, P. and Winkenbach, F.,** Function of lysosomes and lysosomal enzymes in the senescing corolla of the morning glory *(Ipomoea purpurea), J. Exp. Bot.,* 22, 759—771, 1971.
89. **Maxie, E. C., Farnham, D. S., Mitchell, F. G., Sommer, N. F., Parson, R. A., Snyder, R. G., and Rae, H. L.,** Temperature and ethylene effects on cut flowers of carnations (*Dianthus caryophyllus* L.), *J. Am. Soc. Hortic. Sci.,* 98, 568—572, 1973.
90. **Mayak, S. and Dilley, D. R.,** Effect of sucrose on the response of cut carnations to kinetin, ethylene, and abscisic acid, *J. Am. Soc. Hortic. Sci.,* 101, 583—585, 1975.
91. **Mayak, S. and Dilley, D. R.,** Effect of sucrose on response of cut carnation to kinetin, ethylene, and abscisic acid, *J. Am. Soc. Hortic. Sci.,* 101, 583—585, 1976.
92. **Mayak, S. and Dilley, D. R.,** Regulation of senescence in carnation *(Dianthus caryophyllus).* Effect of abscisic acid and carbon dioxide on ethylene production, *Plant Physiol.,* 58, 663—665, 1976.
93. **Mayak, S. and Halevy, A. H.,** Cytokinin activity in rose petals and its relation to senescence, *Plant Physiol.,* 46, 497—499, 1970.
94. **Mayak, S. and Halevy, A. H.,** Interrelationship of ethylene and abscisic acid in the control of rose petal senescence, *Plant Physiol.,* 50, 341—346, 1972.
95. **Mayak, S. and Halevy, A. H.,** The action of kinetin in improving the water balance and delaying senescence processes of cut rose flowers, *Physiol. Plant.,* 32, 330—336, 1974.
96. **Mayak, S., Halevy, A. H., and Kats, M.,** Correlative changes in phytohormones in relation to senescence in rose petals, *Physiol. Plant.,* 27, 1—4, 1972.
97. **Mayak, S. and Kofranek, A. M.,** Altering the sensitivity of carnation flowers (*Dianthus caryophyllus* L.) to ethylene, *J. Am. Soc. Hortic. Sci.,* 101, 503—506, 1976.
98. **Mayak, S., Kofranek, A. M., and Tirosh, T.,** The effect of inorganic salts on the senescence of *Dianthus caryophyllus* flowers, *Physiol. Plant.,* 43, 282—286, 1978.
99. **Mayak, S., Vaadia, Y., and Dilley, D. R.,** Regulation of senescence in carnation *(Dianthus caryophyllus)* by ethylene. Mode of action, *Plant Physiol.,* 59, 591—593, 1977.
100. **McGlasson, W. B.,** The ethylene factor, in *The Biochemistry of Fruits and their Products,* Vol. 1, Hulme, A. C., Ed., Academic Press, London, 1970, 475—519.
101. **Mishra, S. D., Gaur, B. K., Bedeker, V. M., and Singh, B. B.,** Isolation, identification, and significance of free radicals in senescing leaves, *Acta Bot. Ind.,* 4, 131—138, 1976.
102. **Nichols, R.,** The response of carnations *(Dianthus caryophyllus)* to ethylene, *J. Hortic. Sci.,* 43, 335—349, 1968.
103. **Nichols, R.,** Senescence of the cut carnation flower: respiration and sugar status, *J. Hortic. Sci.,* 48, 111—121, 1973.
104. **Nichols, R.,** Senescence and sugar status of the cut flower, *Acta Hortic.,* 4, 21—29, 1975.
105. **Nichols, R.,** Cell enlargement and sugar accumulation in the gynoecium of the glasshouse carnation *(Dianthus caryophyllus)* induced by ethylene, *Planta,* 130, 47—52, 1976.
106. **Nichols, R.,** Sites of ethylene production in the pollinated and unpollinated senescing carnation *(Dianthus caryophyllus)* inflorescence, *Planta,* 135, 155—159, 1977.
107. **Nichols, R. and Ho, L. C.,** Effects of ethylene and sucrose on translocation of dry matter and ^{14}C-sucrose in the cut flower of the glasshouse carnation *(Dianthus caryophyllus)* during senescence, *Ann. Bot. (London),* 39, 286—296, 1975.
108. **Nichols, R. and Ho, L. C.,** An effect of ethylene on the distribution of ^{14}C-sucrose from the petals to other flower parts in the senescence cut inflorescence of *Dianthus caryophyllus, Ann. Bot. (London),* 39, 433—438, 1975.

109. **Oaks, A. and Bidwell, R. G. S.**, Compartmentation of intermediary metabolities, *Annu. Rev. Plant Physiol.*, 21, 43—66, 1970.
110. **Owens, L. D., Lieberman, M., and Kunishi, A.**, Inhibition of ethylene production by rhizobitoxine, *Plant Physiol.*, 48, 1—4, 1971.
111. **Packet, R. C.**, Color changes in flowers of *Lathyrus hirsutus* during senescence, *Nature (London)*, 211, 1215, 1966.
112. **Parups, E. V.**, Disc electrophoresis of proteins of senescing and fresh leaves and petals of certain ornamental plants, *J. Am. Soc. Hortic. Sci.*, 55, 775—781, 1971.
113. **Parups, E. V. and Chan, A. P.**, Extension of vase-life of cut flowers by use of isoascorbate-containing preservative solution, *J. Am. Soc. Hortic. Sci.*, 98, 22—26, 1973.
114. **Paulin, A.**, Influence de la composition de la solution nutritive sur la teneur en divers acides aminés libres et en ammoniac des pétales de fleurs coupées, *Ann. Technol. Agric.*, 20, 283—303, 1971.
115. **Paulin, A.**, Metabolism glucidique et protéique de la fleur d'oecillet alimentée ou non avec une solution de saccharose, *Acta Hortic.*, 71, 241—257, 1977.
116. **Paulin, A. and Ulrich, R.**, Observation physiologique sur les fleurs d'inflorescences isolées d'Iris, *Rev. Gen. Bot.*, 69, 369—388, 1962.
117. **Phan, Chon-Ton.**, Conversaion of various substrates to ethylene by flowers, *Physiol. Plant*, 23, 981—984, 1970.
118. **Porter, N. G.**, The role of abscisic acid in flower abscission of *Lupinus luteus*, *Physiol. Plant.*, 40, 50—54, 1977.
119. **Proctor, J. T. A. and Creasy, L. L.**, An anthocyanin-decolorizing system in florets of *Cichorium intybus*, *Phytochemistry*, 8, 1401—1403, 1969.
120. **Ronen, M.**, The Effect of Abscisic Acid on Cut Carnation Flowers, and Relation to Ethylene, M. S. thesis, The Hebrew University, Rehovot, Israel, 1978.
121. **Rogers, M. N.**, An historical review of post-harvest physiology research on cut flowers, *HortScience*, 8, 189—194, 1973.
122. **Sacalis, J. N.**, Vascular blockage and its inhibition in cut rose flowers, *Acta Hortic.*, 41, 159—164, 1975.
123. **Sacalis, J. N. and Chin, C. K.**, Metabolism of sucrose in cut roses. I. Comparison of sucrose pulse and continuous sucrose uptake, *J. Am. Soc. Hortic. Sci.*, 101, 254—257, 1977.
124. **Sacher, J. A.**, Senescence and postharvest physiology, *Annu. Rev. Plant Physiol.*, 24, 197—310, 1973.
125. **Santarius, K. A.**, The protective effect of sugars on chloroplast membranes during temperature and water stress and its relationship to frost, desiccation, and heat resistance, *Planta*, 113, 105—114, 1973.
126. **Shisa, M. and Takano, T.**, Effects of temperature and light on the coloration of rose flowers, *J. Jpn. Soc. Hort. Sci.*, 33, 140—146, 1964.
127. **Simons, R. K.**, Anatomical changes in abscission of reproductive structures, in *Shedding of Plant Parts*, Kozlowski, T. T., Ed., Academic Press, New York, 1973, 383—434.
127a. **Simon, E. W.**, Types of leaf senescence, *Symp. Soc. Exp. Biol.*, 21, 219—230, 1967.
128. **Simon, E. W.**, Phospholipids and plant membrane permeability, *New Phytol.*, 73, 377—420, 1974.
129. **Simpson, D. J., Baqar, M. R., and Lee, T. H.**, Ultrastructure and carotenoid composition of chromoplasts of the sepals of *Strelitzia* reginae Aiton during floral development, *Ann. Bot. (London)*, 39, 175—183, 1975.
130. **Smith, M. and Butler, R. D.**, Ultrastructural aspects of petal development in *Cucumis sativus* with particular reference to the chromoplasts, *Protoplasma*, 73, 1—13, 1971.
131. **Smith, W. H., Meigh, D. F., and Parker, J. C.**, Effect of damage and fungal infection on the production of ethylene by carnations, *Nature (London)*, 204, 92—93, 1964.
132. **Smith, W. H. and Parker, J. C.**, Prevention of ethylene injury to carnations by low concentrations of carbon dioxide, *Nature (London)*, 211, 100—101, 1966.
133. **Smith, W. H., Parker, J. C., and Freeman, W. W.**, Exposure of cut flowers to ethylene in the presence and absence of carbon dioxide, *Nature (London)*, 211, 99—100, 1966.
134. **Solomos, T.**, Cyanide-resistant respiration in higher plants, *Annu. Rev. Plant Physiol.*, 28, 279—297, 1977.
135. **Stead, A. D. and Moore, K. G.**, Flower development and senescence in *Digitalis purpurea* L., cv. Foxy, *Ann. Bot. (London)*, 41, 283—292, 1977.
136. **Stewart, R. N., Norris, K. H., and Asen, S.**, Microspectrophotometric measurement of pH and pH effect on color of petal epidermal cells, *Phytochemistry*, 14, 937—942, 1975.
137. **Stickland, R. G.**, Changes in anthocyanin, carotenoid, chlorophyll, and protein in developing florets of the chrysanthemum, *Ann. Bot. (London)*, 36, 459—469, 1972.
138. **Trippi, V. S. and Tran Thanh Van, M.**, Changes in the pattern of some isoenzymes of the corolla after pollination in *Phalaenopsis amabilis* Blume, *Plant Physiol.*, 48; 506—508, 1971.

139. **Uota, M.**, Carbon dioxide suppression of ethylene-induced sleepiness of carnation blossom, *J. Am. Soc. Hortic. Sci.*, 94; 598—601, 1969.
140. **Valadon, L. R. G. and Mummery, R. S.**, Changes in carotenoid composition of certain roses with age, *Ann. Bot. (London)*, 33, 671—677, 1969.
141. **Varner, J. E. and Ho, D. T.-H.**, Hormones, in *Plant Biochemistry*, Bonner, J. and Varner, J. E., Eds., Academic Press, New York, 1976, 714—770.
142. **Wagner, G. J. and Siegelman, H. W.**, Large scale isolation of intact vacuoles and isolation of chloroplasts from protoplasts of mature plant tissues, *Science*, 190, 1298—1299, 1975.
143. **Wang, C. Y., Baker, J. E., Hardenberg, R. E., and Lieberman, M.**, Effect of two analogs of rhizobitoxine and sodium benzoate on senescence of snapdragon, *J. Am. Soc. Hortic. Sci.*, 102, 517—520, 1977.
144. **Weinstein, L. H.**, Senescence of roses. I. Chemical changes associated with senescence of cut 'Better Times' roses, *Contrib. Boyce Thompson Inst.*, 19, 33—48, 1951.
145. **Wiemken, A. and Nurse, P.**, Isolation and characterization of the amino acid pools located within the cytoplasm and vacuoles of *Candida utilis*, *Planta*, 109, 293—306, 1973.
146. **Wiemken-Gehring, V., Wiemken, A., and Matile, P.**, Mobilisation von Zellwandstoffen in der welkenden Bluten von *Ipomoea tricolor*(Cav.), *Planta*, 115, 297—307, 1974.
147. **Wiemken, V. and Wiemken, A.**, Dichtemarkierung von β-Glucosidasen in welkenden Bluten von *Impomoea tricolor*(Cav.), *Z. Pflanzenphysiol.*, 75, 186—190, 1975.
148. **Wiemken, V., Wiemken, A., and Matile, P.**, Physiologie der Bluten von *Ipomoea tricolor* (Cav.): Untersuchungen an abgeschnittenen Bluten und gewinnung eines Phloemexudates, *Biochem. Physiol. Pflanz.*, 169, 363—376, 1976.
149. **Winkenbach, F. and Matile, P.**, Evidence for *de novo* synthesis of an invertase inhibitor protein in senescing petals of *Ipomoea*, *Z. Pflanzenphysiol.*, 63, 292—295, 1970.
150. **Woolhouse, H. W. and Batt, T.**, The nature and regulation of senescence in plastids, in *Perspective of Experimental Biology*, Vol. 2, Botany, Sunderland, N., Ed., Pergamon Press, Oxford, 1976, 163—176.
151. **Yazaki, Y.**, Copigmentation and the color change with age in petals of *Fuchsia hybrida*, *Bot. Mag.*, Tokyo, 89, 45—47, 1976.
152. **Zimmerman, P. W., Hitchcock, A. E., and Crocker, W.**, The effect of ethylene and illuminating gas on roses, *Contrib. Boyce Thompson Inst.*, 3, 459—481, 1931.

Chapter 8

THE MATURATION AND RIPENING OF FRUITS

M. J. C. Rhodes

TABLE OF CONTENTS

I. INTRODUCTION

The senescence of a plant organ can be defined as the final stage in its ontogeny during which a series of essentially irreversible events occurs which ultimately leads to cellular breakdown and death.[1] Ripening involves a series of changes occurring during the early stages of the senescence of fruits, in which the structure and composition of the unripe fruit is altered so that it becomes acceptable to eat. The botanical role of the fruit is in the dispersal of the seeds, and the striking colors, sweet and juicy flesh, and the distinctive aroma of the ripe fruit may be related to this function. The present article is concerned with a discussion of the processes involved in the initiation and control of ripening, i.e., the first stages of senescence, in the fleshy fruits of commercial importance; it does not cover the analogous changes occurring in dry fruit such as nuts and grains.

The edible fleshy tissue of the commercially important fruits is derived from a diversity of tissues of the inflorescence.[2] This group of fruits includes both true fruit (such as the tomato) in which only ovular tissue develops into the fruit, and false fruits (such as the strawberry) in which tissue of the receptacle of the flower develops into the bulk of the fruit. The group can be divided into simple fruits (apricot, avocado, banana, cherry, date, grapefruit, mango, and tomato), which develop from a single ovary, and aggregate fruits, which either develop from a number of ovaries of a single flower (strawberry and raspberry) or from a number of ovaries of different flowers in the compound fruits such as pineapple and Figure 4. The fruits of commercial importance are derived from both monocotyledonous and dicotyledonous plants and from a wide range of families within each group.[4] With this diversity of origin, it is perhaps not surprising that there is a range of patterns of development and of ripening among different fruits.

Many fruits show a pattern of growth that follows a single sigmoidal growth curve (see Figure 1 and Table 1). There is an initial phase of slow growth characterized by active cell division, followed by a period of major increase in size, fresh and dry weight, dominated by cell expansion, and a final period of declining growth rate during which ripening is usually initiated.[2] In the apple there are about 2 million cells at anthesis and about 40 million at harvest. This is achieved by about 21 doublings before, but only 4.5 after anthesis,[5] which occur during the first 4 to 6 weeks of growth following full blossom.[6] An exception to this pattern is the avocado, in which cell division can continue throughout the life of the fruit,[7] and in which growth continues all the time the fruit remains attached to the tree.[8] A special feature of the ripening of the avocado is that ripening is inhibited on the tree and only occurs after separation.[8] In fruits that show a single sigmoidal pattern, ripening normally occurs during the final phase of slow growth of the fruit. In fruit ripening on the tree, there is probably no absolute separation in time between ripening and further growth processes and probably some small degree of cell expansion is occurring at least early in ripening. However, in those fruits that store large quantities of reserve materials, and which can ripen normally after detachment from the parent plant, there can be an almost complete separation of senescence from growth phenomena. The factors involved in the maintenance of the unripe mature condition and in the timing of the natural onset of the ripening process will be discussed in detail later in this article.

In addition to those fruits showing a single sigmoidal growth curve, there are fruits showing more complex growth patterns with double[2,4] and even triple sigmoidal patterns.[9] In these there are two or three rapid growth phases interrupted by one or two periods of little or no growth. The double sigmoidal pattern is shown by many fruits including the stone fruits, figs, and grapes (see Table 1 and Figure 1). The two rapid

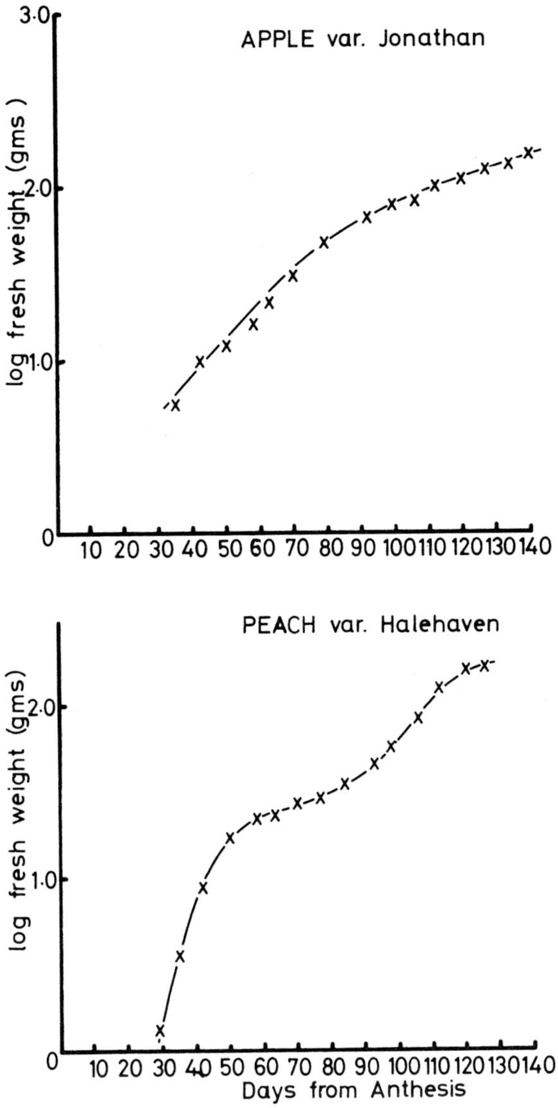

FIGURE 1. Changes in fresh wieght of apple and peach fruits during development. (Re-drawn from data of Figures 1 and 2, Bollard, E. C., in *The Biochemistry of Fruits and their Products,* Vol. I, Hulme, A. C., Ed., Academic Press, London 1970, 387. With permission.)

growth phases are designated periods I and III, while the interrupting slow growth phase is designated period II. Ripening begins during period III and may be towards the end of the period in stone fruits,[4] but near to the beginning of this period in grapes.[10] In the grape about 40% of the total growth is achieved in period III, while in peach over 80% is achieved in this period. It is thus clear that in those fruits showing double sigmoidal growth patterns active cell expansion accompanies some aspects of the ripening process. In the grape, the onset of ripening (veraison) is marked by the start of sugar accumulation, loss of acidity, softening, and skin coloring; these characteristic ripening changes are associated with renewed cell expansion as the fruit makes the transition from period II to III.[10] In stone fruits, the two periods of rapid

TABLE 1

List of Fruits Showing Different Patterns of Growth

Single	Double	Triple
Apple	Apricot	Chinese Gooseberry
Avocado	Black currant	
Banana	Blueberry	
Date	Fig	
Mango	Grape	
Melon	Olive	
Orange	Peach	
Pear	Plum	
Pineapple	Raspberry	
Strawberry	Sour Cherry	
Tomato	Sweet Cherry	

growth are associated with the development of different tissues. For example, in period I the endocarp and seed reach full size, in period II lignification of the endocarp occurs together with the development of the ovary and endosperm, while in period III the expansion of the mesocarp tissue, which ultimately becomes edible, occurs. It is thus the tissues in which the ripening changes occur that are expanding rapidly in the period just prior to, and possibly during, ripening. It is clear that in at least some fruits of this type it is not possible to separate senescence phenomena from those of growth, and this can lead to complications in the interpretation of the observed biochemical changes when both processes occur together. These difficulties can often be overcome by studying the ripening of mature detached fruit, but there are a number of fruits with both simple growth patterns (such as the mature green bell pepper),[11] and those with double sigmoidal patterns (such as the grape),[12] in which ripening does not proceed normally when the fruit is detached from the parent plant.

II. PATTERNS OF RIPENING IN FRUIT

In view of the differences in the origin of the cells that ultimately constitute the edible parts of fruit, and the differences in the botanical origin and in the patterns of development of different fruits, it is perhaps not surprising that there is a range of patterns of behavior during ripening. The processes of ripening generally occur in a relatively coordinated fashion in most fruits under natural conditions. An important factor to consider is that the modern varieties of fruit are often the result of continual selection by man, and they have features that make them very different from their ancestral wild types. Their structure and composition has been largely determined by artificial selection rather than by the rigors of natural selection. The nature of the desirable changes in color, flavor, and texture varies to some extent from fruit to fruit, and the desirable properties of the ripe fruit are characteristic of that particular fruit. The point during the sequence of ripening at which a fruit is ripe is, of course, to some extent a subjective evaluation. These changes in sensory quality responsible for the color and taste of the ripe fruit will be considered in Section 3. Here we will consider some of the metabolic events that underlie these sensory changes, and the study of which has led to much of our understanding of the physiological and biochemical nature of ripening.

A. Climacteric Fruit

The research of Kidd and West[13] in the 1920s made the first major contribution to the study of patterns of ripening behavior. They showed that the onset of the visible ripening changes in apples was marked by an upsurge in the rate of respiration. Kidd and West coined the term "respiration climateric" to describe what they saw as a critical phase in the life of the fruit. In detached apples (see Figure 2), the respiration rate falls to a minimum value (the climacteric minimum) and begins to rise at the time of onset of ripening to a peak (the climateric peak) roughly twice the level of the minimum, and then declines slowly. The fruit reaches optimum eating quality some time after the climacteric peak. This pattern of behavior had been observed in some of the earlier studies of Gerber[14] and Müller-Thurgau and Schneider-Orelli[15] (see review[16]). Kidd and West saw the respiration climacteric as the critical phase separating maturation processes concerned with growth from the onset of the essentially irreversible changes of senescence. In the years following the pioneering efforts of Kidd and West, it became clear that the majority of both temperate and tropical fruits followed a similar climacteric pattern of increase in respiration during ripening. The extent and rapidity of the process varies with temperature and from fruit to fruit. In the banana at a normal ripening temperature the rise may be as much as tenfold,[17] while in the apricot it may only be 30%.[19] In many tropical and sub-tropical fruits, the climacteric peak is only a transitory one with a rapid decline following the peak and this is associated with a more rapid rate of ripening than in the apple. In the tomato the climacteric peak is reached before there is appreciable accumulation of the red pigment, lycopene, and the rate of respiration is declining while the red pigment is accumulating. The climacteric rise in respiration was shown by Kidd and West to occur even when the fruit remains attached to the tree, and this appears to be the case for nearly all of the fruits showing the climacteric pattern of ripening, and which ripen normally on the parent plant. An obvious exception is the avocado, for although it has a normal climacteric rise in respiration in detached fruit, attached fruit do not ripen and do not show a rise in respiration.[8] The respiratory quotient (RQ) during the rise in respiration in apples increases from 1.04 to 1.40,[13] but in bananas there is a fall in RQ from 1.0 in pre-climacteric fruit to about 0.76 during the rise in respiration, returning again to 1.0 at the climacteric peak.[17]

During the period 1920 to 1930, several workers showed that the olefinic gas, ethylene, could promote the ripening of fruits. It was shown that treatment of unripe fruit with ethylene promoted both the onset of the respiration climacteric and the other ripening changes. Such a treatment required a minimum time of exposure of the fruit to a stimulatory concentration of the gas in order to elicit the response in ripening. The early history of the discovery of the properties of ethylene in promoting fruit ripening has been the subject of several reviews.[20-22] Gane[23] showed that ethylene was produced as a natural product of the metabolism of fruit, but it was not until the advent of highly sensitive gas-liquid chromatographic (GLC) methods for measurement of the extremely low levels of ethylene produced by most fruits[24,25] that it was possible to investigate in any great detail the involvement of ethylene in ripening. It is now proved that all fruits showing the climacteric pattern of ripening produce small amounts of ethylene throughout their lives.[20,26] There is usually a peak of ethylene production during flowering, and then a fall to a low level which is maintained in the fruit until the time of onset of ripening. The rate of production then rises dramatically, at or before the time at which the respiration climacteric and the onset of ripening changes are initiated, to reach a peak, and then its production is maintained at a relatively high rate throughout the period of ripening. The peak in ethylene production can, depending on the fruit involved, precede (banana), coincide with (mango, pear, avocado) or follow (apple, Cherimoya, Feijoa, tomato) the peak in respiration.[20] Kidd and West[27] introduced the concept of a threshold concentration of ethylene — which

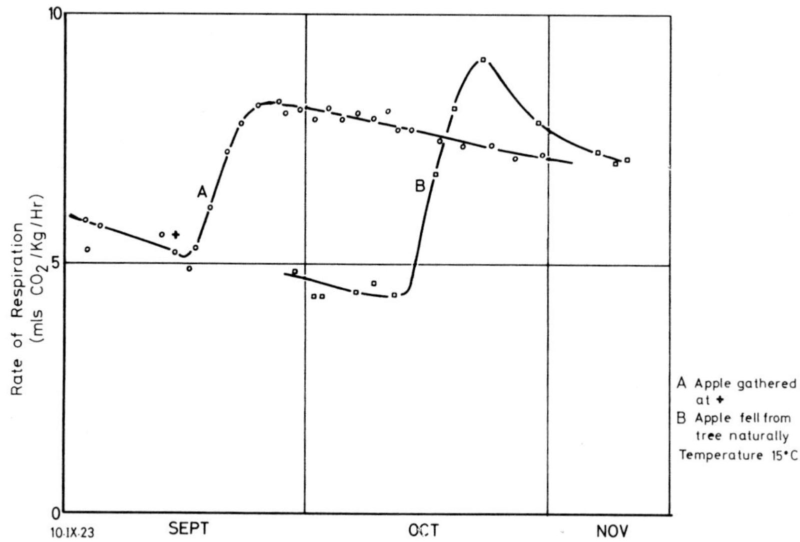

FIGURE 2. Changes in the rate of respiration of apple fruits after detachment from the tree. (Data of Kidd, F. and West, C. Ann. Rep. of Food Invest. Bd. for the year 1924, p. 31, reproduced with permission).

must accumulate in the tissue in order to elicit the ripening response. For a variety of ethylene-dependent responses of plants, it was found that the threshold value was about 0.1 ppm[20] and that concentrations of 1 to 10 ppm were saturating. The levels of ethylene that ultimately accumulate in fruits (see Table 2) can be very much in excess of this minimum level required to stimulate the respiration climacteric and ripening, and can be as much as 10^4 times greater. It was shown by Kidd and West[27] that treatment of apples for short periods with ethylene promotes not only the onset of the respiration climacteric, but also the autogenous production of ethylene by the tissue. This burst of ethylene production follows an essentially logarithmic increase during most of its rise. It is thus said to be produced autocatalytically. This to some extent represents the localized onset of ripening within the fruit, leading to production of the gas and its diffusion throughout the tissues of the fruit, initiating ripening in distant cells and a concomitant increase in ethylene production. The nature of this secondary autocatalytic production of ethylene has been elegantly demonstrated in the experiments of McMurchie et al.[28] who used propylene, a mimic of ethylene action,[26] as a ripening initiator, so that it was possible to study both effects on the initiation of ripening changes (represented by the direct effect of propylene) and on the secondary induction of autocatalysis. They showed that treating bananas with propylene initiated ripening and induced a large secondary production of ethylene. They proposed that two systems were involved in the natural regulation of ripening. In System 1 ethylene biogenesis is initiated by a factor involved in the process of aging in the tissue, and this brings the concentration of ethylene to a sufficient level to induce the respiration climacteric and the associated ripening changes, one of which (System 2) leads to a further large increase in ethylene production. The ability of fruit tissues to respond to the presence of a threshold level of ethylene, either applied externally or developed internally, in producing ethylene autocatalytically is a characteristic of the climacteric type of fruits.

The induction of ripening in mature unripe climacteric fruit by exogenous ethylene underlines the characteristic behavior. Ethylene treatment shortens the period after harvest required for the induction of the climacteric rise without affecting the respira-

TABLE 2

The Respiration Climacteric and Internal Levels of Ethylene in Some Climacteric Fruit

Fruit	Variety	Extent of rise in respiration (°C)	Internal Ethylene Content $\mu\ell/\ell$ (ppm): Pre-climacteric	Internal Ethylene Content $\mu\ell/\ell$ (ppm): Peak of production	Comments	Ref.
1. Fruits with a Single Sigmoidal Growth Curve						
Apple	Cox's Orange Pippin	× 2 at 12°	0.02 — 0.03	~100	Rise in respiration and in ethylene production almost simultaneous	150
Avocado	Choquette	× 2—4 at 15°	0.1 (0.04 at harvest)	300—700	0.5—1.0 ppm at onset of respiration rise	148
Banana	Silk fig	× 4—10 at 20°	0.2	~25	Rise in ethylene production precedes rise in respiration by ~4 hr ~1.0 ppm at onset of respiration climacteric	34
Melon	Cantaloupe	× 2 at 20°	0.05	30—70	About 0.3 ppm at onset of respiration climacteric	246
	Honeydew	× 2 at 20°	0.1—3.0	~12⁺	Rise 0.1—3.0 ppm during 8—10 days before onset of respiration climacteric	21
Mango	Kent, Haden	× 5 at 24°	0.04—0.05	2—3		34
Pear	Anjou	× 2 at 20°	0.08—0.46	40	0.46 ppm at onset of respiration climacteric	127
Persimmon	Yohono	× 3 at 20°	0.44			19
Tomato	VC 243-20, Eurocross BB	× 2 at 20°	0.03—0.08	27	~0.8 ppm at onset of respiration climacteric	247

TABLE 2 (continued)

The Respiration Climacteric and Internal Levels of Ethylene in Some Climacteric Fruit

Fruit	Variety	Extent of rise in respiration (°C)	Internal Ethylene Content $\mu\ell/\ell$ (ppm) Pre-climacteric	Internal Ethylene Content $\mu\ell/\ell$ (ppm) Peak of production	Comments	Ref.
			0.1	20—30		Rhodes unpublished
2. Fruits with a Double Sigmoidal Growth Curve						
Apricot	Moorpark	× 1.3 at 20°	0.1[a]	0.8[a]	—	19
Fig	Mission	~ × 2.0 during development on tree	~0.1	4.6—5.2	Rise in ethylene and respiration occur at about the onset of period III (up to 2 ppm present in period I)	43
Peach	Halehaven	~ × 2.0 during development on tree	~0.02⁺ (in late period II)	42[a]	Rise in ethylene and respiration occur at about onset of period III	192

[a] Estimated from rates of production of the gas using a conversion factor[20]

tion rate at the climacteric peak. As shown in Figure 3, this shortening of the pre-climacteric period bears a roughly logarithmic relationship to the concentration of ethylene applied.[20] Since such ethylene treatment induces a large rise in endogenous ethylene production following the induction of the climacteric, and the level accumulating in the tissues far exceeds the level required to saturate all the ripening responses, it follows that once the climacteric has been initiated further exogenous ethylene treatment has no effect in promoting the ripening processes.

It has become evident that the rise in respiration rate is only one of a number of changes in metabolism that occur at the critical period in fruit development, which is the climacteric. This led Rhodes[29] to attempt to redefine the climacteric period in the light of more recent knowledge of these changes, while still incorporating the essential concept of Kidd and West. He thus defined the climacteric as a "period in the ontogeny of certain fruits during which a series of biochemical changes is initiated by the autocatalytic production of ethylene, marking the change from growth to senescence and involving an increase in respiration and leading to ripening".

B. Non-climacteric Fruit

It became clear in the 1940s that not all fruits had the distinctive and well coordinated pattern of ripening of the typical climacteric fruits, and Biale[30] was able to distinguish a group of fruits (Table 3), which showed neither a rise in respiration nor an accelerated production of ethylene during the ripening of mature fruit. The typical examples of this group are the citrus fruits, whose physiology has been studied in some detail. Citrus fruits are characterized by long growth periods and a very extended ripening process in which changes in chloroplast disorganization and pigment contents are often divorced from other ripening changes.[31] Sound oranges produce small amounts of ethylene (0.02 to 0.06 $\mu\ell$/kg/hr)[20] throughout their development and maturation, but there is no massive rise in the production of ethylene associated with ripening as there is in the climacteric fruits. The respiration of detached fruits declines slowly during storage,[30] However, it was found[32] that oranges and lemons could be induced to produce enhanced amounts of ethylene and to give a rise in respiration if stored in an atmosphere of pure O_2. If green oranges were treated with ethylene,[30] there was a rise in respiration accompanied by an increased rate of loss of green color from the rind, and by the accumulation of reducing sugars.[33] The response in the ultimate maximal respiration rate observed was related to the logarithm of the applied ethylene concentration.[20] This situation clearly differs from the classical climacteric fruit in which the effect of added ethylene is to advance the time of onset of the respiration rise without altering its magnitude (Figure 3). Furthermore, if the exogenous ethylene used to treat the oranges was removed, the respiration rate and the other attendant changes fell back to their initial values and the tissue became susceptible to repeated stimulation by ethylene. McMurchie et al.[28] applied propylene to oranges and found that it induced a rise in respiration and in the rate of degreening without inducing the autocatalytic production of ethylene. Thus in their nomenclature the citrus fruits have system I, but lack system II. This result confirmed the earlier suggestions[34,35] that the essential difference between the classical climacteric fruits and the non-climacteric citrus fruits lies in their relative abilities to produce ethylene autocatalytically in response to the accumulation within the tissues of threshold levels of ethylene.

The low level of ethylene in citrus fruit poses a problem as to whether ethylene plays any role in the endogenous ripening of these fruits. Reid[19] has shown that there was a small but significant rise in ethylene levels from 0.014 to 0.028 ppm during the transition from green to fully ripe mandarin oranges (see Table 3). Apelbaum et al.[36] inves-

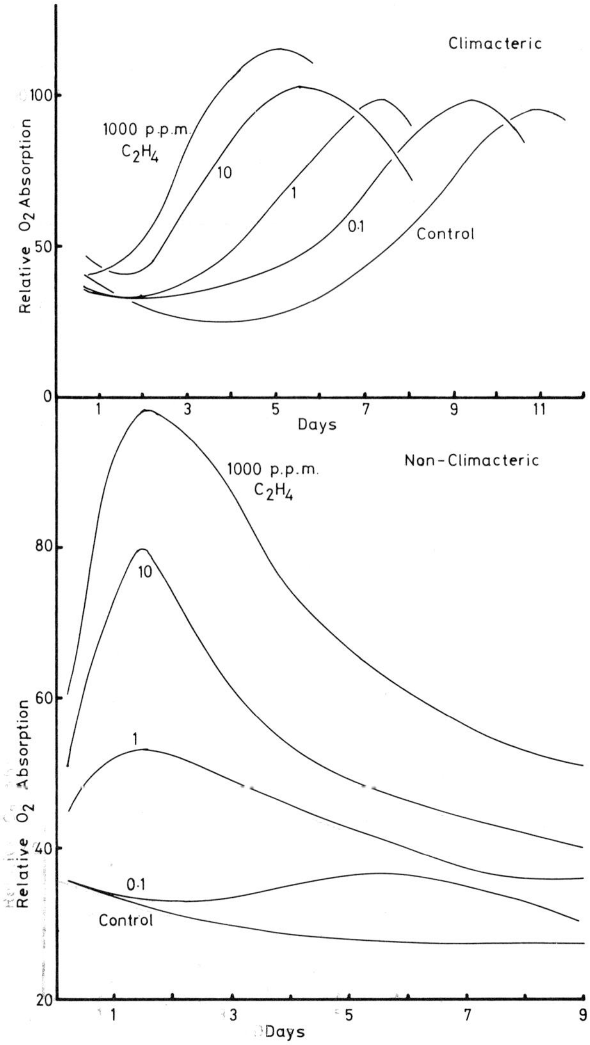

FIGURE 3. Effect of different concentrations of applied ethylene on the respiratory behavior of an unripe climacteric fruit (upper curves) or a non-climacteric fruit (lower curves). (Reproduced with permission from McGlasson, W. B., in *The Biochemistry of Fruits and their Products,* Vol. I., Hulme, A. C., Ed., Academic Press, London, 1970, 496, Fig. 1.)

tigated the storage of Shamouti oranges under sub-atmospheric pressures in which the partial pressure of oxygen was maintained at near the atmospheric level. Such a storage procedure reduced the internal concentration of ethylene in the tissue,[37] but did not alter the rate of color change; the authors concluded that endogenous ethylene may not be the primary inducer of the natural color changes in oranges.

Citrus fruits can, under various conditions, produce quite large amounts of ethylene. For instance, various factors leading to physiological stress such as treatment with chemicals[38] or irradiation[39] can induce citrus fruit to produce ethylene at relatively high rates. Very immature detached fruits can also produce quite large amounts of ethylene; immature oranges and grapefruit harvested at an early stage in development produce considerable amounts of ethylene during subsequent storage,[40,41] while in mature fruit

TABLE 3

The Internal Levels of Ethylene in some Non-Climacteric Fruits

Fruits	Internal ethylene content range of values (ppm) μℓ/ℓ	Notes	Ref.
1. Single Sigmoidal Growth Curve			
Capsicum annuum	0.025—0.040		11
Citrus fruits:			
Oranges	0.04—0.12[a]		34
Lemons	0.11—0.17		34
Mandarins	0.014—0.028	Small increase from green to orange stage	19
Tamarello	Not measureable		19
2. Double Sigmoidal Growth Curve			
Cherry (var. Montmorency)	0.05—0.2[b]		49
Grapes (Doradillo)	0.14—0.02	High in early part of period II, falls to low level during ripening	48
Shiraz	0.1—0.01		

[a] Estimated from rates of production of the gas using a conversion factor[20]
[b] nℓ ethylene/g

the rate of ethylene production remains low and constant. Small oranges and grapefruit weighing about 15 g produce no detectable ethylene for the first 3 days of storage at 20°C but then there is a rise in ethylene production to a maximum rate of about 3 to 5 μℓ/kg/hr, together with a rise in respiration to a peak after about a further 3 days and then a decline. This pseudo climacteric rise does not occur in fruit beyond the half grown stage, in which the respiration falls during subsequent storage and the ethylene production rate remains at less than 0.01 μℓ/kg/hr.[41] The rise in respiration in these very small fruits appears to be directly related to the increased production of ethylene in these tissues, rather than to the onset of ripening.[29,41] It is interesting that in lemons there is a fall in respiration without a burst in ethylene production during storage of fruit harvested at all stages during development.[41] It is clear from these experiments that under various stress conditions, and at various stages in their development, citrus fruits have the capacity to produce sufficient ethylene to stimulate a respiration rise, but that this capacity is not expressed in the normal ripening and, in particular, the mere presence of triggering amounts of ethylene is not in itself capable of stimulating a large autogenous production of ethylene.

C. Variations in the Pattern of Ripening within the Climacteric and Non-climacteric Groups

The typical climacteric fruits (apple, avocado, banana, tomato) are a fairly homogeneous group, which closely fits the pattern of ripening previously described. There are some exceptions such as the mango, in which even in initiated fruit the rise in the internal ethylene concentration is relatively slow, so that it is possible to promote ripening by exogenous ethylene treatment even during the climacteric.[34] In honeydew melons, both the height of the respiration peak and the rate of production of ethylene at the peak are very dependent on the stage of maturity at which the fruit is harvested.[42] Very immature fruit fail to give a respiratory response, but as the fruit matures on the vine the extent of the respiration rise and its capacity to produce ethylene increases.[42] Immature melons respond to ethylene treatment in much the same way as do non-climacteric fruit, in that the height of the respiratory response appears to be related to the ethylene concentration applied and there is no autocatalytic increase in ethylene production. On the other hand, with fruit held on the vine beyond a critical time, the response to ethylene is characteristic of the climacteric fruits, ethylene shortening the time taken for initiation of the climacteric rise and inducing its own production autocatalytically.[42]

The fig is one of a number of fruits that were originally classified as non-climacteric,[30] but which, on more detailed study, have been placed in the climacteric category. Marei and Crane[43] have shown that during ripening on the tree there is a rise in respiration and a burst in ethylene production associated with ripening. The fig is a compound fruit made from the ovaries of a number of different flowers,[4] which shows a double sigmoidal pattern of growth. There is a relatively low concentration of ethylene (less than 0.1 $\mu\ell/\ell$) in the internal atmosphere of the fruit during period II. At the onset of period III, there is a rise in respiration reaching a peak (roughly double the rate at the end of period II) at about halfway through period III.[43,44] During this rise in respiration, the internal ethylene concentration rises from less than 0.1 $\mu\ell/\ell$ to a peak of from 4 to 5 $\mu\ell/\ell$. These changes in the rate of respiration and the internal ethylene level are illustrated in Figure 4. The onset of the increase in the internal ethylene content of the fruit precedes the increase in respiration during period III. Treatment of the fruit on the tree with ethylene had different effects, depending on the time during the growth cycle at which the ethylene was applied.[44] Treatment with ethylene in period I inhibits growth, and this was correlated with the inhibition by ethylene of cell division, which is active at this stage. Treatment of fruit in the early part of period II stimulates some growth, but this is accompanied by the development of an atypical dark reddish color and ultimately by fruit abscission. Treatment with ethylene in the later part of period II stimulates growth, accelerates the onset of the rise in respiration,[43,44] induces autocatalytic production of ethylene, and promotes ripening[45] (See Figure 4). A typical respiration climacteric is not displayed by fruit harvested before the natural initiation of ripening on the tree. However, fruit in which ripening had been initiated on the tree gave a normal rise to the climacteric peak and ripened fully after detachment from the tree. Thus in many respects the fig behaves as a typical climacteric fruit, but as a result of its growth pattern the increases in respiration and ethylene production preceding ripening are occurring during a phase of active cell expansion. The fig achieves a substantial proportion of its total growth during the period in which the climacteric and ripening are proceeding. In such cases, it is extremely difficult to separate growth from senescence processes, and this complicates any interpretation one might wish to make concerning the significance of the increased ethylene production and respiration.

While the climacteric pattern of ripening is a fairly uniform one, there is considerable difference in the ripening patterns of non-climacteric fruits and frequently marked differences in behavior from the pattern of ripening, as typified by the citrus fruits. The grape is a good example of this and in this case, as in the fig, cell expansion and growth occur in parallel with some aspects of ripening, since the grape also shows a double sigmoidal pattern of growth. The onset of period III (veraison) is associated both with changes in color, texture, the loss of acid, and accumulation of sugar, and with the onset of renewed growth.[2,10] This period is also accompanied by a substantial rise in respiration, measured either by CO_2 production or oxygen uptake.[46] The rise in respiration is associated with an increase in RQ from 1.0 to 1.3 or higher,[12] and this is the change that would be predicted if there was a change in respiratory substrate at the time of veraison from carbohydrates towards organic acids.[12] The basis for the classification of grapes as non-climacteric was originally that detached grapes showed

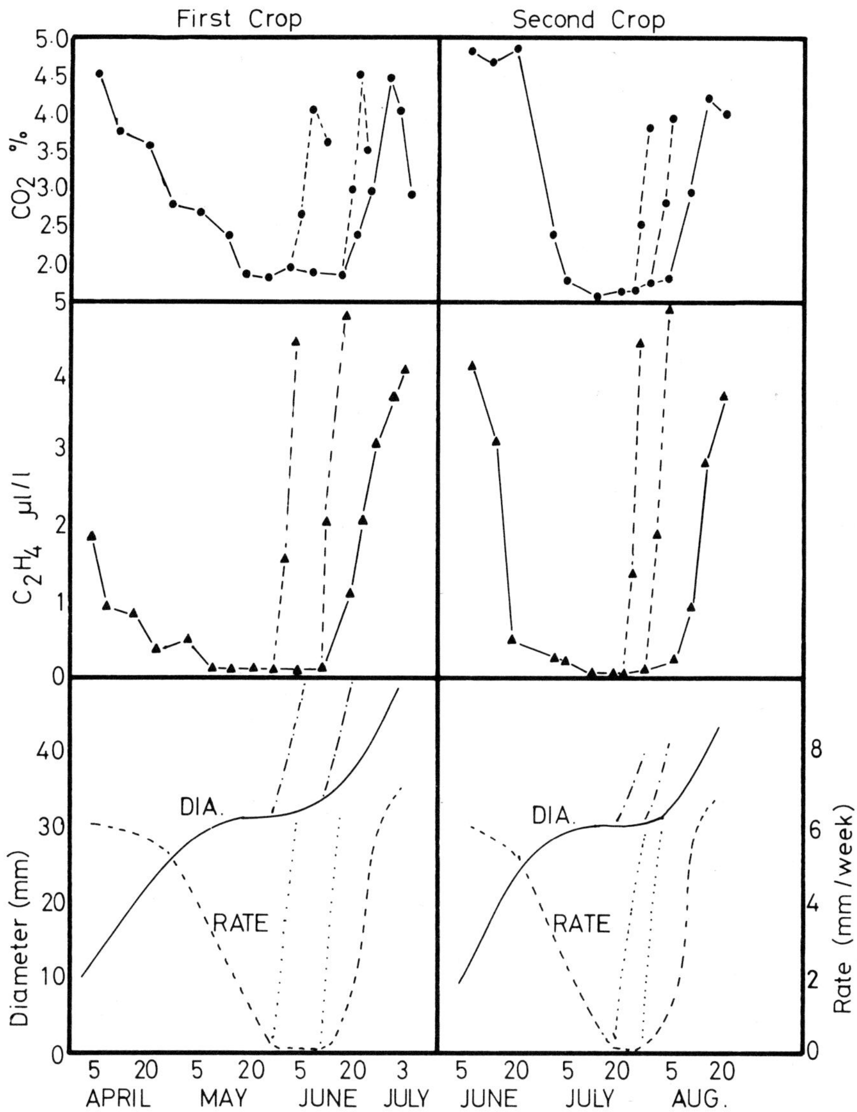

FIGURE 4. Changes in the internal CO_2 (•) and C_2H_4 contents (▲) and in the diameter of two crops of Mission fig fruits developing on the tree (continuous line). At two stages during development the fruit was treated with ethylene and the resultant changes in internal CO_2 and ethylene contents and in the diameter of the fruit are shown as dashed lines. In addition, on the lower graph changes in the rate of growth are shown both with and without ethylene treatment. (Reproduced with some modification from Marei, N. and Crane, J. C., *Plant Physiol.*, 48, 251, 1971, Fig. 4. With permission.)

a slow decline in respiration during storage.[30] However, the significance of this slow downward drift is not clear, since grapes do not continue to ripen properly when detached from the vine.[12]

Ethylene, applied either as a gas or as 2-chloroethyl-phosphonic acid (a compound that breaks down on contact with plant tissues to release ethylene), will delay ripening of grapes if applied in the early part of the slow growth period (II).[47] The effect when applied beyond the mid-point of period II is to shorten the delay before the onset of period III, so that both ripening and renewed growth are advanced. While there is an increase in respiration at veraison there is no corresponding increase in ethylene production. Coombe and Hales[48] showed that the internal ethylene content of Doradillo grapes fell from 0.14 $\mu\ell/\ell$ 10 days before veraison to about 0.08 $\mu\ell/\ell$ at veraison, and then declined slowly during period III. Thus the grape has in common with citrus fruits the ability to respond to ethylene, but lacks the ability to produce endogenous ethylene in response to exogenous treatment with the gas, and in these characteristics behaves like the typical non-climacteric fruits. However, in its response to ethylene at different stages of growth, in its rise in respiration at the start of period II, and in the fact that growth and senescence phenomena occur simultaneously, it resembles the behavior of the fig, a climacteric fruit. The fact that grapes do not produce ethylene autocatalytically seems a good reason for classifying them as non-climacteric.[34,35]

In cherries, either sweet or tart varieties,[49] the ethylene content of the fruit is low throughout the ripening period from the beginning of period III to the fully ripe stage at about 0.1 to 0.2 $\mu\ell/\ell$. The rate of CO_2 production of fruit on the tree declines from the end of pit hardening (period II) to the end of ripening.[50] This confirmed the findings of a number of other workers[50-52] that cherries did not show the climacteric pattern of behavior. Hartman[51] showed that green immature cherries picked early in development show substanial rises in respiration during post harvest storage, but no information as to whether these were associated with an increase in ethylene production was provided. It may be that these changes in immature cherries are analogous to the similar changes occurring in immature citrus fruits.[40,41] In general, it is fairly easy to place the cherry in the non-climacteric category, since ripening is achieved in the absence of increases in either respiration or ethylene production. The ripening does occur following the period of renewed growth, and it is interesting that the increased energy demands required for renewed cell expansion do not, in this case, lead to an increase in the rate of respiration.

Various fruits of the genus *Capsicum,* the peppers, are classified as non-climacteric. These fruits have single sigmoidal growth curves and are examples of fruit in which ripening is often inhibited if fruit is detached at the mature green stage.[11] There are measureable differences in the internal CO_2 and ethylene content of fruit ripening attached to the parent plant, but these are very small. The rise in CO_2 content is only about 10% from an initial value of about 0.6% CO_2, while the ethylene content changes from 0.025 to 0.40 $\mu\ell/\ell$. Treatment of attached fruit with 2-chloroethyl-phosphonic acid increases the red color development and increases the proportion of red fruit at harvest.[11] Treatment of harvested pimiento peppers at or beyond the breaker stage with 2-chloroethyl-phosphonic acid enhanced red color production,[53] while treatment of mature green fruit failed to promote the formation of red pigment.[53,54] Treatment of mature green detached fruit with propylene[11] induced a very small transitory increase in respiration rate, measured as internal CO_2 content, but had no effect in inducing autocatalytic production of ethylene. On this basis the peppers are classified as non-climacteric.

An example of a fruit that was originally classified as non-climacteric,[55] but which on re-examination is classified as climacteric, is the Chinese gooseberry (*Actinidia chinensis*) whose growth curve is characterized by a triple sigmoidal pattern.[9] Pratt and Reid[9] suggested that the reason why earlier workers had failed to detect the climacteric rise in respiration in the fruit was due to the extremely variable time between anthesis and the onset of ripening in this fruit, the earlier workers having used bulk samples of fruit in experiments of relatively short duration. These fruits contain substantial reserves of polysaccharides and ripen only very slowly after detachment, but it was found that all samples eventually ripened; the ripening was accompanied by a rise in respiration which proceeded in two phases with a plateau in between which corresponded with a peak in ethylene production. The rise in respiration was fourfold and the ethylene production rate at the peak was 100 $\mu\ell$/kg/hr, equivalent to an internal ethylene content of approximately 200 $\mu\ell/\ell$.[20] Thus the Chinese gooseberry is typical of the climacteric fruits.

Among the so-called soft fruit many members are considered non-climacteric but, in general, there is a lack of detailed studies. The strawberry produces only very small amounts of ethylene[56] and ripens in the absence of a rise in respiration;[30,56] exogenous ethylene has no effect on the ripening process.[57] However, the raspberry contains signifcant amounts of ethylene in its internal atmosphere at the onset of ripening[49] and this concentration increases as the fruit changes color. Both blueberry and cranberry have been considered non-climacteric, since a number of studies showed that their respiration did not rise during ripening.[58] In the cranberry, exogenous ethylene treatment has no effect on the respiration rate, on the sugar or on acid content of the fruit, but it does stimulate degreening.[59] The status of the blueberry is a matter of current study. It has been shown[59] in a study of seven varieties over five seasons that berries attached to the parent plant undergo a twofold rise in respiration to a peak that corresponds to the green-pink stage in development, and this is accompanied by a 50 to 70% rise in the rate of ethylene production. Similar increases in ethylene production have recently been found during ripening in another blueberry variety.[60] Thus the status of the blueberry has yet to be established.

There is a need for further studies on soft fruit, particularly of the effects of exogenous ethylene on endogenous ethylene production, since these could throw important insights into the ripening mechanisms in these fruits.

D. Non-ripening Mutants

Further evidence for the essential physiological and biochemical differences between the typical climacteric and non-climacteric fruits has come from a study of the non-ripening mutants of a climacteric fruit, the tomato. Studies on a non-ripening mutant, *rin*, have thrown light on the role played by ethylene in the ripening process. The strain *rin* developed as a spontaneous mutant of a normal breeding line (61—37 Fireball × Cornell 54—149).[61] Fruit bearing the *rin* gene (a monogenic recessive characteristic with no known linkage associations except to large sepal size[63]) does not produce the red pigment lycopene even after extended periods at normal ripening temperatures. There is a slight degree of softening[62] and some loss of chlorophyll, leading eventually to fruit with a pale yellow color. Herner and Sink[63] found that detached fruit monitored for up to 120 days after harvest showed little or no change in either CO_2 or ethylene production associated with the gradual color change from green to yellow. The internal ethylene concentration remained at a constant level of about 0.15 to 0.29 $\mu\ell/\ell$ throughout this color change, while that of control fruit rose rapidly from about 0.2 $\mu\ell/\ell$ to about 18 $\mu\ell/\ell$ by the time the fruit had achieved a red color. Treatment of the fruit with propylene induced an increase in respiration, but there was no autocatalytic production of ethylene. In *rin* fruit the respiration is capable of repeated stimulation by ethylene, as in citrus fruit. The *rin* tomatoes are capable of a limited production of lycopene following treatment of fruit on the vine with ethylene,[64] treatment of detached fruit with ethylene in an atmosphere of 100% oxygen,[65] or treatment of immature detached fruit with high concentrations of 2-chloroethyl phosphonic acid in air.[66] Wounding *rin* fruit mechanically has been shown to induce a large production of ethylene.[63] Thus *rin* fruit have the capacity to produce both lycopene and ethylene, but these attributes are not expressed in normal ripening. In their inability to respond to propylene treatment by autocatalytically producing ethylene, *rin* fruit behave like the non-climacteric citrus fruits in lacking system II of McMurchie et al.[28] Thus it seems that the *rin* mutation has led to a change in behavior from the normal strains of tomato to one that clearly resembles that of the citrus fruits, the typical non-climacteric fruit.

The foregoing discussion of the various patterns of ripening in normal and mutant strains of fruit shows that there are two extremes in the pattern of ripening, with the climacteric pattern of the apple, banana, and avocado at one end, and that of the typical non-climacteric citrus fruits at the other. Although a considerable number of fruits correspond closely to the climacteric pattern, there is quite a range of ripening behavior among non-climacteric fruit — and this behavior is often quite different from that of the citrus fruit. In addition, the interpretation of patterns of ripening and of any physiological or biochemical study of the mechanism of ripening is always complicated in those fruits in which a substantial amount of cell expansion accompanies some of the ripening changes. It does seem reasonable to divide fruits on the basis of their pattern of ripening behavior, but a better basis for doing this would be their ability to produce excess ethylene autocatalytically in response to the initiation of ripening, rather than their different patterns of respiratory behavior.

III. CHANGES IN STRUCTURE AND COMPOSITION RESPONSIBLE FOR THE SYMPTOMS OF RIPENING

The visible signs of ripening are those changes in the color, texture, taste, and aroma of the fruit tissue that lead to the state in which the fruit is fit to eat, and these are determined by changes in the structure and metabolism of the fruit tissue. Ripening involves complex changes in metabolism of the fruit in which both anabolic and catabolic processes are involved.[67] New pathways of metabolism are brought into play — which may lead, for instance, to the production of new pigments or new volatile compounds. In addition, degradative processes such as those involved in the conversion of chloroplasts to chromoplasts, and in changes in adhesion between cells leading to tissue softening also occur. These metabolic changes must reflect changes in the in vivo activity of the enzymes involved and the change in activity may involve the synthesis, degradation, release, activation, or inactivation of these enzymes. These changes, at least in the climacteric fruit, have to be considered in relationship to the changes in respiration and in ethylene production. The underlying structural and biochemical changes responsible for the visible changes of ripening will now be considered.

A. Changes in Cell Structure During Ripening

The edible tissue of fleshy fruit is composed usually of parenchymatous cells, frequently of very large size. The cells of ripe fruit are among the largest cells found in plants and are commonly 150 to 700 μm in length.[2] The cells of the tissue surrounding the seeds of the pomegranate can even exceed 2mm.[68] The flesh cells of fruit frequently have polyploid nuclei; in apricots the degree of ploidy has been shown to increase during development and to reach up to 64-ploid in the ripe fruit.[69] The increase in ploidy may be a factor determining the large size of fruit cells.[2] These cells have a high water content and accumulate organic substances such as sugars and organic acids. In addition to the large parenchymatous cells, the flesh of many fruits contains stone cells, which may play a role in the texture of the ripe fruit and which contribute to the taste of some fruits, since they frequently contain tannin-like compounds, leucoanthocyanins, which are responsible for the astringency of these fruits. The extreme example of an astringent fruit is the persimmon, whose cells contain a water soluble tannin, diospyrin (+ leucodelphinidin-3-glucoside).[70] These cells increase in size and number during ripening, and it is frequently necessary to apply post-harvest treatments to decrease the astringency of the fruit in order to make it acceptable to eat.[70]

The study of the ultrastructure of fruit cells during ripening has often proved difficult as a result of their large size and their high content of organic acids and other compounds potentially destructive of their fine structure. The difficulty has been to develop a suitable fixation technique to preserve the fine structure, and to improve the infiltration of the embedding materials to minimize artifacts in sectioning.[71] These problems seem to have been overcome in some recent studies, but caution should be exercised in interpretation of the results of some earlier studies. The major ultrastructural changes occurring during ripening occur in the plastids. Harris and Spurr[73] showed that mature green tomatoes contain chloroplasts with relatively small grana and frequently with one or two vacuoles bearing starch. The first signs of the onset of ripening changes are the appearance of electron dense regions in the granal and intergranal membranes which are thought to represent the initial stages in lycopene deposition, and an increase in the number of plastoglobules. Later the inter-granal region swells and the granal diameter increases markedly. By the time the orange stage is reached, new structures called thylakoid plexes appear; these resemble the loose type of prolamellar body found in plastids in etiolated leaves exposed to light for short periods. Lycopene crystalloids of up to 18 μ in length appear by the orange-red stage, and crystalloids composed of concentrically arranged lamellae are frequently found. In the later stages of ripening there is a reduction in the amount of membranous material in the stroma and a large increase in the size and number of the plastoglobules. The appearance of thylakoid plexes and of concentrically arranged membranes is also a feature of the formation of chromoplasts in ripening *Capsicum* fruits.[74] The accumulation of phytoferritin has been reported in the plastids of ripening tomatoes.[71] In apples during ripening there is a disorganization of the thylakoid system with the appearance of plastoglobules.[75,76] While in pears the formation of chromoplasts is accompanied by the appearance of plastoglobules and at an early stage there appears to be extensive vacuolation of the plastids.[72]

Other cell components show less dramatic changes during ripening. All the non-plastid cytoplasmic structures observed in the cytoplasm of immature tomatoes are also observed in the ripe fruit.[71] These structures include mitochondria, golgi bodies, ribosomes, endoplasmic reticulum, small vacuoles and vesicles, microbodies, and lipid droplets. Bain and Mercer[72] observed that there was a considerable increase in the vacuolation of the cytoplasm during the ripening of pears. However, Simpson et al,[71] using improved fixation procedures, observed no obvious degenerative changes in the cytoplasm of cells from the pericarp of tomatoes — even at the fully red stage. In particular, mitochondria retain their ultrastructure throughout the ripening period, and this structure is retained until late in senescence when degeneration has started in all other cell organelles.

B. Factors Responsible for the Changes in Color During Ripening

A major factor in the color changes of fruit during ripening is the transition from chloroplasts rich in the green pigment, chlorophyll, into chromoplasts rich in red or yellow carotenoid pigments. The nature of the transition and its control by hormonal and environmental factors will be described elsewhere in the volume (see Goldschmidt, Chapter 9). The metabolic significance of the transition is the loss of the potential ability to carry out photosynthetic CO_2 fixation. It has been shown that in apples[76,77] the chloroplasts of the peel are functionally active in photosynthesis, and even the internal tissues of the pulp contain active plastids.[78,79] On a chlorophyll basis the photosynthetic activity, measured by differences in the CO_2 exchange of this fruit between successive light and dark periods, was of the same order as that of leaves.[77] This method of measuring photosynthetic activity does not take into account possible effects of light on respiratory activity and, since fruits contain peroxisome-like bodies,[71] the presumed site of photorespiratory activity,[80] there may be errors in the measured rates. However, the total chlorophyll per unit area is much lower in fruits than in leaves, and generally it seems there is very little net fixation of CO_2 in apples; at most, the photosynthetic activity only just compensates for respiratory CO_2 production.[76,77,81] Experiments[81] using $^{14}CO_2$ showed that in the light CO_2 was fixed into the normal products of photosynthetic CO_2 reduction. During ripening of apples[76,81] the rate of photosynthetic activity was constant during the early part of the climacteric rise in respiration, but decreased during the later stages (see Figure 5). The ratio of respiration to photosynthesis, which had been constant at between 1.1 and 1.4 during growth, rose to 8 to 12 at the late stages in ripening. There was no complete parallelism between the loss of chlorophyll and the loss of photosynthetic activity. Clijsters[76] covered apples developing on the tree with black bags and found that after 80 days in the dark there was a complete disappearance of both chlorophyll and measureable photosynthetic activity in the fruit. Nevertheless, the growth continued in the dark and at harvest was only 6% less than that of control fruit. Development in darkness had no effect on the total malate content of the fruit, but there was less malate in the outer layers of cells compared with control fruit.[76,82] Covering fruit in this way had no significant effect on the time of onset of the respiration climacteric and of ripening.[81] It thus seems clear that the limited photosynthetic activity of fruit has only a small effect in promoting the accumulation of reserve materials such as malate; this is in line with the view that most reserve materials are transported to the fruit from other parts of the tree, principally from the leaves.[4] Changes in the photosynthetic activity, which would be expected to decrease the availability of photochemically generated $NADPH_2$ and ATP, do not appear to play an important role in the regulation of the climacteric and ripening.

There are also changes in the pigments that make an important contribution to the color of fruits, but which are stored outside the chloroplasts. These are the water soluble phenolic pigments, the anthocyanins, which are responsible for the red and blue colors of many fruits. They exist as complex conjugates of the parent aglycones, the anthocyanidins, and are thought to be stored in the vacuole of the cell. The six principal anthocyanidins (see Table 4) that are found in fruit are usually present as the 3-glycosides. Cyanidin-3 galactoside is the principal pigment responsible for the color of red apple varieties, while pelargonidin-3-glucoside is the main pigment of ripe strawberries.[83] In fruits such as grapes, acylated anthocyanins are found in which the organic acid is esterified to a hydroxyl group on the sugar moiety rather than to the aromatic hydroxyl group. The organic acids involved are commonly cinnamic acid derivatives such as p-coumaric, caffeic, and ferulic acids. Frequently, a considerable number of different anthocyanins are present in a single tissue, and they each contribute towards the color of the fruit. The color of anthocyanins is red at acid pH and blue at alkaline pH, but it seems unlikely that localized pH is a major factor in determining the color of anthocyanin-containing fruits.[84] Rather, interaction between anthocyanins and other colorless phenolic pigments (the co-pigmentation effect) and between anthocyanins and metal ions may be important in the blue colors of fruits with an acidic sap.

Anthocyanidins are formed in the cell from simple sugars via the shikimate pathway and phenylalanine in the formation of the so-called B ring (see Table 4), while the A ring is formed by head to tail condensation of acetate units.[83-84] In apples the formation of anthocyanins in the skin is dependent on light; action spectral studies[85] show that light between 600 and 750 nm is necessary for the synthesis of anthocyanins with an action spectrum maximum at 650 nm. The synthesis of anthocyanins has been shown to require a high-energy photoreaction in which the photosynthetic system may

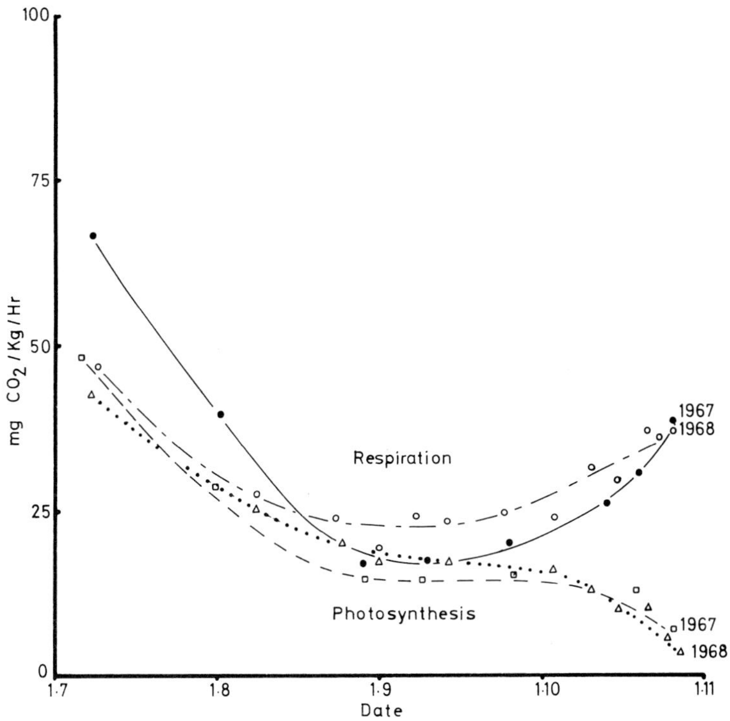

FIGURE 5. Changes in the rates of respiration and of photosynthetic activity of apple fruits developing on the tree. (Reproduced with permission from Clijster, H., *Qual. Plant Mater. Veg.*, XIX 132, 1969, Fig. 4.)

be the receptor, and which may also involve the photomorphogenic pigment, phytochrome.[86] There have been relatively few studies of changes in the level of enzymes of anthocyanin biosynthesis[87] during ripening, but it seems likely that the enzyme, phenylalanine ammonia lyase, may play an important role in the regulation of anthocyanin biosynthesis in the apple.[88]

C. Factors Responsible for the Softening of Fruit During Ripening

Changes in the texture of fruit during ripening result from changes in the structure and composition of their cell walls. The cells of fruit are frequently large parenchymatous cells with large air spaces between them. Changes in the degree of contact between cells, as well as in the structure of the cell walls themselves, will affect texture. It is only in the past few years that knowledge of the composition and the interrelationship between components has reached the stage that it has been possible to propose tentative models for the structure of plant cell walls. Albersheim[89] and co-workers have used highly purified cell wall-degrading enzymes coupled with improved techniques for the separation and analysis of the partial hydrolysis products, in a study of the structure of the primary cell wall of Sycamore cells in cell suspension cultures. Knee[90] applied the same type of analysis to the cell walls of ripening apples. He finds three polysaccharides in the primary wall of the parenchymatous cells (see Figure 6), α-cellulose, a xyloglucan with a β,1-4 linked core of glucose units with xylose units attached by 1 to 6 linkages, and a rhamnogalacturonan consisting of chains of α,1-4 linked galacturonic acid units interrupted by rhamnose units. The rhamnogalacturonan is modified by up to five β,1-4 linked galactose units attached to some of the galacturonic acid residues and a branched chain arabinan attached to the rhamnose residues. The

TABLE 4

Structure of the Anthocyanidin Aglycone Constituents of Anthocyanin Pigments Found in Fruits

(a)	R = R′ = H	Pelargonidin
(b)	R = OH, R′ = H	Cyanidin
(c)	R = OCH_3, R′ = H	Peonidin
(d)	R = R′ = OH	Delphidin
(e)	R = OH, R′ = OCH_3	Petunidin
(f)	R = R′ = OCH_3	Malvidin

galactose side branches of the rhamnogalacturonan are thought to be linked to the xyloglucan moieties, which in turn are hydrogen-bonded to cellulose microfibrils. The middle lamella contains both pectinaceous material, consisting of a linear α,1-4 linked polygalacturonan in which a substantial proportion of the carboxylic groups are esterified with methanol, and a protein rich in hydroxyproline residues, which in the apple are glycosylated with up to 10 arabinose units. Knee[90] found that during ripening in the apple there was a decrease in the galactose content of cell wall hydrolysates, which was interpreted as being due to a rupture of the linkages between the xyloglucan and rhamnogalacturonan leading to a weakening of the structure of the primary wall. In addition there is an increase in water-soluble polyuronide, which has been correlated with cell separation and the softening of the tissue.[90] In apples the polyuronide released has a low content of neutral sugar and is of the type thought to be in the middle lamella. An increased solubility of the cell wall polyuronides is the characteristic cell wall change occurring during softening in a whole range of ripening fruits.[91-93] In some cases, such as the pear, evidence has been presented[94] that suggests that there is extensive degradation of cell wall polyuronides as well as its solubilization. In the peach, there is evidence of a decrease in the molecular weight of the polyuronides solubilized with progressive ripening,[95] and evidence for an increase in the free galacturonic acid content of detached peaches as they ripen.[96] In the apple, however, it appears that the solubilized polyuronide is of high molecular weight and its liberation is thought to be due to subtle changes in the relationships between cell wall components, rather than to an extensive degradation of the polyuronides.[90]

A number of enzymes attacking polyuronides have been detected in ripening fruit. Frequently the activity of such enzymes is low or absent in unripe fruit and increases during the ripening process.[97] The metabolic function of these enzymes is at present uncertain, but in recent years considerable evidence of their properties has been obtained. The enzyme frequently invoked as being involved in the solubilization of pectin is the cell wall-bound polygalacturonase (PG), which acts on de-esterified pectin molecules and attacks linkages between galacturonic acid groups in the polyuronide. Evidence for such a role for PG has come from studies on the *rin* mutant of tomatoes in which the very limited degree of softening has been correlated with a very low content of PG in relation to normal varieties.[98] Two types of PG have been found in fruit; one, an exo-PG, acts by cleaving free galacturonic acid groups from the non-reducing end of the molecule and the other, an endo-PG, which acts by cleaving the chains of polyuronides in a random fashion. In peaches both types of PG are present, and Pressey and Avants[99] found that the endo-PG was able to catalyse the release of water soluble (but 70% ethanol insoluble) polyuronide from isolated washed cell wall preparations from peaches, while the peach exo-PG was inactive in such a solubilization. They proposed that in the peach the endo-PG was responsible for the solubilization of the pectin molecule by randomly attacking internal regions of the long chains of

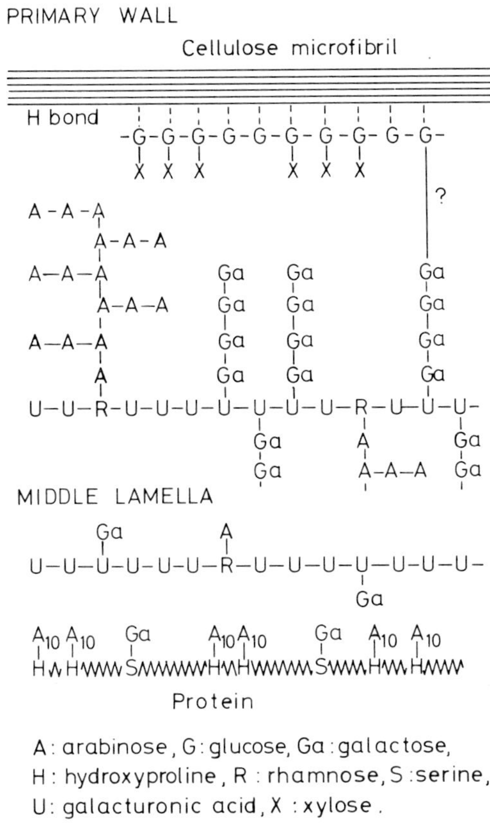

FIGURE 6. Possible structure of cell walls of cortical parenchyma of apple fruit (From Knee, M., in *Facteurs et Régulation de la Maturation des Fruits,* C.N.R.S. Colloq. No. 238, Ulrich R., Ed., Paris 1974; 342, Fig. 1. Reproduced with permission.)

polyuronide and so increasing their solubility. The role of the exo-PG was in the further step by step degradation of the solubilized polyuronide molecules leading to the appearance of free galacturonic acid in ripe peaches. This secondary role of exo-PG is unclear and it is uncertain how important such a further degradation of the solubilized pectic molecules would be in changes in texture of the ripening fruit. In the pear,[100] there are also two PGs which are similar to the enzymes found in peaches, with the exception that the exo-PG acts only effectively with high molecular weight substrates. In the tomato,[101] an endo-PG is present and is capable of solubilizing polyuronides from cell wall preparations in vitro. The apple[102] contains only one PG, an exo-PG, and there is no evidence for the presence of the endo-PG, the enzyme implicated in the solubilization reactions in peaches, pears, and tomatoes. However, the apple exo-PG is capable of degrading apple cell wall preparations, releasing both low molecular weight uronic acid residues and polyuronides. In this situation, it is suggested that the solubilization reaction depends on the rupture of linkages between the polyuronides and other structural components of the cell wall, rather than extensive degradation of the polyuronide chains. If PGs play a role in the increase in the water solubility of pectins during ripening, then it is likely that a second enzyme, pectin methyl esterase (PE) also plays a role. The polyuronides of the middle lamella are highly methylated, while PG is relatively inactive towards methylated polyuronides and requires free carboxyl groups for its action. PE is a wall-bound enzyme that can be readily desorbed from cell walls in vitro by addition of neutral salts at slightly alkaline pH, and its role may be to modify pectin structure prior to PG action. There is some evidence for a decrease in pectin methylation during ripening[97] and it may well be that quite limited PE activity could greatly promote PG activity and softening. PE has been found to increase in activity in the tomato,[103] but earlier reports of an increase in ripening bananas have not been confirmed.[104] The enzymes from bananas[104] and tomatoes[105] have been purified and shown to be similar in their molecular weights (28,000 to 30,000) and in their pI values (pH 8 to 9). PE can exist in multiple forms in ripening fruit.[104,106] In general,[104,105] the potential PE activity far exceeds the net change in methylation observed during ripening, so that even if it is indeed involved in the softening reaction the factors controlling its activity in vivo are far from understood.

Bartley[107] correlated the decrease in the cell wall galactose content during ripening of apples with the appearance of a β-galactosidase. His evidence suggests that the loss of galactose units precedes the release of polyuronide, and that the two processes are not necessarily interrelated. The role of cellulases which catalyse the breakdown of cellulose in fruit ripening is controversial. In avocado, cellulase activity rises during the climacteric to very high values and Awad and Young[107a] have suggested that cellulases may be synthesized or activated very early in the climacteric and play an important role in the early stages of softening. In other fruits, however, cellulases appear to play only a minor role in softening.[107b] It is clear that changes occurring in the softening of fruit tissues during ripening are extremely complex, so it is perhaps not surprising that there is still doubt about the role the various enzymes play in the softening reaction. The enzymes PG and PE appear to be associated with the cell wall in fruit and it is perhaps important to consider the problem of the accessibility of such enzymes to their substrates, which may be bound within a complex matrix structure in the wall. Here the evidence for wall modifying factors, such as those suggested by Karr and Albersheim,[108] may be important.

D. Factors Responsible for the Changes in Flavor of Fruit

In general fruits show a decrease in acidity and an increase in sweetness during ripening. There are, however, some exceptions to this general rule such as the banana, in which the pH falls during ripening from 5.4 to 4.5, and this is associated with an increase in the content of organic acids such as malate and citrate.[109] In fruits such as melons (which have essentially no reserve carbohydrate) there is no increase in sugars during ripening off the vine although, of course, this does increase during ripening on the vine by transport from leaves. However, most fruits have reserve carbohydrates, principally starch, which serve both as substrates for respiration and as sources of sugars that impart sweetness to the ripe fruit. The absolute levels of sugar and acids, as well as the ratio of sugar to acid, play an important role in taste of the ripe fruit. In bananas, unripe green fruit contain 20 to 25% starch and almost all (98 to 99%) of this is converted into simple sugars such as sucrose, fructose, and glucose, which can ultimately constitute 15 to 20% of the dry weight of the ripe fruit.[17] There is a decrease in total carbohydrate of about 2 to 5% during ripening in bananas, which may represent loss of sugars utilized in respiration. Grapes are the highest accumulators of sugars among fruit and levels of more than 25% on a fresh weight basis (80% on a dry weight basis) have been reported.[2] Although loss of water from the flesh cells occurs during ripening, the principal factor in the increase in the sugar content of expanding grape cells is an increase in the import of sugar rather than in the export of water.

The pH of the sap of fruit cells is frequently below pH 7 and can be as low as pH 3.0 in lemon cells. Two acids are most frequently found in fruit cells; these are malate and citrate. Malate predominates in the apple, apricot, banana, cherry, peach, pear, and plum; citrate predominates in the citrus fruits, figs, guava, loganberry, raspberry, strawberry, and pineapple, while a mixture of more or less equal amounts of citrate and malate occurs in the tomato and gooseberry. While these are the major acids, many fuits store a whole range of other acids in relatively small amounts. An exception to this is the grape — in which tartaric acid is a major stored acid and occurs at more or less the same levels as malate. These organic acids are assumed to be stored in the vacuole and there is now some fairly direct evidence for this.[110] The fact that the organic acids are stored in the vacuole means that they are not fully available to the mitochondrial oxidation system, which is thought to be the major site of their ripening in their metabolism during ripening in fruits. In apples there appears to be an additional cytoplasmic malate decarboxylating system, which may also play a role in the decrease in acidity during ripening.[111] Whatever the location or route of oxidation of the malate, it is clear that factors altering the permeability characteristics of the tonoplast membrane can have an important role to play in utilization of organic acids.

In addition to the changes in sugarid balance, an important factor in the flavor of fruits is their aroma, which arises from the production of volatile compounds by the fruit during ripening. Even compounds of low volatility can contribute to the aroma of fruits on account of the extraordinary sensitivity of the olfactory epithelium tissue of the nose to some compounds. An understanding of the contribution of particular compounds to the aroma must stem from a knowledge of the quantity of the compound produced, the quality of the aroma of each compound, and the reaction of the sensitive tissues of the nose to a range of concentrations of that compound. The quantity of total volatiles produced is typically 1 to 20 ppm, although there are exceptions with some varieties of banana in which the volatile content can reach over 300 ppm.[112] In the banana over 200 different compounds have been detected, each present at under a ppm and in some cases at only parts per thousand million. These volatile compounds include many different classes of organic molecules such as organic acids, alcohols, esters, carbonyl compounds, hydrocarbons, terpenoids, and there are wide differences in the contributions that individual compounds make to the resulting aroma, which are not simply related to their rates of production. An example[113] of this is ethyl 2-methyl butyrate which, although it is produced in only very small amounts, plays an important role in the aroma of ripe Delicious apples, since its olfactory threshold level is 0.0001 ppm.

The production of most volatile odor compounds is generally very low in unripe pre-climacteric fruit, but it rises many-fold during ripening. Drawert[114] studied the production of a range of volatiles in ripening apples and bananas. In apples (var. Glockenapfel) the production of methyl-2-butanol, for instance, was about 10 to 20 μg/100g in pre-climacteric fruit and rose, at 17°C, to a peak value of over 250 μg/100g 10 to 20 days after the achievement of the peak of the respiration climacteric. A number of volatile products including hexanol, hexyl hexylate, hexyl butyrate, and isoamyl acetate showed peaks of production following the climacteric. In contrast, the production of methyl-2-propanol was high during the pre-climacteric period, fell to a minimum 10 days after the achievement of the respiration peak, and then rose to a second peak 20 days after the respiration climacteric. Similarly, in bananas (Gros Michel) peaks of production of a number of volatile compounds were achieved about 10 days after the climacteric peak. Bayonove[115] has studied the formation of γ-decalactone[116] during the ripening of Cardinal peaches. Peaches that are harvested immature show practically no formation of γ-decalactone during ripening, while if the fruit is allowed to mature to a certain extent on the tree, it then becomes capable of producing the volatile during subsequent ripening off the tree. This formation of γ-decalactone has been shown to parallel the solubilization of the pectin in the fruit as it ripens.

Attention in the area of volatile production has turned away from cataloguing the composition of the total volatile fraction towards a study of the pathways of production of the individual volatile compounds and of factors that control their appearance during ripening. It is clear that there are at least two major groups of precursors of volatile compounds, the long chain amino acids, leucine, isoleucine, and valine, and the unsaturated fatty acids, linoleic and linolenic acids. Drawert[114] has postulated pathways for the production of aliphatic esters based on labeling studies with ^{14}C amino acid precursors, but there is very little enzymological evidence to complement these studies. Galliard and his group have studied the formation of the volatile compounds, *trans*-2-nonenal, *trans*-2-*cis*-6-nonadienal and *trans*-2-hexenal, which are important constituents of cucumber flavor. These are formed as result of degradation of the endogenous lipid of the tissue following mechanical damage. These important flavor constituents are therefore not present in vivo, but are formed during chewing. The sequence of enzymic steps in the breakdown of linoleic and linolenic acids involves a lipid acyl hydrolase to release the fatty acid from the lipid, oxidation of the fatty acid by lipoxygenase to form the corresponding hydroperoxides, followed by cleavage of the hydroperoxide to yield *cis*-3 enals, and finally their isomerization to produce the *trans* -2 enals. Enzymes responsible for each of these steps have been characterized in extracts of cucumber fruits.[117,118] A similar sequence of enzymically controlled steps is involved in the breakdown of the lipids to yield the volatile products, hexanal and *cis* 3 hexenal in tomatoes.[119,120] Much remains to be done on the enzymology of the endogenous volatile production of ripening fruits; the description of the properties of the individual enzymes and the changes in their activity during ripening will be a necessary prelude to a study of their regulation during ripening.

E. Interrelationships of the Different Aspects of the Ripening Process

The importance of the respiration climacteric in relation to the other ripening changes has been discussed in some detail elsewhere.[29,121] Two views of this relationship exist. One considers the rise in respiration as a response to the increased energy demands made on the fruit by the various ripening changes (see reviews 21, 29). There is now, however, considerable evidence that the many aspects of ripening are not dependent on the respiration rise. For instance, the application of inhibitors of protein synthesis such as cycloheximide to unripe pears[122] and bananas[123] inhibited the production of ethylene, softening, and loss of chlorophyll during subsequent storage without inhibiting the rise in respiration. Quazi and Freebain[124] showed that in bananas, in which ripening had been initiated by a short exposure to ethylene, subsequent storage in a low oxygen (0.5 to 1%) atmosphere containing ethylene accelerated the change in fruit color in the absence of a rise in respiration. Dostal and Leopold[125] showed that treatment of mature green tomatoes with gibberellic acid had no effect on the respiration climacteric, but delayed the onset of color changes during subsequent storage at 20°C. Treatment of such fruit with ethylene did not completely overcome the effect of GA_3 on the color changes. In bananas[126] it was found that short periods of exposure to ethylene, which did not induce the respiration climacteric, did induce an increase in the amino acid incorporation into proteins usually associated with ripening in the fruit (see Section IV.B). The softening of immature pears was initiated by treatment with 0.08 ppm of ethylene without an increase in respiration.[127] Similarly in the honeydew melon, the processes of softening and increased carotenoid synthesis precede the onset of the respiration climacteric.[42,128] These results show that it is possible to isolate many aspects of ripening from the rise in respiration, and they clearly throw doubt on the view of ripening as a highly integrated process. The view has been expressed that the respiration climacteric is neither dependent upon nor integrated with other aspects of ripening.[123]

An alternative view of the role of the respiration climacteric has been proposed by Romani,[129] who views the activity of the mitochondrial system as operating relatively independently of the other ripening changes. The rise in respiration should be regarded, in Romani's view, as just another independent ripening change and this he relates to the quasi-autonomous status of mitochondria in cells. Ultrastructural studies on a wide range of fruits show that, although other organelles such as plastids and endoplasmic reticulum undergo changes in structure during ripening, mitochondria appear to retain their structural integrity until well past the climacteric peak. Studies on the properties of mitochondria isolated from fruit at various stages of ripening suggest that, although there may be changes in the mitochondrial composition such as changes in the ability to metabolize oxaloacetic acid during the climacteric,[130,131] mitochondria remain fully functional throughout the period of the respiratory rise and beyond. Isolated mitochondria oxidize substrates at increasing rates during ripening,[130,132] and retain the ability to phosphorylate ADP with a high respiratory control ratio throughout ripening. Indeed the evidence appears to suggest that there is an increased capacity for phosphorylation as measured by ADP/O ratios as the fruits ripen.[67] The earlier labeling data showing that the rate of esterification of ^{32}Pi increased[67] as the fruit ripens, as well as the data for isolated mitochondria clearly indicate that mitochondria in fruit remain tightly coupled throughout the climacteric rise in respiration.[67] Romani[129] suggested that the increase in respiration is a response to changes occurring in the cytoplasm, the nature of which is unknown, but increased availability of cofactors and substrates or removal of some of the other constraints limiting in vivo activity of mitochondria could be involved. As Solomos[133] points out, mitochondrial activity in vivo is highly constrained and the full mitochondrial potentiality is rarely expressed in vivo in bulky organs. Thus removal of constraints can lead to an increased activity without the need for mitochondrial biogenesis. It is interesting to speculate as to how mitochondria insulate themselves from the degradative changes occurring elsewhere in the cell. As Romani[129] points out, mitochondria have a quasi-autonomous existence in the cell. However, the autonomy of both animal and plant mitochondria is strictly limited. The mitochondrial genome is only capable of coding for about 8 to 10 polypeptide chains, and over 90% of the mitochondrial protein is made in the cytoplasm and transported into the developing mitochondrion.[134,135] The polypeptides (which are made intra-mitochondrially) represent only sub-units of proteins of which the other sub-units are synthesized in the cytoplasm. It is thus clear that the mitochondrial autonomy is very limited, yet mitochondria remain intact and functional at the climacteric peak and beyond, even though the rate of protein synthesis in the cytoplasm has then fallen to very low levels (see Section IV.B). This could be achieved if the rates of turnover of mitochondria and of mitochondrial proteins were very low, but there are, as yet, no published measurements of the rates of turnover of mitochondrial proteins during the period of ripening and senescence. Despite the difficulties of the mechanisms involved, Romani's suggestion that the respiratory rise represents a relatively independent process and that the products of its activity, $NADH_2$ and ATP are not coupled to dependent anabolic reactions, is worth considering. It is clear that many of the changes of ripening are independent of the rise in respiration, but whether there are some changes that are dependent on the increase in supply of energy remains to be seen.

It is clear from the preceding discussion that at least some of the ripening changes can be experimentally dissociated from each other. However, in the field, under normal conditions, ripening does occur as a well-coordinated sequence of changes, and it is interesting to speculate whether factors other than the respiratory rise may play a part in this co-ordination. This will be considered in the next section.

IV. INTERNAL CONTROL OF RIPENING

A. Hormonal Control

1. The Role of Ethylene in Fruit Ripening

The biosynthesis of ethylene in fruits has been the topic of a number of reviews[136-138] and recently progress has been made on the route of biosynthesis from its main precursor, methionine. It has been shown that S-adenosylmethionine[139] (SAM) and 1-amino-cyclopropane-1-carboxylic acid[139a-139b] (ACC) are intermediates in the conversion of methionine to ethylene. There is evidence that the reaction between SAM and ACC may be the rate limiting step in ethylene biosynthesis.[139c] An enzyme to catalyze this step has been separated from tomato tissue[139c,139d] and evidence has been presented that it is a pyridoxal-containing enzyme[139d] with a high affinity for SAM and that it is sensitive to inhibitors of the rhizobitoxine type,[139c] which are powerful inhibitors of ethylene biosynthesis in plants.[144] The nature of the enzyme(s) catalyzing the conversion of ACC to ethylene is unknown and the earlier evidence suggesting a role for peroxidase-like enzymes in ethylene biosynthesis[140] needs to be re-evaluated in the light of this new information. Cyclic pathways for the conservation of sulfur during ethylene biosynthesis from methionine have been proposed for apple[141] and flower tissues.[142]

It is clear from the fact that externally applied ethylene can induce ripening in fruit, and that potentially stimulatory concentrations of the gas are present in fruit at the inception and during ripening, that the gas plays some fairly direct role in ripening. The precise nature of that role has still to be elucidated. It would be highly desirable from an experimental point of view to study ripening in the absence of endogenous ethylene. This has been attempted by either removing the ethylene as it was formed or by blocking its biosynthesis. Burg and Burg[143] used hypobaric conditions as a means of reducing the internal ethylene concentration in bananas stored at one fifth atmospheric pressure at a partial pressure of O_2 of about 0.20. This procedure delayed ripening and this delay could be overcome by addition of ethylene under the hypobaric conditions (see Section 5A). Lieberman[144] showed that when the fungal toxin, Rhizobitoxine (2 amino-4-(2′amino-3′hydroxypropoxy) *trans* -3-butenoic acid), which blocks the conversion of methionine to ethylene,[148] was infiltrated into unripe apples it caused a 72% inhibition of ethylene production and markedly reduced the respiration climacteric, but did not cause significant differences in the texture of the apples compared with untreated control fruit after 11 days of ripening at 23°C. In this case a considerable amount of ethylene was present in the tissues of the fruit (equivalent to 80 ppm[20]) even in the presence of inhibitor. Wang and Mellenthin[146] vacuum infiltrated commercially mature pears of two varieties, Anjou and Bartlett, with the amino-ethoxy analogue of Rhizobitoxine (AAR). Both varieties contained significant levels of ethylene in their internal atmospheres at harvest and were thus initiated into ripening. AAR reduced the ethylene production in both varieties, but had a much greater inhibitory effect on the slow ripening variety, Anjou, compared with the faster ripening Bartlett. The internal concentration of ethylene in the Bartlett pears infiltrated with the inhibitor was equivalent to 30 ppm, compared with less than 1 ppm in Anjou. In Bartlett the toxin had very little effect on any of the ripening changes, while in Anjou it delayed changes in firmness, in the solubilization of the pectic substances, and in the increase in protein nitrogen and promoted the disappearance of malate. It is thus probable that the varietal differences can be directly related to different degrees of inhibition of ethylene production rather than to any direct effect of AAR on the individual processes. The results from these two approaches clearly indicate an essential role for ethylene in the ripening process.

It is likely that in the climacteric fruits ethylene plays at least three distinct roles during ripening. It is a component of the system of opposing hormonal factors responsible for the initiation of the ripening process and the transition from growth to senescence. It is, in its autocatalytic production of supra-optimal concentrations of the gas, a product of the ripening process. The large secondary production of ethylene triggered by the priming of the initiation mechanism may play a role in coordinating the other ripening changes.

There has been considerable debate as to whether an increase in ethylene production to a stimulatory level is a prerequisite for the initiation of the respiration climacteric and ripening. Pratt and Goeschl[21] recognized at least three patterns of relationships between changes in the level of ethylene and the initiation of the respiration climacteric and ripening. In one the level of ethylene increased in the tissue up to a critical threshold concentration, which then initiated the ripening syndrome. The clearest case of this situation is the honeydew melon,[43] in which there is a rise in the internal ethylene concentration from about 0.1 $\mu\ell/\ell$ to about 3 $\mu\ell/\ell$ in the 10-day period prior to the rise in respiration. This concentration is the minimum amount required to promote ripening in immature melons and there is no change in the sensitivity of the fruit to ethylene as it matures on the vine. The honeydew melon is an extreme example of this type of behavior in that: (1) the rise in ethylene production occurs over a long period of time, and (2) the level of ethylene required for initiation of the climacteric is relatively high. In other fruits (such as the banana) the increase in ethylene production may only (see Table 2) precede the onset of the climacteric by a few hours.[34] Both situations leave open the question of the factors responsible for the initial rise in ethylene production, but it does appear that in this case the climacteric is *initiated* by a change in the rate of ethylene production. The other two types of relationship described[20,21] invoke the presence of other factors inhibitory to ripening. These factors may either be endogenous products of the fruit or be produced outside the fruit and be continuously transported to the fruit from other parts of the parent plant. In the avocado variety Fuerte, ripening is completely inhibited if the fruit is left attached to the tree. Gazit and Blumenfeld[147] showed that treatment of avocados on the tree with 50 ppm ethylene for 48 hr failed to elicit a ripening response. Similarly, there was no response in freshly picked fruit treated with 100 ppm for 24 hr. However, if treatment was delayed by 24 hr, ripening was initiated. In the avocado variety Choquette,[148] it was found that the internal ethylene content of fruit on the tree was about 0.04 ppm. Following picking, after a transitory rise and fall, a concentration of about 0.1 ppm was established in the tissue 24 hr before the onset of ripening. It is clear that in the avocado the level of ethylene is not the factor that determines the onset of ripening, and that the loss of a factor supplied to the fruit while attached to the tree and disappearing slowly after harvest is the dominant feature of the initiation system. In the apple,[149] attachment does not inhibit ripening, but it does modify the extent of the respiration climacteric and severely reduces ethylene production. In fruits such as bananas[34] and apples,[150] potentially stimulating levels of ethylene are frequently present in the fruit long before the onset of ripening. It appears, that in these cases, the decrease in another, as yet unidentified factor, by altering the sensitivity of the tissue to the ethylene already present, appears to determine the time of the initiation of ripening. In these fruits, the levels of exogenous ethylene required to promote ripening decrease as the fruit matures.[20,151,152] It was shown that with very immature apples[151] or bananas[153] short ethylene treatments, which did not in themselves immediately promote ripening, shortened the time to the eventual onset of ripening compared with untreated controls. Thus ethylene appears to affect the aging of the tissue and may act in promoting the disappearance of the other factors.

The concentration of ethylene in the interior of bulky organs is determined by the rate of production of the gas and by the structure of the tissue.[154] Burg and Burg[154] showed that the diffusion of ethylene, as well as that of O_2 and CO_2 through a fruit obeys Fick's Law, namely that the rate of diffusion is proportional to the concentration gradient between the inside and outside of the fruit. The proportionality constant of the relationship includes factors for the diffusion coefficient for the gas and a constant for a particular fruit — which relates to the geometry and thickness of the barrier limiting diffusion. In fruits a gaseous gradient of ethylene is established, which is due to the resistance to diffusion offered by the tissues and the relatively low surface area/volume of fruits. In fruits such as the apple, the epidermal layers offer the principal barrier to diffusion. In the apple diffusion is mainly via the lenticels (pores of the skin), while in the tomato it has been estimated that up to 60% of the gaseous diffusion occurs through the calyx scar.[154] The development of stimulatory levels of ethylene inside the tissue can be an important part of the ripening initiation process. Changes in these levels can be dependent on changes in rate of production, but it is possible that changes in the resistance to diffusion of the tissue could have a bearing on the internal ethylene levels. It has been suggested that the deposition of wax on the skin of fruit[155] and the clogging of the air spaces with solutes[156] could lead to an increased resistance to diffusion, but these changes are only likely to affect the gaseous exchange in ripe fruit[20] and would not be likely to interfere with changes involved in the initiation of the ripening process. The phenomenon of auto-inhibition of ethylene production has been suggested by Zeroni et al.[157] as a mechanism for the maintenance of the preclimacteric state. In studies on figs (at the non-ripening stage) and on mature green tomatoes they found that if they allowed ethylene to accumulate around the fruit, ethylene production was stopped once a critical concentration had been reached (about 0.6 ppm), while no such auto-inhibition of ethylene production occurs if the experiment is repeated with figs in the ripening phase, or with ripe tomatoes. The production of ethylene is one factor in the system for the initiation of ripening, and the possible involvement of other growth substances in it will be discussed in Section 2, below.

The second involvement of ethylene in ripening is the autocatalytic production of the gas; this can be considered as a ripening change alongside those of color, texture, and flavor. In many fruits, once ripening is initiated there is a rapid increase in the rate of production of ethylene and the internal concentrations of the gas rises to levels as high as 1000 ppm. Initially, ethylene production increases logarithmically (see Figure 7) to about the time of the climacteric peak and then the rate of increase declines.[150] The mechanism by which ethylene production increases in this way is unclear, but several possible explanations have been advanced. For instance, Yang[158] has proposed that ethylene treatment of immature fruits induces the formation of unique isoenzymes of peroxidase, the enzyme thought to be involved in ethylene production, thus increasing progressively the potential for its synthesis. Kende and Hanson[142] proposed that autocatalysis is due to ethylene induced increase in permeability of the tonoplast which releases more substrate for ethylene biosynthesis. Whatever the intracellular nature of the mechanism of the autocatalytic production of ethylene, there are clearly gradients of ripening within fruits, and the autocatalytic increase in ethylene production may to some extent reflect an increasing number of ethylene-producing cells within the fruit. As shown in Table 2, the internal concentrations can rise to several hundred ppm, a concentration well beyond the level required to saturate all known ethylene effects on plants, and the significance of this over-production is unclear.

It was found that storage of uninitiated bananas in low oxygen atmospheres (1 to 5%) led to an almost complete inhibition of ripening,[159] which could be reversed by application of ethylene in the low oxygen atmosphere. Similarly, in bananas in which

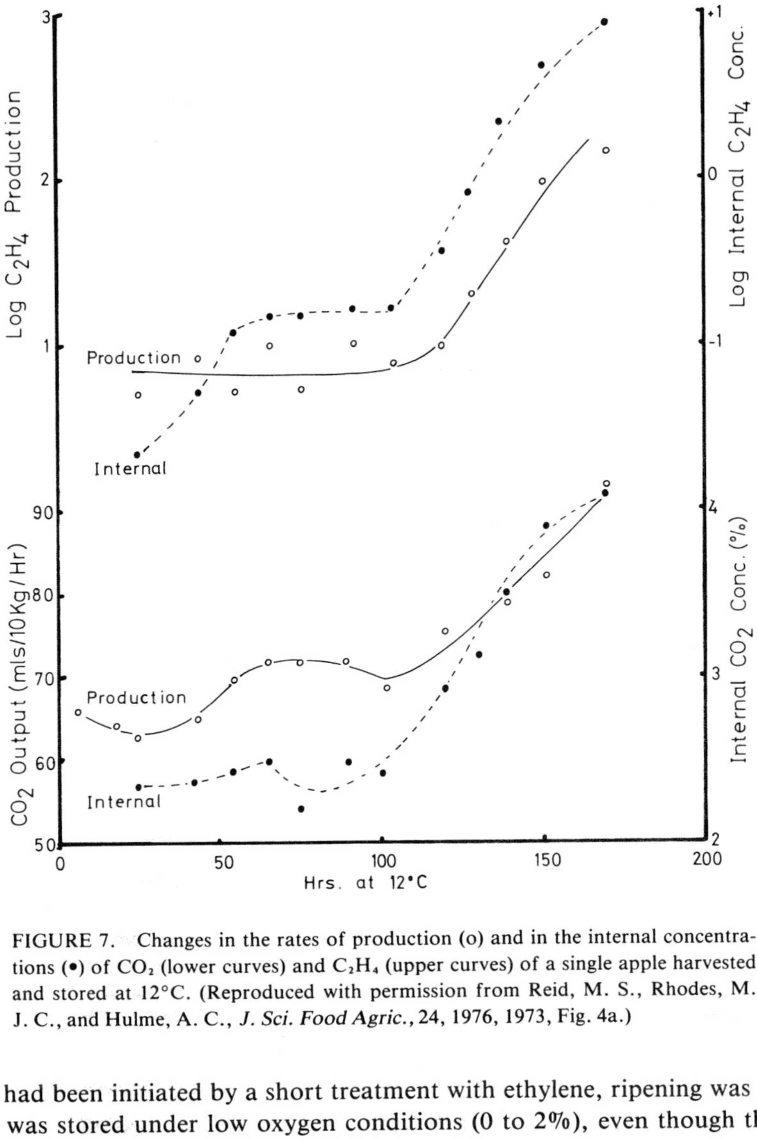

FIGURE 7. Changes in the rates of production (o) and in the internal concentrations (•) of CO_2 (lower curves) and C_2H_4 (upper curves) of a single apple harvested and stored at 12°C. (Reproduced with permission from Reid, M. S., Rhodes, M. J. C., and Hulme, A. C., *J. Sci. Food Agric.*, 24, 1976, 1973, Fig. 4a.)

ripening had been initiated by a short treatment with ethylene, ripening was delayed if the fruit was stored under low oxygen conditions (0 to 2%), even though the control of ripening was less than with uninitiated fruit.[160] However, this effect could be reversed by including ethylene in the low oxygen atmosphere. This has led to the suggestion that ethylene has an important role in promoting directly the progress of some ripening changes, as well as in the initiation of ripening. Although, as shown in Section III.E above, many of the ripening changes can proceed quite independently of the respiration climacteric,[121] in climacteric fruit ripening is inhibited under conditions that inhibit ethylene biosynthesis.[127] There is evidence that ethylene may directly affect particular ripening processes and that activation of some events in ripening may be a direct response to the presence of high levels of ethylene. The best studied example is the relationship between ethylene treatment and the induction of the rise in respiration. A number of studies[35,67,133,161,162] have shown that ethylene treatment of a wide range of plant tissues other than fruits can lead to a rise in respiration and this, of course, occurs without the attendant ripening changes. Among fruit tissues, including the nonclimacteric citrus fruits and the *rin* mutants of tomato, the response in respiration is initiated by ethylene treatment, but requires the continued presence of ethylene for the respiratory response to be sustained. Such a direct relationship between ethylene production in the cytoplasm and mitochondrial activity may explain the relationship between cytoplasm changes and the activity of mitochondria suggested by Romani[129] (see Section III.3). Solomos and Laties[163,164] have correlated this ability to respond to ethylene with the presence of the cyanide-insensitive pathway of mitochondrial respiration.[133] Although neither the details of the electron carriers involved in the cyanide-insensitive pathway, nor the mechanisms by which ethylene could activate the system are known, this correlation does provide an interesting approach for further work and it does indicate a direct relationship between the presence of ethylene and the induced respiration. In the climacteric fruits, once initiated, the autocatalytic production of ethylene leads to the accumulation of ethylene, which in turn could directly induce and sustain a rise in respiration.

There are similar direct relationships between the presence of ethylene and other aspects of ripening. It has been suggested[165] that the presence of high levels of ethylene has an effect on the softening of apples during ripening in store. These workers showed that the presence of an ethylene-absorbing compound in the store, which reduced the level of ethylene from about 1000 ppm to below 10 ppm, had a marked effect in retaining the texture of the fruit after storage. Knee[90] performed similar experiments, but interpreted his data as suggesting that it was the delay in achieving catalytically significant amounts of ethylene (rather than the lowering of the ultimate concentration) which was responsible for the observed inhibition of softening. In the olive, a non-climacteric fruit, ethylene treatment has no effect on the respiration rate or on ripening, as measured by increased softening or development of red color, but it did increase the destruction of chlorophyll.[166] Another example is the effect of the presence of ethylene on protein turnover in the banana,[126] in which short exposure to ethylene induces the acceleration of amino acid incorporation in the absence of the climacteric and ripening.

There is some evidence that the threshold levels for the different ethylene-dependent responses of plant tissue may be different.[167] For instance, the half-maximal concentration for the effect of ethylene on pea epicotyl geotropism was about 95 $n\ell/\ell$ while that for the effect in promoting increase in the diameter of cotton hypocotyls was 1350 $n\ell/\ell$. In fruits there is also some evidence that the various processes of ripening are initiated at different internal concentrations of ethylene (see Table 5). It was shown that the climacteric in honeydew melons required a threshold of ethylene concentration of about 3 $\mu\ell/\ell$[43] and a concentration for half-maximal response of 5 to 10 $\mu\ell/\ell$. In contrast, the softening of this tissue was induced by a threshold concentration of about 100 $n\ell/\ell$ and a half-maximal value of 320 $n\ell/\ell$.[167] The increase in carotenoid biosynthesis in ripening honeydew melons appears to have a similar dependence on ethylene to that of softening.[128] There can thus be approximately a 30-fold difference in the sensitivity of these ripening processes to ethylene. Similarly, in the pear,[127,168] the onset of softening was induced by ethylene concentrations of only 0.08 $\mu\ell/\ell$ while the onset of the respiration climacteric in the same fruit required 0.46 $\mu\ell/\ell$. Knegt, Kramer, and Bruinsma[169] found that the internal concentration of ethylene rose to 5 $\mu\ell/\ell$ in ripening tomatoes before the onset of pectin solubilization, while in the same fruit a concentration of only 0.1 ppm was sufficient to induce the respiration climacteric.[22] Vendrell and McGlasson[170] showed that longer exposure of bananas to ethylene was required to initiate chlorophyll breakdown than to cause starch degradation, these examples show that different ripening processes seem to have different sensitivities to ethylene. It is thus tempting to suggest a role for ethylene in coordinating the sequence of ripening changes, as its concentration rose and individual responses were initiated. It must, however, be borne in mind that similar ripening changes occur in the non-cli-

TABLE 5

Levels of Ethylene Required to Initiate Various Processes Occurring during Ripening

Fruit	Ripening process	Internal concentration for the initiation of process (ppm) $\mu\ell/\ell$	Ref.
Melon	Softening	~0.1	167
	Carotenoid biosynthesis	~0.1	239
	Respiration rise	3.0	43
Pear	Softening	0.08	168
	Respiration rise	0.46	127
Tomato	Respiration rise	0.1—0.2	22, 63, 248
	Solubilization of pectins	5.0	

macteric fruit in the absence of increased ethylene production and the role of ethylene in the climacteric fruit may be to accelerate processes that proceed slowly in its absence.

Burg and Burg,[26] in a study of the concentration dependence of a number of different responses of plant tissue to ethylene, found that the concentration for apparent half-maximal biological activity was 6×10^{-10} *M*. From a study of the relative activity of related hydrocarbons on the pea stem section assay, Burg and Burg[26] defined the molecular requirements for biological activity. Activity was only found with a range of unsaturated compounds (see Table 6), and the relative activities of the various compounds tested closely corresponded with the stability constants of the silver complexes of these olefinic compounds; Burg and Burg[26] thus proposed the presence of a metal binding site of ethylene. In view of the requirement of O_2 for ethylene action, they proposed that O_2 was necessary for ethylene to bind to the site. Goeschl and Kay[167] have suggested that, since different ethylene responses show different concentration dependencies on the gas, there may be more than one ethylene binding site within the cell, each perhaps specific for a single response or group of responses. A criticism of the Burgs' approach is that they analyzed their data by the method of enzyme kinetics derived from the analysis of the initial rates of enzymic reactions, and these may not be strictly applicable to relatively long term physiological responses to the gas.

If one accepts the hypothesis that ethylene binding sites exist in the cell, and that binding to that site is a prerequisite for ethylene action, and there is gathering evidence that such sites exist for other plant growth substances,[171,172] then it follows that the ethylene, which reaches the gaseous phase, is excess or waste ethylene. Ethylene is produced and causes its various responses within the cells of the tissue, and hence gaseous ethylene which disappears into the vapour phase is irrelevant to its action and merely a consequence of its gaseous nature. If the binding of ethylene to the site is the first step in the elicitation of the response, then clearly the concentration of ethylene in the proximity of the site will determine the extent of binding that occurs. Abeles[58] has calculated, on the basis of the presence of a single species of acceptor in the cell and a response with a half-maximal activity at 0.1 $\mu\ell/\ell$ that there would be about 500 ethylene attachment sites/cell. It is clear that the binding to the site must be transitory and not involve ionic or hydrogen bonding between ethylene and the attachment site. Using deuterated ethylene[173,174] it has been shown that there is no hydrogen exchange between ethylene and the cell, and that *cis-trans* isomerization does not occur during the interaction between the ethylene and cell tissue. It seems likely that extremely weak Van der Waal's forces may be involved in the binding of ethylene.[173]

Since the nature of the binding site is unknown, it is unclear whether ethylene acts in the aqueous or hydrophobic regions of the cell, but ethylene has an appreciable

TABLE 6

Biological Activity of Ethylene and Other Unsaturated Compounds as Determined by Assay with Pea-stem Sections

Compound	K_A relative to ethylene[a]
Ethylene	1
Propylene	130
Vinyl chloride	2,370
Carbon monoxide	2,900
Vinyl fluoride	7,100
Acetylene	12,500
Allene	14,000
Methyl acetylene	45,500
1-Butane	140,000
Vinyl bromide	220,000
Ethyl acetylene	765,000
Vinyl methyl ether	1,175,000
Butadiene	1,200,000
1,1-Difluoro ethylene	2,060,000
Vinyl ethyl ether	5,440,000

[a] The Michaelis-Menten constant (K_A) is the molarity in solution required for half-maximum biological activity. Its value for ethylene at 24°C in the presence of an infinite concentration of O_2 is 6×10^{-10} *M*

(Reproduced from Burg, S. P. and Burg, E. A., *Science*, 148, 1190, 1965. With permission.)

solubility in both water and lipids.[59] The concentration around the active site is dependent on the rate of production, the solubility of the gas, its access to the active sites, and the dissipation of the gas into the surrounding atmosphere. In view of the gaseous nature of ethylene and the ease by which it is lost, it is perhaps surprising to find reports that ethylene gas is metabolized by tissues. Early studies of this problem were confounded by doubts about the chemical and radiochemical purity of the ^{14}C ethylene used to detect such metabolism (see review 22). Beyer[175] has devised methods for producing highly purified ^{14}C ethylene using GLC techniques. He applied highly purified ^{14}C ethylene to two plant systems; pea seedlings[176-178] and carnation flowers,[179] and found evidence for two separate systems for metabolizing ethylene; one, the oxidation of ^{14}C ethylene to $^{14}CO_2$ and the other the incorporation of ^{14}C ethylene principally into water soluble components by a pathway unique to ethylene metabolism. CO_2, a known inhibitor of ethylene action, markedly inhibits $^{14}C_2H_4$ conversion to $^{14}CO_2$ without affecting the incorporation into soluble components.[176] Giaquinta and Beyer[178] studied the labeled products of the incorporation into soluble products and found five as yet unidentified radioactive products. More recently Blomstrom and Beyer[179a] have tentatively identified two of the water soluble metabolites of ethylene as ethylene glycol and its sugar conjugate. In cotyledons of broad bean, Jerie and Hall[179b] have identified ethylene oxide as the major metabolite of ethylene with between 85 to 95% of the applied radioactive ethylene being converted to ethylene oxide in a 2 hr incubation period. The significance of this massive conversion of ethylene to ethylene oxide in relation to the metabolism of ethylene to ethylene glycol is unclear. Beyer and his associates[178] were able to rule out the possibility of CO_2 acting as a precursor of these water soluble metabolites and concluded that at least two unique systems for metabolising ethylene exist in plant tissues. The significance of these two routes of metabolism on the regulation of ethylene action is as yet unclear, but Beyer[176] has suggested that there is a close relationship between the metabolism of ethylene and its biological activity. The nature of the changes that result from the binding of ethylene to the attachment site will be discussed in Section IV.B.

2. *The Role of Other Hormonal Factors in Fruit Ripening*

Studies on the involvement of ethylene have in the past dominated the study of the hormonal control of fruit ripening. This stems from the early discovery of the importance of the gas in fruit ripening, and also from the ease with which its concentration within the tissues can be measured. It is clear from the foregoing discussion that ethylene plays a unique role in ripening, but there is clear evidence that it is not the only factor involved and, as information becomes available, the importance of the other factors becomes increasingly apparent. Particularly in the systems controlling the initiation of ripening, factors other than ethylene are important and can be dominant. The hormonal control of some aspects of ripening will be discussed elsewhere in this volume (Goldschmidt, Chap. 9), and here I will attempt only to outline the progress that has so far been made in attempting to assign roles to other growth substances in ripening.

In the past few years there have been a number of studies of changes in levels of growth substances in fruits during ripening. These studies, touched on also in Chapter 10, are at present far from complete, but they are beginning to throw light on the role of these substances in ripening. There have been numerous studies of the effects of applied growth substances and these have frequently yielded ambiguous and confusing data. The problems generally relate to the difficulties of the uptake and distribution throughout the tissue of growth substances applied to bulky fruit tissues. An example of this is provided by the work of Vendrell,[180] who showed that treatment of banana slices by vacuum infiltration with auxin solutions delayed ripening, while dipping whole bananas in solutions of auxins advanced ripening. This was shown to be due to the uneven distribution of the auxin in the whole fruit, which caused a localized production of ethylene in the peel, which in turn promoted ripening in the pulp. The peel cells, although they produced ethylene in the presence of high amounts of auxins, were retarded in ripening compared with the pulp tissue, which received very little auxin, but was subjected to the stimulatory effect of the ethylene diffusing from the peel tissue.

The control of maturation and the initiation of ripening is thought to be due to the interaction and balance between opposing promotory and inhibitory factors. Ethylene is one important promoting factor, ABA appears to be another. Externally applied ABA will hasten ripening in a climacteric fruit such as the tomato,[181,182] and will advance ripening in grapes if applied during stage II of the growth cycle.[2] Absiscic acid (ABA) has been shown to stimulate ethylene production in pre-climacteric apples[183] and tomatoes.[184] There have now been a number of studies of changes in the level of ABA in tissues during ripening. In a number of climacteric fruits, for instance the pear[185] and the avocado,[186] the level of ABA is constant during maturation, but rises during ripening.[187,188] These rises in ABA accompany the rise in ethylene production during ripening. Adato et al.[188] showed that there was a threefold rise in free ABA level during ripening of detached avocado at 19°C to a maximum level of 700 μg/100g fresh weight. As shown in Figure 8, the peak in ABA corresponded in time to the

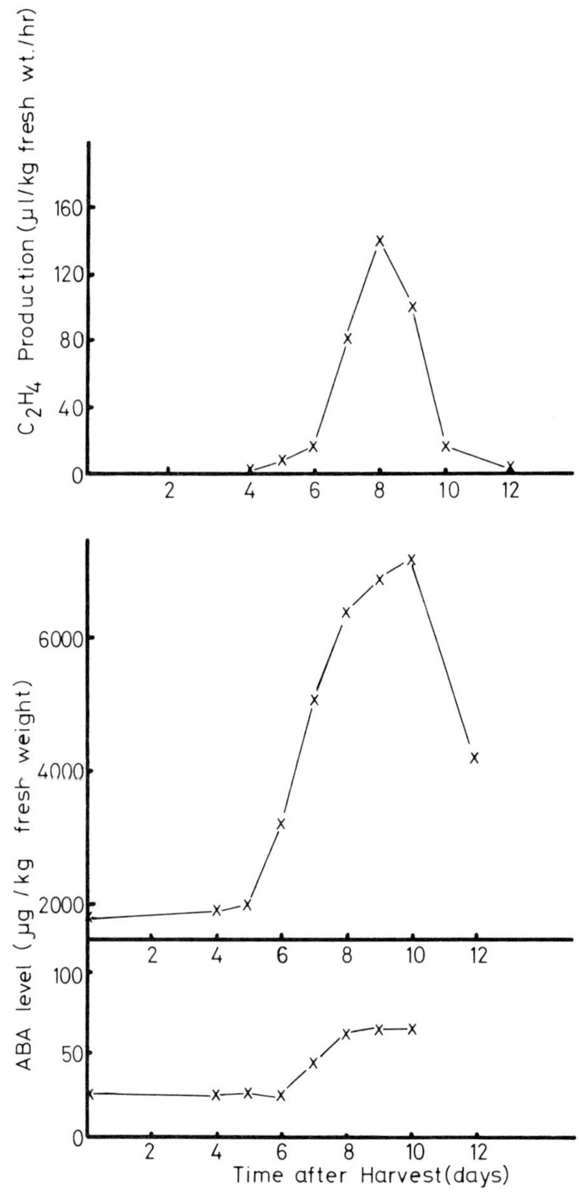

FIGURE 8. Changes in the rate of ethylene production (upper curve) and in the levels of free (middle curve) and bound (lower curve) ABA in Fuerte avocado fruits during ripening. (Redrawn from Figures 1 and 2 of Adato, I., Gazit, S., and Blumenfeld, A., *Aust. J. Biol. Sci.*, 3, 555—558, 1976. Reproduced with permission.)

peaks of ethylene production and of the respiration, and preceded the softening of the tissue. A similar pattern of change was found in the bound ABA fraction, although the peak occurred about one day later than that in free ABA. The ratio of free/bound ABA was about 10:1 throughout ripening, and thus the increase in ABA must represent net synthesis rather than the release of the bound form.

Increase in ABA level during ripening occurs in a wide range of fruits of different ripening pattern. For instance, Rudnicki[189] has shown a threefold increase in ABA levels in the strawberry, in which ripening occurs in the presence of little or no ethylene. Similarly, there is a rise in ABA in citrus fruits, which also ripen in the presence of only very low levels of ethylene.[190,191] ABA level increases during ripening of some climacteric (peach)[192] and non-climacteric (grape,[49] cherry[193]) fruit that show a three-phase growth cycle. In the grape the ABA level rises steeply at veraison (onset of period III), while in the peach the ABA level increased just before and during period III,[192] and this increase tends to precede the increased ethylene production associated with ripening.[192] External treatments that delay or advance ripening also delay or advance the changes in ABA. In the pear, cold treatments that have been shown to promote ripening also advance the increase in ABA.[194] It was shown[187] that the ABA level in pears increased by up to fourfold during ripening, and that the use of CA conditions to delay ripening also delayed the rise in ABA. Similarly, in grapes Coombe and Hales[48] found that treatment of the fruit during stage II of growth with 2-chloro-ethyl-phosphonic acid, which degrades on contact with the tissue to yield ethylene, promoted ripening and advanced the onset of the rise in ABA, while treatment with benzothiazole, 2-oxy acetic acid, which delayed ripening, also delayed the onset of increased ABA accumulation. These results clearly show an association between ABA and ripening, but they do not elucidate its role. McGlasson and Adato[195] have made a detailed study of the changes in ABA level during the growth of tomatoes of both normally ripening varieties and ones showing ripening defects. They found in all these varieties an increase in ABA level during the major period of growth of the fruit, reaching a peak shortly before the time of onset of the respiration climacteric and increased ethylene production; this confirmed the results of an earlier study.[184] The increase in ABA appears to be due to synthesis rather than release of the bound form, which only represented a small portion of the total ABA. The peak in ABA occurs in advance of the symptoms of ripening in all the strains and then declines. Throughout the period of ABA accumulation the ethylene level is low, and during the period of rapid rise in ethylene production in normal fruit the ABA level is falling. Clearly, the pattern of changes in ABA in the tomato is very different from that found in the other fruit discussed and further work needs to be done before the differences can be appreciated. However, it is perhaps significant that the rise during development occurred in strains such as *rin*, *nor* and *Nr*, which show markedly curtailed ripening, so its significance to ripening is at least questionable. The facts remain that in all fruits studied ABA is present in the unripe condition and can play a role in interacting with other factors in promoting ripening. In addition, rather like ethylene, ABA synthesis appears in some cases to be stimulated by the initiation of the ripening process. The analogy with the role of ethylene in ripening can be taken a step further if one speculates that the increased tissue content of ABA may have a role in regulating some ripening changes once the ripening process has been initiated. It is interesting to point out that the increase in ABA cuts across the boundary of the climacteric and non-climacteric fruits, and it may be that ABA is the dominant ripening promoter in some non-climacteric fruits.

The nature of the factors that antagonize the ripening promoters is still a matter for speculation. There is good evidence that tree fruits receive ripening inhibitors from the parent tree, and there is also good evidence to relate the presence of ripening inhibitors in fruits to the changing sensitivity of maturing climacteric fruits to ethylene. There is certainly little information on the role that cytokinins and gibberellins may play in the ripening processes. There are few studies on changes in their concentrations during ripening, and most of the available information is based on studies of the effects of externally applied compounds. What information is available suggests that they may affect color changes in fruit (see Goldschmidt, Chap. 9), but not significantly other aspects of ripening. For instance, GA_3 will delay the loss of green color from citrus fruits and the appearance of the red pigment lycopene during ripening of tomatoes.[125] Vendrell[196] showed that gibberellic acid applied either to whole green bananas or to slices of unripe bananas delayed the yellowing of the fruit, while other aspects of ripening proceeded normally. Similarly, the cytokinin kinetin[197] applied to banana slices immediately before exposure to 60 ppm ethylene for 16 hr had no effect on the development of the respiration climacteric or on the accumulation of ethanol-soluble carbohydrates, but it did markedly retard the de-greening of the peel. It thus appears that, although gibberellins and cytokinins may play a role in regulating chloroplast senescence, they do not play an important role in controlling ripening as a whole. However, these conclusions are based on a minimum of evidence and, in view of the diversity of the chemical nature of the various gibberellins and cytokinins, further developments must be awaited before the role of these compounds can be fully assessed.

A possible endogenous hormonal inhibitor of ripening is indole-acetic acid (IAA). IAA acts both as an inhibitor of ripening[180] and as a promoter of ethylene biosynthesis.[198] Frenkel[199] has shown that infiltration of pre-climacteric pears with either IAA or 2,4-dichlorophenoxy-acetic acid inhibited ripening, measured either as chlorophyll degradation or as softening, even though it simultaneously stimulated ethylene production. Thus auxins are able to overcome the stimulatory effect of added ethylene on these processes. Auxin antagonists such as α-p-chlorophenoxy-isobutyric acid relieved the inhibition of ripening, and this occurred whether or not the experiment was carried out at atmospheric or 0.1 atmospheric pressure, a treatment designed to reduce the internal concentration of ethylene.[37] It now seems likely that IAA acts in vivo as an inhibitor of ripening. There are very few studies of changes in IAA level during the period immediately before and immediately after the onset of ripening in fruits. Varga and Bruinsma[200] studied changes in IAA level in tomatoes using a fluorometric method for the determination of IAA, and showed that the IAA in tomatoes was mainly confined to the seeds, although concentrations of up to about 3ng/g fresh weight ($\sim 10^{-8}$ *M*) were found in the pericarp. There was a peak in the IAA content of the seeds at about the 50% development stage and then a fall to low levels by the later stages of ripening. A rather similar sequence of changes occurred in the pericarp tissue, even though the absolute level of IAA was always very low. The significance of the interaction between IAA and the stimulation of increased ethylene biosynthesis is hard to assess. It may be that it is not significant at the very low physiological levels of IAA (10^{-8} *M*), but it is potentially an interesting control system with a direct interaction between the levels of inhibitor and activator. Frenkel[199] has suggested that products of IAA oxidation such as 3-methylene-oxindole may act as ripening promoters. He envisages a slow decrease in IAA during the pre-climacteric period under the influence of IAA oxidase, which has been shown to increase in activity during the early stages of ripening,[201] leading to the accumulation of oxidation products such as methylene oxindole. Frenkel et al.[202] showed that purified methylene-oxindole (MO) infiltrated into unripe pears, which were subsequently stored at either atmospheric or one half atmospheric pressure, promoted softening even under hypobaric conditions. The authors argued that their results indicated that IAA acted directly on the ripening initiation system rather than indirectly via ethylene production, although this possibility could not be entirely ruled out.

Thus there is increasing evidence that, at least in some fruits, both ABA and ethylene are important endogenous promoters of ripening. Information on changes in the levels of potential ripening inhibitors at the time of the natural initiation of ripening is very limited at present, and much more evidence will be needed before an understanding of the role they play in the mechanisms controlling the initiation and regulation of ripening can be obtained.

B. Biochemical Control

While there is uncertainty about the mechanism by which ripening is initiated, there is equal uncertainty about the nature of the underlying biochemical events that result from this initiation and that lead to the visible symptoms of ripening. In the climacteric fruits, ripening is seen as the final phase in development of the organ and, as such, is under genetic control. There is a sharp transition in metabolism at the time of the onset of the climacteric and this involves the promotion of both anabolic and catabolic processes. Two theories have been advanced to explain the underlying biochemical control of the levels of enzyme activity responsible for the promotion of these processes. In one, the important feature is the *de novo* synthesis of a small group of specific enzymes, which catalyze critical steps in the changes leading to ripening, but which are not present before the climacteric. Inherent in this theory is the synthesis of new messenger RNA species for these enzyme proteins. The alternative theory suggests that the activation or release of pre-existing enzymes is the dominant feature and this is linked with changes in the compartmentation of the cell, which could lead to the release of hydrolytic enzymes from lipoprotein-bound cell organelles and to changes in the permeability of cell membranes, leading to increased availability of substrates, cofactors or activators to the active site of the enzymes. No conclusive evidence for either theory has yet been advanced, but it seems likely that both types of change can operate in the various ripening processes.

Evidence for the first theory was presented by the work of Hulme[203] who showed that in detached apples the protein nitrogen content rose during ripening. This finding has since been confirmed for pears[204] and cantaloupe melons,[205] but it is not a feature of the ripening of other fruit such as avocados[1] and bananas.[206] However, for a number of climacteric fruits such as apples,[207] avocados,[208] bananas,[209] pears,[122] and tomatoes[210] there is evidence for a period of enhanced incorporation of amino acids into proteins during the early stages of the climacteric. By the time the climacteric peak is reached the rate of incorporation is usually relatively low.[207,208,210] An increase in total RNA content has been found during the ripening of apples,[211] but such a rise does not occur in either avocados[1] or bananas.[206] As in the case of proteins, so with RNA there is evidence for a period of enhanced incorporation of RNA precursors such as uridine into RNA at the early stages of the climacteric in apples,[207,212] avocados,[213] pears,[214] and tomatoes.[210] Richmond and Biale[213] showed that the increased labeling was mainly into ribosomal RNA and into a fraction eluting from an MAK column shortly after the RNA peak, which may represent a messenger-RNA fraction. Romani[214,215] found evidence for the presence of polyribosomes in ripening pears at the time of the climacteric minimum, and that there was a decrease in the ratio of polyribosomes to ribosomes as ripening proceeded. There is thus good evidence for a period early in ripening in which both RNA and protein synthesis are active.

Further evidence for the importance of protein synthesis in ripening has come from the use of inhibitors of protein synthesis. It has been shown that infiltration of preclimacteric pears[122] and bananas[123] with the inhibitor of protein synthesis, cycloheximide, inhibited all the processes of ripening except the rise in respiration. These experiments may be interpreted as showing that protein synthesis plays a vital role in several responses of ripening but, as the specificity of cycloheximide is now doubted,[216] this conclusion may not necessarily be valid. There are many reports of increases in the in vitro activity of enzymes extracted from fruit at various stages of ripening,[1] but often it is unclear whether these result from changes in the extractability of enzymes as the structure and composition of fruit changes as it ripens, from changes in the form and activity of the isolated enzymes or from genuine changes in the tissue concentration of the enzyme. Skakoun and Daussant[217] have shown, using immunochemical techniques, that alcohol dehydrogenase, which may play a role in the production of volatile aldehydes and alcohols, increases in extractable activity during ripening of bananas, and that this increase is due to an increase in alcohol dehydrogenase protein rather than to activation. Similarly, Iki et al.,[218] also using immunochemical techniques, have demonstrated the *de novo* synthesis of β-fructo-furanosidase in ripening tomatoes.

Although an increased rate of amino acid incorporation is a feature of the early stages in ripening in a number of climacteric fruits, there are few reports of studies on the nature of the protein products of the incorporation. It was, however, shown[122] that the increased incorporation of phenylalanine into protein during the ripening of pears was associated with an increased labeling of malic enzyme, an enzyme thought to play a role in malate decarboxylation during the ripening of pome fruits.[111] Brady and O'Connell[126] reinvestigated the activation of amino acid incorporation in bananas. It had previously been shown[209] that ripening in the banana is associated with small but significant changes in the patterns of proteins, as revealed by polyacrylamide gel electrophoresis. Brady and O'Connell[126] showed that the incorporation of amino acids into proteins during the early stages of ripening represented the turnover of a substantial proportion (25 to 50% of the total protein in a 24 hr period) of the proteins of the pulp cells, and that the incorporation was into a wide range of soluble proteins rather than into a few novel proteins specifically involved in ripening. They further showed that the general turnover of proteins is a direct response to the presence of ethylene and not necessarily related to ripening changes, since short treatments with ethylene of slices of unripe fruit elicited the response in enhanced protein turnover without initiating ripening. On the other hand, Rattanapanone et al.[219] found that there was a peak in the incorporation of tritiated uridine into RNA just before the onset of color changes in ripening tomatoes. They isolated from the total RNA a polydisperse RNA fraction, rich in polyadenylic acid, which had the properties of messenger-RNA, since it was able to direct protein synthesis in an in vitro wheat germ protein translation system. Although the ratio of messenger to total RNA did not change during the ripening period, there were distinct differences in the nature of the protein products of translation when m-RNA extracted from mature green tomatoes was compared with that from tomatoes that had commenced ripening.[220] These differences were in only a few proteins, and it was concluded that at about the time of onset of ripening there was a change in the composition of the m-RNA fraction which could be involved in directing the synthesis of a few novel enzymes involved in ripening.

These conflicting reports leave the significance of the role of RNA and protein synthesis in ripening unclear. The reports of Rattanapanone et al.[219,220] are the clearest evidence yet for the synthesis of new m-RNA and protein species during ripening, but it is clear that the changes in the pattern of m-RNA production at the time of the initiation of ripening are small, perhaps only involving one or two proteins. Brady and O'Connell's work indicates that the increases in amino acid incorporation observed in a number of fruits probably reflect ethylene-induced changes in the rate of turnover of a wide range of proteins, but it is still possible that underlying the general turnover there is the production of a few novel species. More work on the nature of the translation products of the new m-RNA species produced during ripening is necessary before the significance of protein synthesis in ripening can be assessed.

Blackman and Parija[221] proposed long ago that the respiration climacteric and the attendant ripening changes resulted from a change in the "organizational resistance" of the cells. Bain and Mercer[72] observed large increases in the vacuolation of the cytoplasm of pears during ripening and found a decrease in the electrical impedance of the tissue just before the start of the climacteric; they concluded that there was a change in the permeability properties of the cell membranes during ripening. Such changes in permeability could clearly have a profound influence on the activity of enzymes by making substrates, cofactors, activators or inhibitors more freely available to them. There are considerable difficulties in making suitable measurements of the permeability of cell membranes, and some of the earlier studies[1] may have overestimated the permeability changes in fruit cells since they were carried out in hypotonic media, which can lead to membrane damage.[222]

Changes in membrane permeability have been estimated by measurements of changes in apparent free space (AFS) to organic molecules and to inorganic ions, and by measurements of changes in the leakage of solutes from tissue slices. All the methods necessarily involve the use of tissue slices, but with all the metabolic changes that slicing induces in plant tissues this is not a desirable situation. The AFS of the pulp cells of bananas[223,224] has been shown to start to increase just before the onset of the climacteric rise in respiration, and Sacher[223] suggested that the increase in free space was a causative factor in the onset of the respiration rise and of ripening. In green bananas the free space is about 20% of the total fruit volume and this rises to about 40 to 60% at the climacteric peak. Similar increases in AFS accompany the ripening of avocados.[225] An increase in the leakage of solutions accompanies the ripening of apples,[226] avocados,[225] and bananas.[224] The significance of these apparent changes in permeability was questioned by Brady et al.[224] who found that, in banana slices ripening in the absence of ethylene, an increase in amino acid leakage and in AFS to various ions and organic molecules preceded the onset of the climacteric rise in respiration ripening by 3 to 4 days. However, if such slices were treated with ethylene the respiration climacteric was induced without changes in either amino acid leakage or AFS. Such permeability changes *followed* the onset of the climacteric by 32 hr and are thus not a prerequisite for its onset. It is clear that, while permeability changes may not be causative of ripening, they appear to accompany it and probably affect its course.

The measurements of changes in AFS and of solute leakage are relatively crude measures of changes in the permeability characteristics of the cell. They give no information on changes in compartmentation within the cytoplasm, but give some measurement of tonoplast and/or plasmalemma membrane permeability. Vickery and Bruinsma[227] have attempted to obtain estimates of the permeability coefficients of particular cell membranes in a computer analysis of the progress curves of efflux of K^+ from slices of tomato pericarp taken from fruit at various stages of ripening. They concluded that the permeability of both the tonoplast and the plasmalemma to the passive efflux of K^+ did not change during development of tomatoes up to the fully ripe stage. They did, however, find that the fraction of the total K^+ found in the cytoplasm increased during ripening. They concluded that this rise resulted from the active transport of K^+ under the full metabolic control of the cell, rather than from changes in the passive permeability of the membrane, and that this metabolic control was maintained even in the ripe fruit. Vickery and Bruinsma[227] suggested that earlier studies showing increasing rates of efflux of ions[226] from tissues during ripening, which had been interpreted as indicating a change in the permeability characteristics of membranes, may have been due to a rise in the cytoplasmic level of ions rather than to membrane changes. These studies show that measurement of changes in membrane permeability is fraught with difficulties, but the notion that there may be changes in compartmentation of solutes, which might be brought about by active transport under metabolic control, is an important one for further study. This should involve a study of factors controlling the level and activity of the proteins involved in the active transport.

It is of interest to know how such permeability changes, whether they involve active or passive transport, might affect the activity of enzymes involved in ripening. An example of this has been proposed by Chalmers and Rowan[228] in a study of the activity of the enzyme phosphofructokinase (PFK), which is the first committed enzyme of the pathway of glycolysis. PFK is a complex regulatory enzyme whose activity is affected by a wide range of soluble metabolites of very different stereochemistry from its substrates, fructose-6-phosphate and Mg ATP. The activity of phosphofructokinase measured in vitro does not increase during the ripening of either bananas[229,230] or tomatoes,[228,231,232] yet there is good evidence from the accumulation of its product, fructose 1:6 diphosphate, during ripening of bananas[229,230,238] and tomatoes[228,231,232] that glycolysis is activated during the climacteric rise in respiration, and that the in vivo activity of PFK increases markedly during this period. Chalmers and Rowan[228] showed that the climacteric peak respiration rate of tomatoes was positively correlated with their P_i contents, and proposed that leakage of P_i from the vacuole was responsible for the increase in PFK activity during ripening. Issac[232] has shown, using PFK purified from tomatoes by affinity chromatography, that P_i stimulates PFK activity, and that this stimulation was much greater at pH 7.5 than at pH 8.0, the pH optimum of the enzyme. Tomato PFK is inhibited by citrate, and to a lesser extent by malate, the two principal organic acids accumulated in the tomato. P_i overcomes completely the inhibition by both citrate and malate. These results show how it is possible that the increased leakage of an ion such as P_i may regulate glycolysis during ripening. Such a mechanism can only be speculative in a situation in which very little is known about either the cellular distribution of ions (such as P_i) and organic acids (such as malate and citrate) or about their concentrations in the cell compartments housing the regulatory enzymes.

This discussion of the biochemical control of ripening shows that there are several important issues still be to resolved. It seems likely that ripening is a differentiation process that is ultimately determined at the genome level. The process by which this control is expressed ultimately as changed activities of enzymes and pathways is unknown. The extent to which the increased enzyme activities responsible for these changes are due either to *de novo* synthesis, activation, or release of enzymes from membrane-bound structures cannot be assessed, although all three processes may be involved. It is clear that increased turnover of proteins occurs, at least in bananas, as a direct response to ethylene and this may obscure the formation of new enzymic proteins at this stage. In some cases, at least, membrane permeability changes appear to occur during ripening, and we know very little of the particular membranes that are affected and can only speculate on the possible metabolic significance of such changes.

V. EXTERNAL CONTROL OF RIPENING

The use of external factors in an attempt to control the ripening process has important implications in the storage of fresh commodities prior to marketing, and this control of the ripening process must be achieved while minimizing the deleterious effects of water loss, microbial infection, and the development of physiological disorders in storage.

A. Effect of Temperature

Ripening occurs in only a relatively narrow range of temperatures. At temperatures below a certain critical temperature, many fruits of tropical and sub-tropical origin develop a disorder known as chilling injury, the symptoms of which include an inhibition of ripening. This critical temperature[234] is between 10 and 13°C for many common fruits such as bananas and tomatoes, while in some temperate fruits such as the Cox's Orange Pippin variety of apples, it can be as low as 3°C. The resulting limitations on the use of low temperatures is a major factor limiting the storage of such fruits. There is also an upper temperature limit above which ripening does not occur. Biale and Young[8] showed that at temperatures above 25°C the extent of the respiration climacteric in avocados decreased. The respiratory rise between the initial pre-climacteric level and the climacteric peak was 250% at 25°C, but only 30% at 30°C. Color changes associated with ripening in the tomato are inhibited in storage at temperatures above 30°C.[235] It is generally true that many fruits fail to ripen properly beyond 30 to 35°C and that in most cases this inhibition is reversible. Above about 30°C bananas fail to ripen normally, the pulp becoming soft and watery.[20] Ethylene synthesis is inhibited at higher termperatures, but this, as with the inhibition of color changes in the tomato, is reversible within a relatively short period after return to lower temperatures. Within the normal ripening range of temperatures respiration rates and ripening behave in the predictable fashion with a Q_{10} over the range 10 to 25°C generally about 3.[30]

B. Gaseous Composition of Storage Atmosphere

The use of modified atmospheres to delay ripening stems from the studies by Kidd and West on English apples. They found that raising the CO_2 level and decreasing that of O_2 in the storage atmosphere delayed ripening, and such a modified atmosphere, coupled with the use of temperatures just above the limit for the onset of low temperature disorders, is the basis for the widely used method of controlled atmosphere storage that has proved so successful with a wide range of fruits.[236]

1. Effect of O_2 Tension

Low levels of O_2 (between 1 and 5%) have been shown by a large number of workers to delay the onset of ripening in a number of different fruits. The role of oxygen in inhibiting ripening does not appear to be related entirely to an inhibition of respiratory activity, although low levels of O_2 do decrease the net respiration rates of bulky fruit tissues.[237] It appears that O_2 has its principal effect in interfering with the initiation process that determines the onset of ripening. This was demonstrated by Mapson and Robinson,[159] who showed that low oxygen (1 to 5%) atmospheres, even at ambient temperatures, almost completely prevented ripening of bananas as measured by the onset of endogenous ethylene production, the respiration climacteric, and changes in color index. The inclusion of low levels of ethylene in the low oxygen atmosphere reversed the inhibition of these ripening changes and ultimately normal ripe fruit were produced, showing that the ripening as such was not inhibited by low O_2. It was shown[160] that, with bananas that had been initiated by ethylene treatment, low oxygen atmospheres had a diminished effect in delaying ripening.

It was proposed by Mapson and Robinson[159] and Burg[20] that the inhibition of ripening by low oxygen atmospheres was due principally to its effects on the involvement of ethylene in the initiation of ripening. There are a number of sites at which O_2 can affect ethylene synthesis and action. The biosynthesis of ethylene from methionine is an aerobic process[136-138] and is completely inhibited in the absence of oxygen. Burg[26] showed that oxygen is required for ethylene action and he suggested that the oxidation of a metal at the attachment site of the ethylene binding protein was a necessary prerequisite for ethylene binding and for the resulting ethylene action. In addition, it has been shown that during the pre-climacteric period in a number of fruits, the sensitivity of the tissue to exogenous ethylene changes, and this may be interpreted as being due to the disappearance of an ethylene antagonist. Frenkel[199] has proposed that in some fruits IAA may play a role in antagonizing ethylene effects, and that an oxidative mechanism is involved in the disappearance of IAA and the appearance of a ripening promoter, methylene-oxyindole. In this case, the low oxygen atmosphere could directly influence the hormonal balance of the tissue and its sensitivity to its own production of ethylene by slowing the disappearance of IAA. Thus oxygen can act in this system by affecting ethylene biosynthesis, ethylene action, and the sensitivity of the tissue to ethylene. The effect of oxygen on fruits other than the banana appears more complex and here respiratory effects may be relatively more important. In tomatoes, for instance (Rhodes, unpublished data[248]), storage of mature green fruits in 3% O_2 completely prevents the onset of ripening, measured by respiratory and color changes, but the inclusion of ethylene in the low oxygen mixture does not overcome this inhibition as it does in the banana. Similarly, with apples, Hulme et al.[238] showed that inclusion of even relatively high levels of ethylene (270 $\mu\ell/\ell$) in 3% O_2 did not overcome the inhibition of the respiration climacteric and ripening by 3% O_2 alone. It is clear that at least these fruits require a higher level of O_2 for normal ripening, even in the presence of ethylene, than does the banana. In conclusion, it seems very likely that the principal effect of low oxygen atmospheres have two effects in promoting strong life. One is due to the inhibitory effect of low O_2 tensions on respiratory metabolism while the other is due to its effect on ethylene action and the initiation of ripening processes. The relative importance of these two types of inhibitory action may vary from fruit to fruit.

Oxygen tensions greater than atmospheric tend to promote ethylene biosynthesis and consequently to promote ripening in a number of fruits.[20] This effect is significant in citrus fruits, which, if placed in atmospheres of 34 to 99% O_2 will have a normal climacteric rise,[30] and their production of ethylene is markedly increased.[30]

2. Effect of CO_2 Tension

Increasing CO_2 concentrations in the atmosphere around fruits delay the onset of the climacteric. Concentrations of CO_2 from 3 to 10% are generally effective in this, although higher CO_2 levels can, in some fruits (such as apples) lead to physiological disorders.[236] The effect of CO_2 in delaying ripening has been related to its effect on ethylene action. Burg[26] has proposed that CO_2 acts as a competitive inhibitor for ethylene action, presumably by competing with ethylene for the attachment site. There is ample evidence to show that high levels of CO_2 will counteract many of the known ethylene effects,[20,26] although there are some exceptions to this.[239] Burg[26] has calculated that the relative affinity of the active site for ethylene and CO_2 is about 100,000:1. The internal concentration of CO_2 in apples in pre-climacteric fruit is about 2%, while the internal ethylene content is about 0.02 ppm,[150] and it is interesting to speculate whether the internal level of CO_2 plays any role during ripening. CO_2 will also inhibit respiration in bulky organs[237] and, although this effect has not been studied in detail, it is thought to be due to the inhibition of succinic dehydrogenase by high levels of CO_2. In contrast to this, CO_2 stimulates the respiration of lemons,[240] but this is due to a stimulation of dark fixation of CO_2 and the resultant accumulation in the tissue of organic acids, which act as respiratory substrates.

C. The Effect of Atmospheric Pressure

Burg and Burg[143] showed that the ripening of bananas was completely inhibited if the fruit was stored at one fifth atmospheric pressure in pure oxygen, so that the atmospheric tension of oxygen was maintained. The inclusion of small amounts of ethylene in the atmosphere overcame the inhibitory effect of the reduction in pressure.

Burg[37] showed that the reduction in pressure led to an increase in the diffusivity of the gas and reduced internal concentrations. The fact that added ethylene reversed the inhibition showed that the removal of ethylene was the factor responsible for the delay in ripening. This system forms the basis of hypobaric storage systems designed by Burg.[241] The system as used allows a slow bleed of humidified air to enter the low pressure storage chamber and thus the partial pressure of O_2 is below the atmospheric level. Thus the hypobaric storage utilizes the delaying effects of a low oxygen atmosphere, as well as the beneficial effects of lowering internal ethylene content. Burg[37] has shown that the storage life of a wide range of fruits is enhanced under hypobaric conditions. Salunkhe and Wu[242] showed that mature green tomatoes could be stored for up to 100 days at 102 mm Hg at 55°F and ripened normally on return to atmospheric pressure. Avocados can be stored for up to 70 days at 6°C at 60 mm Hg and ripened normally on return to atmospheric pressures.[243] This confirmed Burg's earlier work,[143] but conflicts with the findings of other workers[244] who found that avocados could only be stored successfully under low pressure conditions if the partial pressure of CO_2 was maintained at 10%.

Stenvers and Bruinsma[245] showed that, with mature green tomatoes, storage at 19 cm Hg pressure at a partial pressure of oxygen of 0.20 did not delay ripening compared to similar fruit stored under normal atmospheric conditions. However, storage under reduced pressures at lower O_2 tensions did delay ripening. There was considerable accumulation of ethylene in the low pressure vessels in these experiments, but even the addition of an ethylene absorbing compound (which reduced the ethylene level significantly) did not affect the time to ripeness. Thus, with the tomato at least, the value of low pressure storage is in producing a low oxygen atmosphere, rather than in removal of ethylene. This clearly conflicts with the data for banana, and it would be interesting to know the relative importance of low O_2 and of removal of volatiles in delaying ripening during the low pressure storage of other commodities.

VI. CONCLUSION

This review illustrates the great number of questions concerning the hormonal and biochemical control of ripening that are still left unanswered. It is clear, however, that ripening can no longer be considered as a highly coordinated series of interdependent processes; rather it involves a number of relatively independent changes that normally occur at the same time. It seems probable that the accumulation of high concentrations of ethylene in the tissue plays a role in stimulating some of these processes directly, and this may be responsible for the apparent coordination of ripening under natural conditions. Ethylene also is important in the system for the initiation of ripening, but it is increasingly apparent that other factors (such as ABA and IAA) may also be important in the timing of the onset of the sequence of ripening changes.

REFERENCES

1. **Sacher, J. A.,** *Annu. Rev. Plant Physiol.*, 24, 197—224, 1973.
2. **Coombe, B. G.,** *Annu. Rev. Plant Physiol.*, 27, 207—227, 1976.
3. **Hulme, A. C.,** *The Biochemistry of Fruits and their Products,* Vol. I and II, Academic Press, London, 1970, 1971.

4. **Bollard, E. G.,** *The Biochemistry of Fruits and their Products,* Vol. 1, Academic Press, London, 1970, 387—427.
5. **Pearson, J. A., and Robertson, R. N.,** *Aust. J. Biol. Sci.,* 6, 1—20, 1953.
6. **Denne, M. P.,** *Ann. Bot. (London),* 24, 397—406, 1960.
7. **Schroeder, C. A.,** *Proc. Am. Soc. Hortic. Sci.,* 61, 103—109, 1953.
8. **Biale, J. B. and Young, R. A.,** *The Biochemistry of Fruits and their Products,* Vol. II, Academic Press, London, 1971, 1—63.
9. **Pratt, H. K. and Reid, M. S.,** *J. Sci. Food Agric.,* 25, 747—757, 1974.
10. **Coombe, B. G.,** *Acta Hortic.,* 34, 261—274, 1973.
11. **Saltveit, H. E., Jr.,** *J. Am. Soc. Hortic. Sci.,* 102, 523—525, 1977.
12. **Peynaud, E. and Ribereau-Gayon, P.,** *The Biochemistry of Fruits and their Products,* Vol. II, Academic Press, London, 1971, 171—205.
13. **Kidd, F. and West, C. A.,** Report Food Invest. Bd. for 1924, 27—32 1925.
14. **Gerber, C.,** *Ann. Sci. Nat. Bot. Biol. Veg., (Paris), Sēr. 8,* Botan 4, 1—280, 1896.
15. **Müller-Thurgau, H. and Schneider-Orelli, O.,** *Landwirtsch. Jahrb. Schweiz,* 22, 760—774, 1908.
16. **Wolf, J.,** *Qual. Plant. Mater. Veg.,* 20, 355—365, 1971.
17. **Palmer, J. K.,** *The Biochemistry of Fruits and their Products,* Vol. II, Academic Press, London, 1971, 65.
18. **Ulrich, R.,** *Facteurs et Rēgulation de la Maturation des Fruits,* Ûlrich, R., Ed., *Colloq. Int. C.N.R.S,* Collōq No. 238, Paris 1975.
19. **Reid, M. S.,** Facteurs et Rēgulation de la Maturation des Fruits, Ulrich, R., Ed., *Colloq. Int. C.N.R.S., (Paris)* no. 238, 177—182, 1975.
20. **Burg, S. P.,** *Annu. Rev. Plant Physiol.,* 13, 265—302, 1962.
21. **Pratt, H. K. and Goeschl, J. D.,** *Annu. Rev. Plant Physiol.,* 20, 54—584, 1969.
22. **McGlasson, W. B.,** *The Biochemistry of Fruits and their Products,* Vol. I, Academic Press, London, 1970, 475—519.
23. **Gane, R.,** *Nature (London),* 134, 1008, 1934.
24. **Burg, S. P. and Thimann, K. V.,** *Proc. Natl. Acad. Sci. U.S.A.,* 45, 335—344, 1959.
25. **Meigh, D. F.,** *J. Sci. Food Agric.,* 11, 381—385, 1960.
26. **Burg, S. A. and Burg, E. A.,** *Science,* 148, 1190—1196, 1965.
27. **Kidd, F. and West, C. A.,** *Plant Physiol.,* 20, 467—504, 1945.
28. **McMurchie, E. J., McGlasson, W. B., and Eaks, I. L,.** *Nature (London),* 237, 235—236, 1972.
29. **Rhodes, M. J. C.,** *The Biochemistry of Fruits and their Products,* Vol. I, Academic Press, London, 1970, 521—533.
30. **Biale, J. B.,** *Handbuch der Pflanzenphysiologie,* Vol. 12, Springer-Verlag, Berlin, 1960, 536—592.
31. **Goldschmidt, E. E.,** in *Plant Growth Substances 1973,* Hirokawa Publishing Co., Tokyo, 1974, 1027—1033.
32. **Biale, J. B. and Young, R. E.,** *Am. J. Bot.,* 34, 301—309, 1947.
33. **Goldschmidt, E. E., Aharoni, Y., Eilati, S. K., Riov, J. M., and Monselius, S. P.,** *Plant Physiol.,* 59, 193—195, 1977.
34. **Burg, S. P. and Burg, E. A.,** *Plant Physiol.,* 37, 179—189, 1962.
35. **Reid, M. S. and Pratt, H. K.,** *Nature (London),* 226, 976—977, 1970.
36. **Apelbaum, A., Goldschmidt, E. E., and Ben Yehoshua, S.,** *Plant Physiol.,* 57, 836—838, 1976.
37. **Burg, S. P.,** in *Post-Harvest Biology and Handling of Fruit and Vegetables,* Haard, N. F. and Salunkhe, D. K., Eds., Avi Publishing Co., Westport, Conn., 1975, 172—188.
38. **Cooper, W. C., Rasmussen, G. K., Rogers, B. J., Preece, P. C., and Henry, W. H.,** *Plant Physiol.,* 43, 1560—1576, 1968.
39. **Maxie, E. C., Eaks, I. L., Somner, N. F., Rae, H. L., and El-Batal, S.,** *Plant Physiol.,* 40, 407—409, 1965.
40. **Aharoni, Y.,** *Plant Physiol.,* 43, 99—102, 1968.
41. **Eaks, I. L.,** *Plant Physiol.,* 45, 334—338, 1970.
42. **Pratt, H. K. and Goeschl, J. D.,** *Biochemistry and Physiology of Plant Growth Substances,* Wightman, F. and Setterfield, G., Eds., The Runge Press, Ottawa, Canada, 1968, 1295—1302.
43. **Marei, N. and Crane, J. C.,** *Plant Physiol.,* 48, 249—254, 1971.
44. **Maxie, E. C. and Crane, J. C.,** *Proc. Am. Soc. Hortic. Sci.,* 92, 255—267, 1968.
45. **Puech, A. A., Rebeiz, C. A., and Crane, J. C.,** *Plant Physiol.,* 57, 504—509, 1976.
46. **Saulnier-Blache, P. and Bruzeau, F.,** *Ann. Physiol. Veg.,* 9, 179—196, 1967.
47. **Hale, C. R., Coombe, B. G., and Hawker, J. S.,** *Plant Physiol.,* 45, 620—623, 1970.
48. **Coombe, B. G. and Hales, C. R.,** *Plant Physiol.,* 51, 629—634, 1973.
49. **Blanpied, G. D.,** *Plant Physiol.,* 49, 627—630, 1972.
50. **Pollack, R. L., Hoban, N., and Hills, C. H.,** *Proc. Am. Soc. Hortic. Sci.,* 78, 86—95, 1971.

51. **Hartmann, C.,** *C. R. Acad. Sci. Paris,* 273, 1570—1572, 1971.
52. **Romani, R. J. and Jennings, W. G.,** *The Biochemistry of Fruits and their Products,* Vol. II, Academic Press, London, 1971, 411—436.
53. **Lockwood, D. and Vines, H. M.,** *J. Am. Soc. Hortic. Sci.,* 97, 192—197, 1972.
54. **Knavel, D. E. and Kemp, T. R.,** *HortScience,* 8, 403—404, 1973.
55. **Wright, H. B. and Heatherbell, D. A.,** *N. Z. J. Agric. Res.,* 10, 405—414, 1967.
56. **Knee, M., Sargent, J. A., and Osborne, D. J.,** *J. Exp. Bot.,* 28, 377—396, 1977.
57. **Hoad, V., Andersson, H. M., Guttridge, C. G., and Sparks, T. R.,** Long Ashton Res. Stn. Ann. Rpt. 1970, 33—34.
58. **Abeles, F.,** *Ethylene in Plant Biology,* Academic Press, New York, 1973.
59. **Windus, N. D., Shutak, V. G., and Gough, R. E.,** *HortScience,* 11, 515—517, 1976.
60. **Lipe, J. A.,** *J. Am. Soc. Hortic. Sci.,* 103, 76—77, 1978.
61. **Robinson, R. W. and Tomes, M. L.,** Tomato Genetics Co-op Rpt. 18, 36—37, 1968.
62. **Buescher, R. W., Sistrunk, W. A., Tigchelaar, E. C., and Ng, T. J.,** *HortScience,* 11, 603—604, 1976.
63. **Herner, R. C. and Sink, K. C., Jr.,** *Plant Physiol.,* 52, 38—42, 1973.
64. **Mizrahi, Y., Dostal, H. C., and Cherry, J. H.,** *HortScience,* 10, 414—415, 1975.
65. **Frenkel, C. and Garrison, S. A.,** *HortScience,* 11, 20—21, 1976.
66. **Buescher, R. W.,** *HortScience,* 12, 315—316, 1977.
67. **Biale, J. B.,** *Post-harvest Biology and Handling of Fruit and Vegetables,* Haard, N. F. and Salunkhe, D. K., Eds., Avi Publishing Co., Westport, Conn., 1975, 5—18.
68. **Winton, A. L. and Winton, K. B.,** *The Structure and Composition of Foods,* John Wiley, New York, 1935.
69. **Bradley, M. V. and Crane, J. C.,** *Am. J. Bot.,* 42, 273—281, 1955.
70. **Ito, S.,** *The Biochemistry of Fruits and their Products,* Vol. II, Academic Press, London, 1971, 281—301.
71. **Simpson, D. J., Baquar, M. R., McGlasson, W. B., and Lee, T. H.,** *Aust. J. Plant Physiol.,* 3, 575—587, 1976.
72. **Bain, J. M. and Mercer, F. V.,** *Aust. J. Biol. Sci.,* 17, 78—85, 1964.
73. **Harris, W. M. and Spurr, A. R.,** *Am. J. Bot.,* 56, 380—389, 1969.
74. **Spurr, A. R. and Harris, W. M.,** *Am. J. Bot.,* 55, 1210—1224, 1968.
75. **Rhodes, M. J. C. and Wooltorton, L. S. C.,** *Phytochemistry,* 6, 1—12, 1967.
76. **Clijsters, H.,** *Qual. Plant. Mater. Veg.,* 19, 129—140, 1969.
77. **Clijsters, H.,** *Prog. Photosynthesis Res.,* 1, 388—395, 1969.
78. **Phan, C. T.,** *Plant Cell Physiol.,* 11, 823—825, 1970.
79. **Phan, C. T.,** *Facteurs et Régulation de la Maturation des Fruits,* Ulrich, R., Ed., Colloq. Int. C.N.R.S., no. 238, 49—56, 1975.
80. **Tolbert, N. E.,** *Annu. Rev. Plant Physiol.,* 22, 45—74, 1971.
81. **Clijsters, H.,** *Facteurs et Régulation de la Maturation des Fruits,* Ulrich, R., Ed., Colloq. Int. C.N.R.S., no. 238, 1975, 41—47.
82. **Clijsters, H. and Perrin, M. A.,** *Qual. Plant. Mater. Veg.,* 23, 379—393, 1974.
83. **van Buren, J.,** *The Biochemistry of Fruits and their Products,* Vol. I, Academic Press, London, 1970, 269—304.
84. **Ribereau-Gayon, P.,** *Plant Phenolics,* Oliver and Boyd, Edinburgh, 1972.
85. **Siegelman, H. W. and Hendricks, S. B.,** *Plant Physiol.,* 33, 185—190, 1958.
86. **Downs, R. J., Siegelman, H. W., Butler, W. L., and Hendricks, S. B.,** *Nature (London),* 205, 909—910, 1965.
87. **Aoki, S., Araki, C., Kaneo, K., and Katayama, O.,** *J. Food Sci., Technol.,* (Tokyo) 17, 507—511, 1970.
88. **Faragher, J. D. and Chalmers, D. J.,** *Aust. J. Plant Physiol.,* 4, 133—141, 1977.
89. **Albersheim, P.,** *Annu. Proc. Phytochem. Soc.,* 10, 145—164, 1974.
90. **Knee, M.,** *Facteurs et Régulation de la Maturation des Fruits,* Ulrich, R., Ed. Colloq. Int. C.N.R.S., no. 238, 1975, 341—345.
91. **Dolendo, A. L., Luh, B. S., and Pratt, H. K.,** *J. Food Sci.,* 31, 332—336, 1966.
92. **Shewfelt, A. L.,** *J. Food Sci.,* 30, 573—576, 1965.
93. **Knee, M.,** *Phytochemistry,* 12, 1543—1549, 1973.
94. **McCready, R. M. and McCombe, E. A.,** *Food Res.,* 19, 530—535, 1954.
95. **Pressey, R., Hinton, D. M., and Avant, J. K.,** *J. Food Sci.,* 36, 1070—1073, 1971.
96. **Ash, A. S. F. and Reynolds, T. M.,** *Aust. J. Biol. Sci.,* 7, 435—443, 1954.
97. **Pilnik, W. and Voragen, A. G. J.,** *The Biochemistry of Fruits and their Products,* Vol. I, Academic Press, London, 1970, 53—87.
98. **Buescher, R. W. and Tigchelaar, E. C.,** *Hortic. Sci.,* 10, 624—625, 1975.

99. Pressey, R. and Avants, J. K., *Plant Physiol.*, 52, 252—256, 1973.
100. Pressey, R. and Avants, J. K., *Phytochemistry*, 15, 1349—1351, 1976.
101. Wallner, S. J. and Bloom, H. L., *Plant Physiol.*, 60, 207—210, 1977.
102. Bartley, I. M. *Phytochemistry*, 17, 213—216, 1978.
103. Hobson, G. E., *Biochem. J.*, 86, 358—365, 1963.
104. Brady, C. J., *Aust. J. Plant Physiol.*, 3, 163—172, 1976.
105. Lee, M. and Macmillan, J. D., *Biochemistry*, 7, 4005—4010, 1968.
106. Pressey, R. and Avants, J. K., *Phytochemistry*, 11, 3139—3142, 1972.
107. Bartley, I. M., *Phytochemistry*, 13, 2107—2111, 1974.
107a. Awad, M. and Young, R. E., *Plant Physiol.*, 64, 306—308, 1979.
107b. Poovaiah, B. W. and Nukaya, A., *Plant Physiol.*, 64, 534—547, 1979.
108. Karr, A. and Albersheim, P., *Plant Physiol.*, 46, 69—80, 1970.
109. Wyman, H. and Palmer, J. K., *Plant Physiol.*, 39, 630—633, 1964.
110. Buser, C. and Matile, P., *Z. Pflanzenphysiol.*, 82, 462—466, 1977.
111. Rhodes, M. J. C., Wooltorton, L. S. C., and Hulme, A. C., *Qual. Plant. Mater. Veg.*, XIX, 1—3, 167—183, 1969.
112. Nursten, H., *The Biochemistry of Fruits and their Products*, Vol. I, Academic Press, London, 1970, 239—268.
113. Flath, R., Black, D. R., Guadagni, D. G., McFadden, W. H., and Schultz, T. H., *J. Agric. Food Chem.*, 15, 29—35, 1967.
114. Drawert, F., *Facteur et Régulation de la Maturation des Fuits*, Ulrich, R., Ed., 1974, 309—320.
115. Bayonove, C., *Facteurs et Régulation de la Maturation des Fruits*, Ulrich, R., Ed., 1974, 327—339.
116. Jennings, W. R. and Stevenants, M. R., *J. Food Sci.*, 29, 796—801, 1964.
117. Galliard, T., Matthew, J. A., Fishwick, M. J., and Wright, A. J., *Phytochemistry*, 15, 1647—1650, 1976.
118. Phillips, D. R. and Galliard, T., *Phytochemistry*, 17, 355—358, 1978.
119. Galliard, T. and Matthew, J. A., *Phytochemistry*, 16, 339—343, 1977.
120. Galliard, T., Matthew, J. A., Wright, A. J., and Fishwick, M. J., *J. Sci. Food Agric.*, 28, 863—868, 1977.
121. Rhodes, M. J. C., *The Biochemistry of Plants*, Vol. 2, Davies, D. D., Ed., (Series edited by Stumpf, P. and Conn, E. E.), Academic Press, New York, in press.
122. Frenkel, C., Klein, I., and Dilley, D. R., *Plant Physiol.*, 43, 1146—1153, 1968.
123. McGlasson, W. B., Palmer, J. K., Vendrell, M., and Brady, C. J., *Aust. J. Biol. Sci.*, 24, 1103—1114, 1971.
124. Quazi, M. H. and Freebairn, H. T., *Bot. Gaz. (Chicago)*, 131, 5—14, 1970.
125. Dostal, H. C. and Leopold, A. C., *Science*, 158, 1579—1580, 1967.
126. Brady, C. J. and O'Connell, P. B. H., *Aust. J. Plant Physiol.*, 3, 301—310, 1976.
127. Wang, C. Y. and Mellenthin, W. H., *Plant Physiol.*, 50, 311—312, 1972.
128. Pratt, H. K., *The Biochemistry of Fruits and their Products*, Vol. II, Academic Press, London, 1971, 207—232.
129. Romani, R. J., *Facteurs et Régulation de la Maturation des Fruits*, Ulrich, R., Ed., 1975, 229—233.
130. Lance, C., Hobson, G. E., Young, R. E., and Biale, J. B., *Plant Physiol.*, 40, 1116-1123, 1965.
131. Hulme, A. C., Rhodes, M. J. C., and Wooltorton, L. S. C., *Phytochemistry*, 6, 1343—1351, 1967.
132. Hulme, A. C., Jones, J. D., and Wooltorton, L. S. C., *Proc. R. London Soc. Sec. B.*, 158, 514—535, 1963.
133. Solomos, T., *Annu. Rev. Plant Physiol.*, 28, 279—297, 1977.
134. Schatz, G. and Mason, T. L., *Annu. Rev. Biochem.*, 43, 51—88, 1974.
135. Leaver, C. J., *The Chemistry and Biochemistry of Plant Proteins, Phytochemical Society Symp. Series No. 11, Harborne, J. B. and van Sumere, C. F., Eds., 1975, 137—166.*
136. Yang, S. F., *Facteurs et Régulation de la Maturation des Fruits*, Ulrich, R., Ed., *Colloq. Int. C.N.R.S.*, no. 238, 1975, 171—175.
137. Mapson, L. W. and Wardale, D. A., *Phytochemistry*, 11, 1371—1387, 1972.
138. Lieberman, M., *Physiol. Vég.*, 13, 489—499, 1975.
139. Adams, D. O. and Yang, S. F., *Plant Physiol.*, 60, 892—896, 1977.
139a. Adams, D. O. and Yang, S. F., *Proc. Natl. Acad. Sci. (USA)*, 76, 170—174, 1979.
139b. Boller, T., Herner, R. C., and Kende, H., *Planta*, 145, 293—303, 1979.
139c. Yu, Y-B, Adams, D. O., and Yang, S. F., *Archives Biochem. Biophys.*, 198, 280—286, 1979.
140. Mapson, L. W. and Wardale, D. A., *Phytochemistry*, 10, 29—39, 1971.
141. Baur, A. H. and Yang, S. F., *Phytochemistry*, 11, 3207—3214, 1972.
142. Hanson, A. D. and Kende, H., *Plant Physiol.*, 57, 528—537, 1976.
143. Burg, S. P. and Burg, E. A., *Science*, 153, 314—315, 1966.
144. Lieberman, M., Kunishi, A. T., and Owens, L. D., *Facteurs et Régulation de la Maturation des Fruits*, Ulrich, R., Ed., *Colloq. Int. C.N.R.S.*, no. 238, 1974, 161—170.

145. Owens, L. D., Lieberman, M., and Kunishi, A. T., *Plant Physiol.*, 48, 1—4, 1971.
146. Wang, C. Y. and Mellenthin, W. H., *Plant Physiol.*, 59, 546—549, 1977.
147. Gazit, S. and Blumenfeld, A., *J. Am. Soc. Hortic. Sci.*, 95, 229—231, 1970.
148. Burg, S. P. and Burg, E. A., *Nature (London)*, 194, 398—399, 1962.
149. Meigh, D. F., Jones, D., and Hulme, A. C., *Phytochemistry*, 6, 1507—1515, 1967.
150. Reid, M. S., Rhodes, M. J. C., and Hulme, A. C., *J. Sci. Food Agric.*, 24, 971—979, 1973.
151. Harkett, P. J., Hulme, A. C., Rhodes, M. J. C., and Wooltorton, L. S. C., *J. Food Technol.*, 6, 39—45, 1971.
152. Wang, C. Y., Mellenthin, W. H., and Hansen, E., *J. Am. Soc. Hortic. Sci.*, 97, 9—12, 1972.
153. Peacock, B. C., *Queensl. J. Agric. Anim. Sci.*, 29, 137—145, 1972.
154. Burg, S. P. and Burg, E. A., *Physiol. Plant.*, 18, 870—884, 1965.
155. Hall, E. G., Huelin, F. E., Hackney, F. M. V., and Bain, J. M., Congr. Int. Bot. 8th Congr. Paris, Section 11/12, 1954, 405—406.
156. Ben Yehoshua, S., Robertson, R. N., and Biale, J. B., *Plant Physiol.*, 38, 194—201, 1963.
157. Zeroni, M., Galil, J., and Ben Yehoshua, S., *Plant Physiol.*, 57, 647—650, 1976.

157. Zeroni, M., Galil, J., and Ben Yehoshua, S., *Plant Physiol.*, 57, 647—650, 1976.

158. Ku, H. S., Yang, S. F., and Pratt, H. K., *Plant Cell Physiol.*, 11, 241—246, 1970.
159. Mapson, L. W. and Robinson, J. E., *J. Food Technol.*, 1, 215—225, 1966.
160. Hesselman, C. W. and Freebairn, H. T., *J. Am. Soc. Hortic. Sci.*, 94, 635—637, 1969.
161. Solomos, T. and Biale, J. B., *Facteurs et Régulation de la Maturation des Fruits*, Ulrich, R., Ed., *Colloq. Int., C. N. R. S.*, 221—228, 1975.
162. Rhodes, M. J. C. and Wooltorton, L. S. C., *Phytochemistry*, 10, 1989—1997, 1971.
163. Solomos, T. and Laties, G. G., *Biochem. Biophys. Res. Commun.*, 70, 663—671, 1976.
164. Solomos, T. and Laties, G. G., *Plant Physiol.*, 58, 47—50, 1976.
165. Forsyth, F. R., Eaves, C. A., and Lightford, H. J., *Can. J. Plant Sci.*, 49, 567—572, 1969.
166. Maxie, E. C., Catlin, P. B., and Hartmann, H. T., *Proc. Am. Soc. Hortic. Sci.*, 75, 275—291, 1960.
167. Goeschl, J. D. and Kays, S. J., *Plant Physiol.*, 55, 670—677, 1975.
168. Wang, C. Y. and Hansen, E., *J. Am. Soc. Hortic. Sci.*, 95, 314—316, 1970.
169. Knegt, EW., Kramer, S. J., and Bruinsma, J., *Facteurs et Régulation de la Maturation des Fruits*, Ulrich, R., Ed., *Colloq. Int. C.N.R.S.*, no. 238, 1974, 355—358.
170. Vendrell, M. and McGlasson, W. B., *Aust. J. Biol. Sci.*, 24, 885—895, 1971.
171. Venis, M. A., Proc. 9th Int. Conf. Plant Growth Substances, Pilet, P. E., Ed., Lausanne, 1976, 27—34.
172. Fox, J. E. and Erion, J., Proc. 9th Int. Conf. Plant Growth Substances, Pilet, P. E., Ed., 1976, 139—146.
173. Abeles, F. B., Ruth, J. H., Forrence, L. E., and Leather, G. R., *Plant Physiol.*, 49, 669—671, 1972.
174. Beyer, E., Jr., *Plant Physiol.*, 49, 672—675, 1972.
175. Beyer, E., Jr., *Plant Physiol.*, 55, 845—848, 1975.
176. Beyer, E., Jr., *Plant Physiol.*, 56, 273—278, 1975.
177. Beyer, E., Jr., *Nature (London)*, 255, 144—147, 1975.
178. Giaquinta, R. and Beyer, E., Jr., *Plant Cell Physiol.*, 18, 141—148, 1977.
179. Beyer, E., Jr., *Plant Physiol.*, 60, 203—206, 1977.
179a. Blomstrom, D. C., and Beyer, E. M., Jr., *Plant Physiol.*, 63 (Suppl.), 67, 1979.
179b. Jerie, P. H. and Hall, M. A., *Proc. Roy. Soc. (London) Ser. B*, 200, 87—94, 1978.
180. Vendrell, M., *Aust. J. Biol. Sci.*, 23, 1133—1142, 1970.
181. Kader, A. A., Morris, L. L., and Lyons, J. M., *HortScience*, 8, 264, 1973.
182. Mizrahi, Y., Dostal, H. C., McGlasson, W. B., and Cherry, J. H., *Plant Physiol.*, 56, 544—546, 1975.
183. Lieberman, M., Baker, J. E., and Sloger, M., *Plant Physiol.*, 60, 214—217, 1977.
184. Khudairi, A. K., *Am. Sci.*, 60, 696—707, 1972.
185. Gil, G. F., Martin, G. C., and Griggs, W. H., *J. Am. Soc. Hortic. Sci.*, 97, 731—735, 1972.
186. Gazit, S. and Blumenfeld, A., *Physiol. Plant.*, 27, 77—82, 1972.
187. Rudnicki, R., Machnik, J., and Pieniazek, J., *Bull. Acad. Pol. Sci. Ser. Sci. Biol.*, 16, 509—512, 1968.
188. Adato, I., Gazit, S., and Blumenfeld, A., *Aust. J. Plant Physiol.*, 3, 555—558, 1976.
189. Rudnicki, R. and Pieniazek, J., *Bull. Acad. Pol. Sci. Ser. Sci. Biol.*, 19, 421—423, 1971.
190. Goldschmidt, E. E., Goren, R., Even-Chen, Z., and Bittner, S., *Plant Physiol.*, 51, 879—882, 1973.
191. Rasmussen, G. K., *Plant Physiol.*, 56, 765—767, 1975.
192. Looney, N. E., McGlasson, W. B., and Coombe, B. G., *Aust. J. Plant Physiol.*, 1, 77—86, 1974.
193. Davison, R. M., Rudnicki, R., and Bukovac, M. J., *J. Am. Soc. Hortic. Sci.*, 101, 519—523, 1976.
194. Wang, C. Y., Wang, S. Y., and Mellethin, W. M., *J. Agric. Food Chem.*, 20, 451—453, 453, 1972.
195. McGlasson, W. B. and Adato, I., *Aust. J. Biol. Sci.*, 3, 809—817, 1976.

196. **Vendrell, M.,** *Aust. J. Biol. Sci.*, 23, 553—559, 1970.
197. **Wade, N. L. and Brady, C. J.,** *Aust. J. Biol. Sci.*, 24, 165—167, 1971.
198. **Mapson, L. W. and Hulme, A. C.,** *Prog. Phytochem.*, 2, 343—384, 1970.
199. **Frenkel, C.,** *Facteurs et Régulation de la Maturation des Fruits*, Ulrich, R., Ed., *Colloq. Int. C.N.R.S.*, no. 238, 1975, 201—209.
200. **Varga, A. and Bruinsma, J.,** *Z. Pflanzenphysiol.*, 80, 95—104, 1976.
201. **Frenkel, C.,** *Plant Physiol.*, 49, 757—763, 1972.
202. **Frenkel, C., Haddon, V. R., and Smallherr, J. M,.** *Plant Physiol.*, 56, 647—649, 1975.
203. **Hulme, A. C.,** Rep. Food Invest. Bd. London 1932, 1937, 126—128.
204. **Li, P. H. and Hansen, E.,** *Proc. Am. Soc. Hortic. Sci.*, 85, 100—111, 1964.
205. **Rowan, K. S., McGlasson, W. B., and PQRATT, H. K.,** *J. Exp. Bot.*, 20, 145—155, 1969.
206. **Wade, N. L., O'Connell, P. B. H., and Brady, C. J.,** *Phytochemistry*, 11, 975-979, 1972.
207. **Hulme, A. C., Rhodes, M. J. C., Galliard, T., and Wooltorton, L. S. C.,** *Plant Physiol.*, 43, 1154—1161, 1968.
208. **Richmond, A. and Biale, J. B.,** *Plant Physiol.*, 41, 1247—1253, 1966.
209. **Brady, C. J., Palmer, J. K., O'Connell, P. B. H., and Smillie, R. M.,** *Phytochemistry*, 9, 1037—1047, 1970.
210. **De Swardt, G. H., Swanepoel, J. H., and Duvenage, A. J.,** *Z. Pflanzenphysiol.*, 70, 358—363, 1973.
211. **Looney, N. E. and Patterson, M. E.,** *Phytochemistry*, 6, 1517—1520, 1967.
212. **Hulme, A. C., Rhodes, M. J.C., and Wooltorton, L. S. C.,** *Phytochemistry*, 10, 749—756, 1971.
213. **Richmond, A. and Biale, J. B.,** *Biochim. Biophys. Acta*, 138, 625—627, 1967.
214. **Ku, L. L. and Romani, R. J.,** *Plant Physiol.*, 45, 401—407, 1970.
215. **Romani, R. J.,** *Facteurs et Régulation de la Maturation des Fruits*, Ulrich, R., Ed., *Colloq. Int., C. N. R. S.*, no. 238, 1975, 291—298.
216. **Ellis, R. J. and MacDonald, I. R.,** *Plant Physiol.*, 46, 227—232, 1970.
217. **Skakoun, A. and Daussant, J.,** *Facteurs et Régulation de la Maturation des Fruits*, Ulrich, R., Ed., *Colloq. Int. C.N.R.S.*, no. 238, 1975, 281—289.
218. **Iki, K., Sekiguchi, K., Kurata, K., Tada, T., Nakagawa, H., Ogwa, N., and Takehana, H.,** *Phytochemistry*, 17, 311—312, 1978.
219. **Rattanapanone, N., Grierson, D., and Stein, M.,** *Phytochemistry*, 16, 629—633, 1977.
220. **Rattanapanone, N., Spiers, J., and Grierson D.,** *Phytochemistry*, 17, 1485—1486, 1978.
221. **Blackman, F. F. and Parija, P.,** *Proc. R. Soc. (London)*, 103, 412—445, 1928.
222. **Burg, S. A., Burg, E. A., and Marks, R.,** *Plant Physiol.*, 39, 185—195, 1964.
223. **Sacher, J. A.,** *Symp. Soc. Exp. Biol.*, 21, 269—303, 1967.
224. **Brady, C. J., O'Connell, P. B. H., Smydzuk, J., and Wade, N. L.,** *Aust. J. Biol. Sci.*, 23, 1143—1152, 1970.
225. **Ben Yehoshua, S.,** *Physiol. Plant*, 17, 71—80, 1964.
226. **Lewis, T. L. and Martin D.,** *Aust. J. Biol. Sci.*, 22, 1577—1580, 1969.
227. **Vickery, R. S. and Bruinsma, J.,** *J. Exp. Bot.*, 24, 1261—1270, 1973.
228. **Chalmers, D. J. and Rowan, K. S.,** *Plant Physiol.*, 48, 235—240, 1971.
229. **Salminen, S. O. and Young, R. E.,** *Nature (London)*, 247, 389, 1974.
230. **Salminen, S. O. and Young, R. E.,** *Plant Physiol.*, 55, 45—50, 1975.
231. **Issac, J. and Rhodes, M. J. C.,** unpublished data.
232. **Issac, J.,** Ph.D. thesis, University of East Anglia, Norwich, England, 1978.
233. **Barker, J. and Solomos, T.,** *Nature (London)*, 196, 189—191, 1962.
234. **Lyons, J. M.,** *Annu. Rev. Plant Physiol.*, 24, 445—466, 1973.
235. **Goodwin, T. W. and Jamikorn, M.,** *Nature (London)*, 170, 104—105, 1952.
236. **Fidler, J. C., Wilkinson, B. G., Edney, K. L., and Sharples, R. O.,** The Biology of Apple and Pear Storage, Res. Rev. Commonw. Bur. Hortic. Plant. Crops No. 3, 1973, 235 pp.
237. **Forward, D. F.,** in *Plant Physiology, A Treatise*, Steward, F. C., Ed., Vol. IVA, Academic Press, New York, 1965, 311—376.
238. **Hulme, A. C., Rhodes, M. J. C., and Wooltorton, L. S. C.,** *Phytochemistry*, 10, 1315—1323, 1971.
239. **Pratt, H. K.,** *Facteurs et Régulation de la maturation des fruits*, Ulrich, R., Ed., *Colloq. Int. C.N.R.S.*, no. 238, 1975, 153—160.
240. **Young, R. E., Romani, R. J., and Biale, J. B.,** *Plant Physiol.*, 37, 416—422, 1962.
241. **Burg, S. P.,** U.S. Patent 3,333,967, 1963.
242. **Salunkhe, D. K. and Wu, M. T.,** in *Post-harvest Biology and Handling of Fruit and Vegetables*, Haard N. F. and Salunkhe, D. K., Eds., Avi Publishing Co., Westport, Conn., 1975, 153—171.
243. **Apelbaum, A., Zauberman, G., and Fuchs, Y.,** *Hortic. Sci.*, 12, 115—117, 1977.
244. **Spalding, D. H. and Reeder, W. F.,** *Hortic. Sci.*, 11, 491—492, 1976.
245. **Stenvers, N. and Bruinsma, J.,** *Nature (London)*, 253, 532—533, 1975.

246. **Lyons, J. M., McGlasson, W. B., and Pratt, H. K.,** *Plant Physiol.*, 37, 31—36, 1962.
247. **Lyons, J. M. and Pratt, H. K.,** *Proc. Am. Soc. Hortic. Sci.*, 84, 491—500, 1964.
248. **Rhodes, M. J. C.,** Unpublished data.

Chapter 9

PIGMENT CHANGES ASSOCIATED WITH FRUIT MATURATION AND THEIR CONTROL

Eliezer E. Goldschmidt

TABLE OF CONTENTS

I. INTRODUCTION

Fruit maturation and ripening constitute a special type of senescence, in addition to the numerous types of leaf senescence classified by Simon.[1] Fruits are similar to leaves in that they begin their development as green, photosynthetic tissues, at least in their exterior layers. The similarity is maintained to some extent during maturation and senescence — disappearance of chlorophyll and accumulation of yellow-red pigments is typical of both leaves and fruits. Nevertheless, the pigment changes associated with fruit maturation are generally more intense than those occurring during leaf senescence. The structural and biochemical differentiation of chromoplasts, for example, is usually more advanced and complete in mature fruits than in senescent leaves of the same species. The same can be said with regard to anthocyanins, which reach very high concentrations in certain mature fruits.

Although color changes cannot be considered as the central events in fruit ripening, they often serve as important criteria for the visual evaluation of fruit maturity, hence their agricultural and commercial significance.

The area of pigment changes during senescence has been reviewed, mainly from a biochemical standpoint, by Chichester and Nakayama,[2] Chichester and McFeeters,[3] and by Simpson et al.[3a] Pigment changes have been discussed also in other chapters of the treatise edited by Hulme.[4] In the present review, emphasis will be laid upon the environmental and hormonal control mechanisms and not so much on the biochemistry of pigment changes. Notwithstanding the peculiarities of individual fruit species we hope to arrive at a general, unifying concept of the regulation of pigment changes in maturing fruit.

II. THE COURSE OF PIGMENT CHANGES

A. Chlorophylls

Investigators of pigment changes in senescing fruits and leaves diverted most of their efforts towards carotenoids and anthocyanins, leaving the in vivo degradation of chlorophylls least understood. Most investigators were unable to find any colored degradation products, except for diminishing amounts of chlorophyll a and b. Schanderl and Lynn[5] reported the presence of unidentified chlorophyll derivatives in extracts from ripening *Capsicum frutescens* fruits.[5] However, careful studies by Bacon[6] and Bacon and Holden[7] made it clear that "changed chlorophylls" may very well arise as artifacts from reactions with inorganic chromatographic adsorbents and other factors. Holden[8] examined senescing leaves of 30 plant species and could not detect any visible degradation products. The same was confirmed by Durand and Laval-Martin for maturing tomato fruits.[9] The seemingly inevitable conclusion was that chlorophyll undergoes rapid, probably enzymatic cleavage into small, colorless fragments.[2] This is somewhat surprising in view of the well-known breakdown of other porphyrins, such as hemoglobin, into large, pigmented intermediates. Morris and McKinney found that autumn leaves contained high levels of phytol long after the amino-N containing fragments had disappeared.[11a] Removal of the phytol by chlorophyllase had been suggested to be the primary step in chlorophyll degradation, but chlorophyllase activity often correlates better with chlorophyll synthesis than with its degradation.[3,11a] A marked rise in chlorophyllase activity has, nevertheless, been observed recently during ethylene-induced degreening of citrus fruits.[10b,11] The precise role of chlorophyllase in vivo awaits further clarification.

Chlorophyll a/b ratios have been found, with few exceptions[12] to decrease slightly during leaf senescence suggesting a somewhat faster degradation of chlorophyll a.[13]

This pattern may become meaningful if one considers the distribution of chlorophylls a and b in the various chlorophyll-protein complexes.[14] The fate of these complexes during senescence has not yet been determined.

In vivo absorption spectra taken by Goldthwaite[15] did not reveal qualitative differences between green and senescent *Rumex* leaves.

The presence of pigmented degradation products of chlorophyll has, nevertheless, been indicated recently by spectrofluorimetric techniques.[15a]

In summary, the in vivo destruction of chlorophyll is still an open, intriguing problem awaiting elucidation.

B. Carotenoids

The biochemistry of this diversified group has been comprehensively reviewed on several occasions, and the review of Goodwin and Goad[16] deals specifically with carotenoids in fruits.

Carotenoids are normal constituents of the photosynthetic apparatus and are, therefore, present in green fruit tissues prior to maturation. Maturation does not always involve an accumulation of carotenoids — many varieties of apples, pears, grapes, and olives belong to this group. The "yellowing" of mature bananas also results from preexisting carotenoids unmasked due to the disappearance of chlorophyll.[17]

On the other hand, numerous fruits accumulate large amounts of carotenoids during maturation. The ultrastructural, biochemical, and genetical aspects of this process, known as chloroplast-chromoplast transformation, have been studied in great detail in tomatoes, *Capsicum,* and *Citrus* fruits. Young fruits contain mainly foliar carotenoids like lutein and neoxanthin, which decline towards maturation simultaneously with chlorophyll.[18,19] Orange peel carotenoids were found by Eilati et al.[19] to pass through a minimum level during the "color break". A similar minimum in xanthophylls has been found during the mature-green and breaker stages of tomato.[20]

Carotenoids are usually biosynthesized during the later stages of maturation, but the complement of carotenoid pigments differs greatly from one species to the other. The situation may be regarded as relatively simple in tomato, which is normally dominated by the carotenes lycopene and β-carotene. Mature citrus fruits, on the other hand, contain about 115 different carotenoids, which is about one third of the total carotenoids known to occur in nature.[21] In addition to common carotenes and xanthophylls containing 40 carbon atoms, citrus peel contains C-30 carotenoids such as β-citraurin, which is believed to be the main contributor to the bright orange and red color of oranges and tangerines.[21] Along with the increase of total xanthophylls in orange peel, there is also an increase in their esterification — up to 60% of the xanthophylls of colored fruit were found to be esterified.[22] Chichester and Nakayama concluded that "The pathway of either breakdown or the interconversion of carotenoids cannot be, at present, developed into a single scheme which fits all plant tissue".[2] This conclusion appears to be valid even today, 13 years later.

C. Anthocyanins

Anthocyanin pigments belong to the larger family of phenolic substances, and their biochemistry has been studied in considerable detail,[23,24] particularly in grapes.[25,26]

The normal forms of anthocyanins in fruits are the β-glycosides of six principal aglycones (= anthocyanidins). Esterification of the sugar hydroxyls with organic acids yields acylated anthocyanins, which appear in significant amounts in grapes. Other phenols, such as tannins and additional flavanoids, may interact with anthocyanins in a variety of ways to yield effects of copigmentation or discoloration.[27] Phenols and their condensation products appear also as pigments in their own right in dates.[28]

The external expression of anthocyanin pigmentation also involves pH effects and complexes with metals. At low pH levels anthocyanin solutions are red, at pH levels near and above neutrality they become blue or purple. The vacuoles in which many anthocyanins are contained are believed to be at low pH levels. Many fruits having a blue color are at low pH suitable for redness and, indeed, when crushed these fruits yield red juice. Their blue appearance must therefore be attributed to metal complexes and copigmentation.[23]

The biosynthesis of anthocyanins during maturation appears to be intimately related to carbohydrate accumulation.[26,29] Sucrose can partially replace light and CO_2 in promoting anthocyanin synthesis.[29] The link between carbohydrates and anthocyanins may be via the pentose phosphate pathway, which leads through shikimate to anthocyanins.[25] Peynaud and Ribéreau-Gayon proposed that shikimic acid synthesized in leaves is translocated into the grapes, where anthocyanin synthesis occurs.[25] Cluster-grafting experiments have shown that anthocyanin synthesis is nonetheless entirely independent of the leaves of the particular variety, and is controlled by the grape.[26]

Studies with apples[30] and grapes[25] have shown that anthocyanins continue to accumulate throughout fruit maturation. A recent study with Shiraz grapes in Australia has shown, however, that maximum anthocyanin content was obtained at 20 to 30 days after veraison when Brix readings were 21 to 24°. Subsequently anthocyanin levels declined, although Brix readings continued to increase (Figure 1).[31]

III. ENVIRONMENTAL FACTORS

Much horticultural information has accumulated over the years with regard to light and temperature influences on fruit pigmentation. A critical examination of these environmental factors became possible only recently, however, with the introduction of controlled growth chambers. The number of such studies has been quite small until now, and they are still difficult to conduct with fruit trees.

A. Light

Light delays the loss of chlorophyll in detached leaf systems, but the pigment system involved in this light effect has not been determined with certainty (see Chapter 5). The loss of chlorophyll was enhanced by Red light in detached tomatoes.[32] The accumulation of lycopene in tomato could also be induced by Red light and reversed with Far Red.[33,34] β-Carotene, on the other hand, attained similar levels with Red, Far Red, and dark treatments.[34] Phytochrome appears, therefore, to be involved in the control of the biosynthetic route leading to lycopene, but not of the one leading to β-carotene.

Light is required for optimum fruit coloration. Citrus fruits bagged in black polyethylene contained low levels of chlorophyll and carotenoids in their peel,[35] probably due to dark repression of chloroplast development. The light intensities required to saturate the lycopene system in tomato are quite low.[36]

Exposure to light is necessary also for anthocyanin biosynthesis in apple,[37] grape,[26] and other fruits. Anthocyanin synthesis in apple peel discs was one of the first systems used to demonstrate the action of phytochrome — Red promotes, Far Red suppresses.[38]

Light intensity is also important. For full color development in grapes, 54% of sunlight intensity was sufficient, but further reduction caused proportional decrease in anthocyanin levels down to an absence of anthocyanins in darkened clusters.[26] The need for high light intensities points to mechanisms other than phytochrome in at least some fruits. Photosynthetic systems might be involved, since anthocyanin biosynthesis seems to be dependent upon carbohydrate availability, as it is in other tissues.[38a]

B. Temperature

Whereas fruit development and maturation require (in most cases) a relatively high range of temperatures, the changes in pigmentation often favor lower temperatures. This is certainly true for fruits attached to the tree, but much less so for fruits changing their color after harvest, in storage. The differences in optimum temperature between attached and detached fruit suggest that tree influences are superimposed over the biochemical optima for the processes involved in pigment changes.

Loss of chlorophyll in orange fruits maturing on the tree is dependent upon day temperatures below 20°C.[39,40] Citrus fruits from the tropics remain green long after internal maturation has been attained, because of the high temperatures prevailing all year round.[41]

Degreening of detached citrus fruits is performed commercially with ethylene under high humidity at 25 to 30°C.[42,42a] A recent study by Wheaton and Stewart[43] has shown, however, that at 15°C satisfactory degreening may be obtained, even without ethylene.

Numerous studies have approached, since the turn of the century, the effects of temperature on carotenoid accumulation in tomato. A temperature of 25°C seems to be optimal for the biosynthesis of lycopene. Temperatures above 30°C suppress the biosynthesis of lycopene, whereas the biosynthesis of β-carotene remains unchanged.[44]

The distinct temperature (and light, see Section 3A) requirements indicate that lycopene and β-carotene are produced via separate biochemical routes.[34,44]

The picture may be still more complex in citrus fruits that contain a large number of carotenoids. In pink grapefruits, unlike tomatoes, temperatures of 30 to 35°C favored the accumulation of lycopene, but not of carotene.[45] The reverse happened at 5 to 15°C, which suppressed lycopene and promoted carotene formation.[45] Satisfactory color was obtained with detached oranges and tangerines at 15°C, but not at 25°C, due to insufficient biosynthesis of β-citraurin. Other carotenoids such as cryptoxanthin were indifferent to storage temperatures.[46]

Accumulation of anthocyanins is favored by low temperatures; this has been demonstrated for grapes,[26] blood oranges,[45] apples,[30] and other fruits. In apple, more anthocyanins were produced in cooler temperatures (in the range of 21 to 10°C). Attached and detached fruit behaved alike; the effects on chlorophyll destruction were not so clear.[30] Notwithstanding the existence of many anthocyanin derivatives, they all seem to belong to a single biosynthetic route and are therefore subject to a common control mechanism.

IV. EFFECTS OF GROWTH REGULATORS AND OTHER CHEMICALS

Growth regulators have been used and are being used as powerful tools to improve fruit coloration. Of special interest are cases where growth regulators affect pigment changes without having the corresponding effects on other processes of maturation and ripening. It is practically impossible (and perhaps also unnecessary) to relate in detail each of the vast number of reports in this area. Instead, we shall briefly discuss the major effects and provide key references.

A. Ethylene

Since the discovery of its role as a "ripening hormone", ethylene has been used routinely for the degreening of harvested citrus and banana fruits. The introduction in the late sixties of the chemical chloroethylenephosphonic acid (CEPA), which releases ethylene upon entering plant cells, opened the possiblity of field applications of ethylene. The activity of CEPA has been examined during the past decade with numerous fruit species. CEPA treatments result almost invariably in enhancement of most ripening parameters, including color changes. In citrus fruits ethylene has almost no effect on internal pigmentation and overall fruit quality, but prolonged exposure to high doses of ethylene leads nevertheless to rapid deterioration in storage.

Ethylene accelerates the degradation of chlorophyll, as well as the biosynthesis of carotenoids[43,47,48] and anthocyanins.[49-51] Olives are an exception — CEPA delayed the appearance of anthocyanins while having the usual enhancing effect on chlorophyll degradation.[52]

B. Auxins

Considerable significance has been related in the past to the effects of certain synthetic auxins, particularly 2,4,5-trichlorophenoxypropionic and 2,4,5-trichlorophenoxyacetic acid, which improved the anthocyanin coloration of apples along with other ripening processes.[37] According to current views these effects of auxins should be attributed to ethylene, which is known to be evolved from auxin-treated tissues.[52-54]

On the other hand, auxins sometimes act as senescence delaying agents, like cytokinins.[55] The synthetic auxin 2,4-dichlorophenoxyacetic acid delays yellowing of lemons, and is being used commercially to delay aging of citrus fruits after harvest.[56] Conflicting results have been obtained with synthetic auxins in bananas, with enhancement or delay of ripening depending upon the mode of application.[57]

C. Gibberellins

The role of gibberellins as senescence delaying agents has been demonstrated more widely with fruits than with leaves, of which only 3 species have shown a clear-cut response so far (see Chapter 5). Gibberellins interfere with the degradation of chlorophyll, as well as with the biosynthesis of carotenoids and anthocyanins.[35,48,51,58-60] Pigment changes in tomato and citrus appear to be more sensitive to gibberellins than other processes of ripening, such as the climacteric or ethylene-induced rise in respiration.[48,60] Concentrations as low as 0.1 mg GA_3/ℓ effectively delay the degreening of detached citrus fruits, and when sprayed on the tree the effect lasts for several months.[62] Gibberellins also enhance the "regreening" of Valencia oranges,[63] indicating that gibberellins act as inducers of green chloroplast development and not merely as inhibitors of senescence (Figure 2).[64]

D. Cytokinins

The well-known senescence delaying effect of cytokinins (see Chapter 5) has been obtained also with fruits. Cytokinins, like gibberellins, interfere preferentially with certain senescence phenomena, mainly with those related to chloroplasts.[58,60,61,65] Olives are again an exception; cytokinins enhance the accumulation of anthocyanins in this fruit.[66]

E. ABA (Abscisic acid)

There is considerable evidence suggesting that ABA plays an important regulatory role in fruit senescence (see Chapter 8). ABA enhanced the coloration of tomato[33] and grape,[67] but it has proved difficult to obtain an effect of ABA on the coloration of citrus.[64] The reason may lie in the rapid metabolism of ABA in plant tissues. Repeated applications may be required to ensure an effect, as demonstrated previously with leaf disc systems.[68]

F. SADH (Succinic acid-2,2-dimethyl hydrazide)

The growth retarding activities of this synthetic regulator were known long before its effects on fruit coloration were recognized. SADH enhances the biosynthesis of

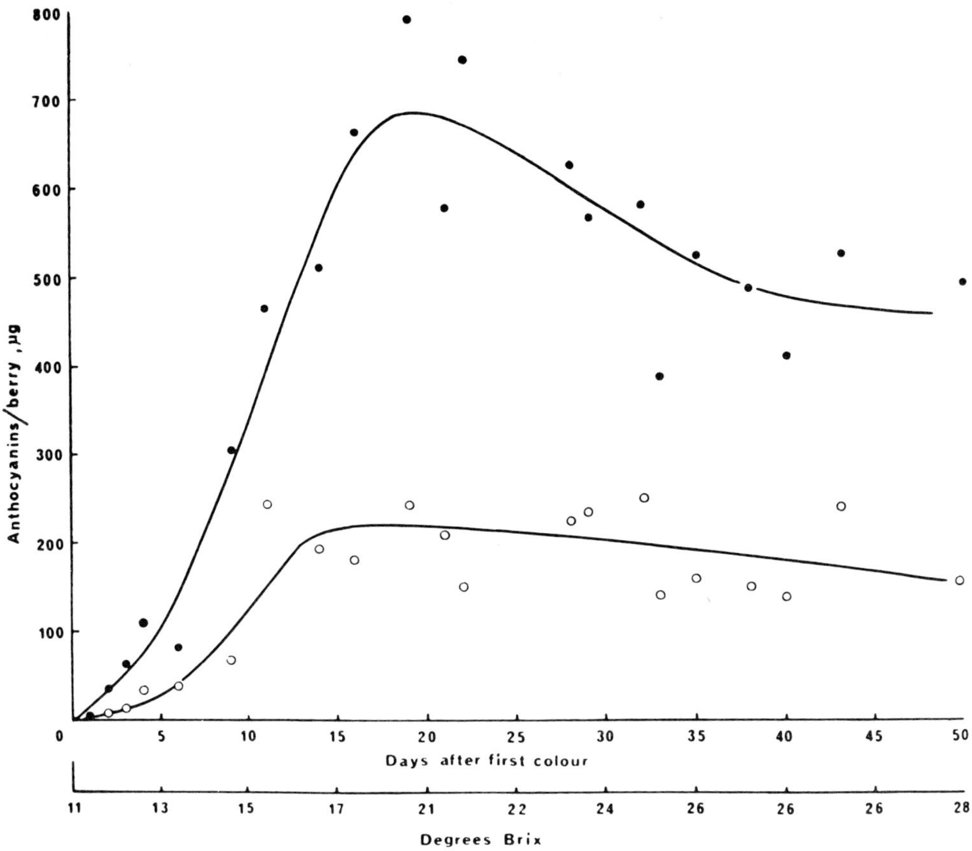

FIGURE 1. Estimates of anthocyanin contents per berry during ripening. o = acylated anthocyanidin -3-glucosides; • = anthocyanidin-3-glucosides. (From Somers, T. C., *Vitis,* 14, 269—277, 1976. With permission.)

carotenoids and anthocyanins in peach, apple, and additional Rosaceae, but its effects on other ripening parameters are rather complex. Following the discussions of Looney[53] and Faust,[54] it would appear that in stone fruits (peach, cherry) SADH increases ethylene production, thereby leading to an overall acceleration of ripening. In pome fruits (apple, pear), on the other hand, SADH reduces ethylene production and delays maturation, except for its striking promotion of reddening in apple. The effects of SADH on fruit coloration are not necessarily related to its activity as a growth retardant, the mechanism of which is also still under debate.

G. Other Chemicals

The nutritional status of trees can have an influence on fruit coloration. Excess N has been shown to have adverse effects on coloration of apples,[37] grapes,[26,69] and citrus fruits.[70]

A large number of chemicals has been tried on apples with meager success.[71]

A mention should be made here of 2-(4-chlorophenylthio)triethylamine hydrochloride (CPTA) and its derivatives. These chemicals interfere with the normal course of carotenoid biosynthesis, causing an accumulation of lycopene, even in fruits that do not normally contain this pigment.[72] CPTA and its derivatives are presently of little practical importance, but they can serve as tools for the elucidation of carotenoid biosynthetic pathways.

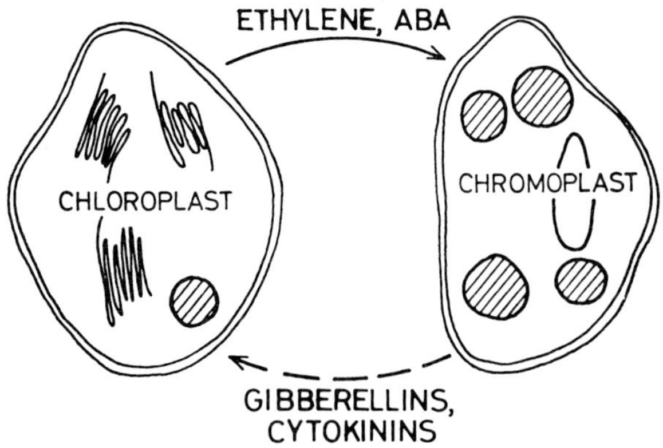

FIGURE 2. A scheme illustrating the partly reversible process of chroroplast — chromoplast transformation in citrus fruit peel, as believed to be regulated by plant growth substances. (Modified after Goldschmidt, E. E., *Plant Growth Substances 1973*, 1974, 1027—1033.)

V. THE ROLE OF ENDOGENOUS PLANT HORMONES

Much of the following will be based on conjecture, rather than on hard evidence. Nevertheless, some kind of an overall picture can be drawn out of the indirect and fragmentary evidence which is presently available.

It was recognized long ago that fruit ripening is set into motion or, at least, accelerated by the detachment from the tree. The avocado is an extreme example, since it does not complete its ripening as long as the fruit remains on the tree. It seems clear, therefore, that the tree provides the fruit with substance(s) that prevent or delay maturation and ripening.[73,74] Mapson thought of auxins as likely agents for this task.[74] Indeed, in detached avocado fruit vacuum infiltration of 1 to 10 μM 1AA delayed ripening.[74a] Auxins are not known, however, to have a general delaying effect on fruit maturation (see section IV.B). Gibberellins and cytokinins, on the other hand, delay maturation over a broad range of fruits and they may be considered as likely candidates for the role of the "tree factor". Both hormones are known to be produced by root systems and transported to the canopy via the stream of transpiration, whereby they can easily reach the fruit. Fruit maturation (and the degreening process in particular) are hampered in many fruits by conditions that favor active growth. These, in turn, can be related back to high levels of gibberellins and to strong influence of root systems. Fruit borne on young, semi-juvenile trees or on vigorously growing water sprouts often fail to mature, as do also citrus fruits in the tropics, where high temperatures permit uninterrupted root growth.

The requirement of low temperatures for optimal fruit coloration may be just the opposite side of the same coin. Suppression of root growth may decrease the supply of gibberellins and cytokinins to the fruit, thus permitting the onset of maturation and ripening.

Coming back to the fruit itself, very little is known about the fate of native auxins, gibberellins, and cytokinins, although it has been demonstrated that their levels in maturing fruits are relatively low.[75-78] The oxidative turnover of auxins has been related to the onset of ripening in pear.[79] The level of gibberellins decreased rapidly in the peel of ethylene-treated orange fruit,[80] but nothing is known so far about the mode of degradation. Even less is known about cytokinins. In olives, where cytokinins induced blackening, it was found indeed that black fruit contained a higher level of cytokinins than green fruit.[81]

There is much more information available on ethylene and ABA. Although it seems difficult to tell which comes first, both seem to accumulate within the fruit towards maturation, suggesting them as participants, if not as primary inducers of fruit maturation. (See Chapter 8 for detailed discussion of the roles of ethylene and ABA).

While considering the simultaneous decrease in growth promotors (auxins, gibberellins, cytokinins), and increase in inhibitors (ethylene, ABA) during maturation, the concept of the hormonal balance creeps back into the mind. Quantitative antagonism between exogenous ethylene and gibberellins or cytokinins in the control of citrus degreening has indeed been demonstrated.[61] The idea of a hormonal balance contributes little, however, to real understanding, since our knowledge about specific sites of action and intracellular compartmentation of plant hormones is still in its infancy.

VI. CONCLUDING REMARKS

Pigment changes appear to be controlled separately, in certain ways, from other processes of maturation and ripening (sugar accumulation, softening, respiratory behavior). This is evident from the beneficial effect of cool temperatures on fruit coloration and from the inhibitory effects of gibberellins and cytokinins, while other processes are affected to a lesser extent.

Notwithstanding the great diversity in fruit morphology and biochemistry, there is a striking similarity in the hormonal control of pigment changes. Pigment changes occurring within plastids (chlorophyll, carotenoids) and in the vacuole (anthocyanins) are induced alike by ethylene and arrested alike by gibberellins and cytokinins. Pigment changes seem therefore to be deeply embedded in the genetic programming of fruit senescence.

Pigment changes of maturing fruits are certainly related to patterns of leaf senescence. Nevertheless, leaves and fruits developing simultaneously on the same plant do not senesce at the same time. Fruits seem to obey a set of environmental and internal signals that dictate their maturation, while the close-by leaves may still continue for a long time with normal photosynthetic activity. Ethylene and ABA, which accumulate in high concentrations within fruits approaching maturity, may very well counteract the tree factors and determine the rate and timing of fruit senescence.

REFERENCES

1. **Simon, E. W.,** *Symp. Soc. Exp. Biol.,* 21, 215—230, 1967.
2. **Chichester, C. O. and Nakayama, T. O. M.,** in *Chemistry and Biochemistry of Plant Pigments,* Goodwin, T. W., Ed., Academic Press, New York, 1965, 439—457.
3. **Chichester, C. O. and McFeeters, R.,** *The Biochemistry of Fruits and their Products,* Vol. 2, Academic Press, New York, 1971, 707—719.

3a. **Simpson, K. L., Lee, T. C., Rodriguez, D. B., and Chicester, C. O.,** *Chemistry and Biochemistry of Plant Pigments,* Goodwin, T. W., Ed., 2nd ed., Academic Press, New York, 1976, 780-842.

4. **Hulme, A. C., Ed.,** *The Biochemistry of Fruits and their Products,* Vols. 1 and 2, Academic Press, New York, 1970—1971.
5. **Schanderl, S. H. and Lynn, D. Y. C.,** *J. Food Sci.,* 31, 141—145, 1966.
6. **Bacon, M. F.,** *Biochem. J.,* 101, 34—36c, 1966.
7. **Bacon, M. F. and Holden, M.,** *Phytochemistry,* 6, 193—210, 1967.
8. **Holden, M.,** *Rep. Rothamsted Exp. Stn.,* 310—319, 1967.

9. Durand, H. and Laval-Martin, D., *J. Chromatogr.*, 97, 92—98, 1974.
10. Park, Y., Morris, M. M., and Mackinney, G., *J. Agric. Food Chem.*, 21, 279—281, 1973.
10a. Aljuburi, H., Huff, A., and Hsieh, M. L., *Plant Physiol.*, 63 (Suppl.), 73, 1979.
10b. Barmore, C. R., *Hortscience*, 10, 595-596, 1975.
11. Shimokawa, K., Shimada, S., and Yaeo, K., *Sci. Hortic.*, 8, 129—135, 1978.
12. Goldthwaite, J. J. and Laetsch, W. M., *Plant Physiol.*, 43, 1855—1858, 1968.
13. Wolf, F. T., *Am. J. Bot.*, 43, 714—718, 1956.
14. Thornber, J. P., *Annu. Rev. Plant Physiol.*, 26, 127—153, 1975.
15. Goldthwaite, J. J., in *Plant Growth Substances 1970*, Carr, D. J., Ed., Springer Verlag, Basel, 1972, 581—588.
15a. Bazzaz, M. B. and Rebeiz, C. A., *Biochem. Biophys. Acta*, 504, 310-323, 1978.
16. Goodwin, T. W. and Goad, L. J., *The Biochemistry of Fruits and their Products*, Vol. 1, Academic Press, New York, 1970, 305—368.
17. Palmer, J. K., *The Biochemistry of Fruits and their Products*, Vol. 2, Academic Press, New York, 1971, 65—105.
18. Laval-Martin, D., Quennemèt, J., and Monéger, R., *Phytochemistry*, 14, 2357—2362, 1975.
19. Eilati, S. K., Budowski, P., and Monselise, S. P., *J. Exp. Bot.*, 26, 624—632, 1975.
20. Rabinowitch, H. D., Budowski, P., and Kedar, N., *Planta*, 122, 91—97, 1975.
21. Stewart, I., in *Proc. Int. Soc. Citriculture*, Grierson, W., Ed., 1, 308—311, 1977.
22. Eilati, S. K., Budowski, P., and Monselise, S. P., *Plant Cell Physiol.*, 13, 741—746, 1972.
23. Van Buren, J., *The Biochemistry of Fruits and their Products*, Vol. 1, 269—304, 1970.
24. Jurd, L., in *The Chemistry of Plant Pigments*, Chichester, C. O., Ed., *Adv. Food Res. Suppl.*, 3, 123—142, 1972.
25. Peynaud, E. and Ribéreau-Gayon, P., *The Biochemistry of Fruits and their Products*, Vol. 2, Academic Press, New York, 171—205, 1970.
26. Winkler, A. J., Cook, J. A., Kliewer, W. M., and Lider, L. A., *General Viticulture*, 2nd ed., University of California Press, Berkeley, 1974, chap. 8.
27. Singleton, V. L., *The Chemistry of Plant Pigments*, Chichester, C. O., Ed., *Adv. Food Res. Suppl.*, 3, 143—191, 1972.
28. Maier, V. P. and Metzler, D. M., *J. Food Sci.*, 30, 747 —752, 1965.
29. Creasy, L. L., *Phytochemistry*, 7, 1743—1749, 1968.
30. Creasy, L. L., *Proc. Am. Soc. Hort. Sci.*, 93, 716—724, 1968.
31. Somers, T. C., *Vitis*, 14, 269—277, 1976.
32. Jen, J. J., *J. Food Sci.*, 39, 907—909, 1974.
33. Khudairi, A. K. and Arboleda, O. P., *Physiol. Plant.*, 24, 18—22, 1971.
34. Thomas, R. L. and Jen, J. J., *Plant Physiol.*, 56, 452—453, 1975.
35. Lewis, L. N. and Coggins, C. W., *Plant Cell Physiol.*, 5, 457—463, 1964.
36. Thomas, R. L. and Jen, J. J., *J. Food Sci.*, 40, 566—568, 1975.
37. Walter, T. E., *Rep. East Malling Exp. Stn.*, 70—82, 1967.
38. Siegelman, H. W. and Hendricks, S. B., *Plant Physiol.*, 33, 185—190, 1958.
38a. Thimann, K. V. and Radner, B. S., *Arch. Biochem. Biophys.*, 74, 209-223, 1958 (and literature there cited).
39. Young, L. B. and Erickson, L. C., *Proc. Am. Soc. Hort. Sci.*, 78, 197—200, 1961.
40. Young, R., Meredith, F., and Purcell, A., *J. Am. Soc. Hort. Sci.*, 94, 672—674, 1969.
41. Reuther, W. and Rio-Castano, D., Proc. 1st Int. Citrus Symp., Chapman, H. D., Ed., University of California, Riverside, 1969, p. 277—300.
42. Grierson, W. and Newhall, W. F., *Proc. Fla. State Hort. Soc.*, 66, 42—46, 1953.
42a. Cohen, E., *J. Hort. Sci.*, 53, 143-146, 1978.
43. Wheaton, T. A. and Stewart, I., *J. Am. Soc. Hort. Sci.*, 98, 337—340, 1973.
44. Tomes, M. L., *Bot. Gaz. (Chicago)*, 124, 180—185, 1963.
45. Meredith, F. I. and Young, R. H., Proc. 1st Int. Citrus Symp. Chapman, H. D., Ed., University of California, Riverside, 1969, 271—276.
46. Stewart, I. and Wheaton, T. A., *Proc. Fla. State Hort. Soc.*, 84, 264—266, 1971.
47. Boe, A. A. and Salunkhe, D. K., *Econ. Bot.*, 21, 312—319, 1967.
48. Dostal, H. C. and Leopold, A. C., *Science*, 158, 1579—1580, 1967.
49. Hale, C. R., Coombe, B. G., and Hawker, J. S., *Plant Physiol.*, 45, 620—623, 1970.
50. Devlin, R. M. and Demoranville, E., *Physiol. Plant.*, 23, 1139—1143, 1970.
51. Proebsting, E. L., Jr. and Mills, H. H., *J. Am. Soc. Hort. Sci.*, 94, 443—446, 1969.
52. Shulman, Y., Erez, A., and Lavee, S., *Sci. Hortic.*, 2, 21—27, 1974.
53. Looney, N. E., *Acta Hortic.*, 34, 397—405, 1973.
54. Faust, M., *Acta Hortic.*, 34, 407—420, 1973.
55. Osborne, D. J., *Symp. Soc. Exp. Biol.*, 21, 305—321, 1967.

56. **Coggins, C. W., Jr. and Hield, H. Z.**, in *The Citrus Industry,* Vol. 2, Reuther, W., Batchelor, L. D., and Webber, H. J., Eds., University of California, Berkeley, Press, 1968, 371—389.
57. **Vendrell, M.,** *Aust. J. Biol. Sci.,* 23, 1133—1142, 1970.
58. **Abdel-Kader, A. S., Morris, L. L., and Maxie, E. C.,** *Hort. Science,* 1, 90—91, 1966.
59. **Vendrell, M.,** *Aust. J. Biol. Sci.,* 23, 553—559, 1970.
60. **Abdel-Gawad, H., and Romani, R. J.,** *Physiol. Plant.,* 32, 161—165, 1974.
61. **Goldschmidt, E. E., Aharoni, Y., Eilati, S. K., Riov, J., and Monselise, S. P.,** *Plant Physiol.,* 59, 193—195, 1977.
62. **Goldschmidt, E. E. and Eilati, S. K.,** *Bot. Gaz. (Chicago),* 131, 116—122, 1970.
63. **Coggins, C. W., Jr. and Lewis, L. N.,** *Plant Physiol.,* 37, 625—627, 1962.
64. **Goldschmidt, E. E.,** in *Plant Growth Substances 1973,* Hirokawa Publishing Co., Tokyo, 1974, 1027—1033.
65. **Wade, N. L. and Brady, C. J.,** *Aust. J. Biol. Sci.,* 24, 165—167, 1971.
66. **Shulman, Y. and Lavee, S.,** *J. Exp. Bot.,* 24, 655—661, 1973.
67. **Coombe, B. G. and Hale, C. R.,** *Plant Physiol.,* 51, 629—634, 1973.
68. **Manos, P. J. and Goldthwaite, J. J.,** *Plant Physiol.,* 55, 192—198, 1975.
69. **Kliewer, W. M.,** *Am. J. Enol. Vitic.,* 28, 96—103, 1977.
70. **Chapman, H. D.,** *The Citrus Industry,* Vol. 2, Reuther, W. Batchelor, L. D., and Webber, H. J., Eds., University of California Press, Berkeley, 1968, 127—209.
71. **Smock, R. M.,** *J. Am. Soc. Hort. Sci.,* 94, 49—51, 1969.
72. **Yokoyama, H., Coggins, C. W., Jr., and Henning, G. L.,** *Phytochemistry,* 10, 1831—1834, 1971.
73. **Burg, S. P. and Burg, E. A.,** *Science,* 148, 1190—1196, 1965.
74. **Mapson, L. W.,** *Endeavour,* 106, 29—33, 1970.
74a. **Tingwa, P. O. and Young, R. E.,** *Plant Physiol.,* 55, 937-940, 1975.
75. **Nitsch, J. P.,** *The Biochemistry of Fruits and their Products,* Vol. 1, 1970, 427—472.
76. **Jackson, D. I.,** *Aust. J. Biol. Sci.,* 21, 209—215, 1968.
77. **Gazit, S. and Blumenfeld, A.,** *Plant Physiol.,* 46, 334—336, 1970.
78. **Erner, Y., Goren, R., and Monselise, S. P.,** *J. Hort. Sci.,* 51, 367—374, 1976.
79. **Frenkel, C.,** *Plant Physiol.,* 55, 480—484, 1975.
80. **Goldschmidt, E. E. and Galily, D.,** *Plant Cell Physiol.,* 15, 485—491, 1974.
81. **Shulman, Y. and Lavee, S.,** *Plant Physiol.,* 57, 490—492, 1976.

Chapter 10

SENESCENCE IN THE WHOLE PLANT

Larry D. Noodén

TABLE OF CONTENTS

I. INTRODUCTION

During the course of their lives, all multicellular organisms and their organs and tissues reach a peak in terms of their physiological function, and then they decline until they die. This process of decline leading to death has been termed senescence, and a distinction has been made between degenerative changes that lead to death (senescence), and those that do not necessarily cause death even though they accumulate with age (aging).[194,225,259] At the present time, our understanding of the biochemistry of these processes is not adequate to define senescence more precisely, or even to say which degenerative changes are senescence and which are aging. It is, however, implied that senescence may involve deliberate or programmed changes that cause death. Although this chapter is concerned more with whole plants than with organs or cells, some consideration will be given to the degeneration of plant parts, because they can tell us much about the way whole plants senesce.

In looking at the biochemical changes that seem to be part of the senescence process in cells and organs, one notices a considerable amount of pattern and sequence, not a chaotic breakdown. At the level of the whole plant, these changes take on a deliberateness and look like an orderly retreat. This is particularly true where the plant's productive capacity and resources are diverted into seed production. Among these biochemical or structural changes, investigators must select criteria for measuring senescence, even though the exact mechanism (or the key process) is yet uncertain. Losses of chlorophyll and protein both seem to be key processes, but they should not be the only acceptable measures. In fact, it may be foolish to allow only one parameter to the exclusion of others; senescence is almost certainly a complex of many sequential and parallel steps. Cessation of vegetative growth, particularly stem elongation or senescence of the apical meristem, and the transition from juvenile to adult phase, though interesting phenomena, may not be the best measures of whole plant senescence (see Section III A). Similarly, photosynthetic rate, especially in attached leaves, may have some limitations as a measure of senescence, because it is determined by sink demand and may decrease to the basal metabolic rate required to maintain life as the vegetative and reproductive growth sinks decline (see Sections IV and V). Likewise, it does not seem necessary to invoke only irreversible processes as components of senescence; reversible steps or steps occurring before the point of no return may be very important components of the senescence process.

As we examine whole plant senescence, the complexity quickly becomes apparent; it can be seen that different things are happening in different parts of the plant, and they influence each other. Thus our review starts with a consideration of the whole plant and how different parts are involved. From here, we turn to the decline in growth, thence to various specific aspects of physiological degeneration, then to allocation of critical nutrient resources, and finally to hormonal controls.

II. SENESCENCE AS A WHOLE PLANT PHENOMENON

A. Monocarpy vs. Polycarpy

In considering the senescence of whole plants, most species can be classified into one of two general groups: (1) plants that have one reproductive phase (flowering, fruiting, or sporulation) followed by death, and (2) plants that normally have more than one reproductive phase before death. These have been termed monocarpic* and polycarpic, respectively.[147,157] An attempt[59] has been made to introduce the terms semelparous and interoparous to designate single and repeated reproductive phases, respectively, but it seems unnecessary to displace the older and better established terminology. In either case, these terms refer primarily to reproductive patterns and not senescence. Extensive listings of examples are compiled or discussed elsewhere.[136,147,235,390] Several generalizations can be made regarding the occurrence of monocarpy. First, monocarpy occurs in a wide range of plant taxa and evidently has evolved independently numeroas times. Secondly, while many annuals and biennials are monocarpic, some perennials (e.g., bamboo, century plant) are also monocarpic. Thirdly, monocarpy is very common in cultivated plants, especially field crops, where the abbreviated life span and complete desiccation prior to harvest and/or synchronous production have apparently been selected.[325] Monocarpic field crops in fact put on the world's most dramatic displays of senescence, for vast fields of individuals undergo senescence and die essentially simultaneously. It is convenient to call this degeneration following the reproductive phase, monocarpic senescence.

B. Internal vs. External Determinants of Longevity

The very pronounced and relatively brief senescence phase in monocarpic plants (curve C in Figure 2 of Chapter 1) distinguishes them from polycarpic species and suggests an internal (genetic) programming of the degenerative sequence. As implied in the definition, monocarpic senescence is related to flower and/or fruit development, and since the bulk of the information available on the physiology of whole plant senescence relates to monocarpy, this will be emphasized almost to the exclusion of polycarpic plants. Although externally-imposed disasters (disease, predation) may play a greater role in terminating the lives of polycarpic plants, longevity is still clearly determined by the genetic constitution of the organism.[235] Whether the longevity-determining genes control a senescence program or simply the susceptibility to (or ability to withstand) life-terminating disasters remains uncertain. Although the early loss of young polycarpic plants may be due largely to accidents as in curve A, the final attrition of established polycarpic plants in their natural environment is not as sharp as that of monocarpic plants and may be closer to curve B.[136,390] In the past, the death of monocarpic annuals has been attributed to adverse environmental factors, e.g., drought, cold; however, careful observations show that many of these plants senesce before the adverse conditions arrive. In this connection, it is of interest to note that *Viola tricolor* and a number of other species die following flowering, but they can overwinter in the vegetative state if they have not flowered.[136,390] This reinforces the idea that they die mainly as a result of flowering instead of adverse weather.

Setting aside such life-terminating disasters as predation and disease, most speculations on external controls of degeneration with age are variations of the "wear and tear" theory, which holds that senescence consists of an accumulation of environmentally-induced lesions, metabolic, genetic, structural, etc., and these finally reach a le-

* Monocarpic (derived from the Greek words for one fruit) should not be confused with monocarpellate meaning one carpel.

thal level or greatly increase susceptibility to accidents. Although some effort has been made to understand these processes and their significance in animals (see for example Rosenfeld[312]), very little has been done on this aspect of plant senescence (or aging), except in seeds (see Chapters 2 and 3). Yet plants provide some interesting data that bear on the general question of the significance of "wear and tear". First, it does not necessarily follow that suboptimal or even stressful environments (which would be expected to cause more "wear and tear") decrease longevity. For example, bristlecone pines and other conifers may attain their greatest age in the harsher environments.[187,324] In addition, both lowered temperature (20°C) and nitrogen deficiency prolong the life of daughter fronds of the duckweed, *Lemna minor.*[391,392] Secondly, one popular variant of the "wear and tear" theory invokes a gradual, time-dependent accumulation of genetic lesions in somatic tissues as a cause of death of the whole organism. Though there is a basis for this view in seeds (Chapter 2), for whole plants an argument against the idea is the remarkable stability of certain clonal lines of woody plants, e.g., winter Pearmain apple, 700 years, and Reine Claude plum, about 400 years.[398] For these and numerous other varieties, parts of individuals have been excised and propagated vegetatively to form new individuals which are notably vigorous and outlive the donor, but possess the special characteristics of that variety. It should be noted, however, that a decline in vigor does occur in some clones, e.g., sugar cane,[385] but in at least some of these cases, the decline is due to viruses. Thirdly, it is well known that the seeds of many species, especially annuals, far outlast their growing siblings if they are kept from germinating (in the soil or in storage).[390] Again, senescence of the whole organism is not merely time-dependent but is tied to an internal developmental program.

C. Is Senescence a Cellular or an Organismic Phenomenon?

Does a cell undergo senescence as an individual or as a part of a larger unit that is senescing? As mentioned above, a part of an individual can be excised and induced to form a new individual, which may flourish long after the donor plant has died. Likewise, the life of individual organs, e.g., leaves[235,390] or roots[390] can be prolonged by excision from a plant if they are maintained in the right conditions. In fact, roots[367,401] or callus tissue[106,402] subcultured on the right nutrient media seem to be capable of indefinite growth; however, for roots, this seems to require successive replacements of the terminal apex with subterminal apices.[367] Thus, it seems that whether or not an individual cell senesces depends more on its position within an organism than on factors intrinsic to that cell. This does not, however, deny the importance of the cellular and biochemical events in senescing cells. These changes play an integral role, but, when the cell is still in the plant, they are initiated by the cells' surroundings.

D. Senescence as a Correlative Phenomenon

The limitation of a cell's life as a result of its membership in a larger, organized community of cells indicates that correlative influences play an important role. The best examples, however, derive from monocarpic plants where the flowers or fruit seem to direct the senescence of the individual bearing them (a curious phenomenon of destruction or matricide). As in the case of clonal stability, horticulturists during a period exceeding 150 years[235,309] have unwittingly provided enormous documentation by demonstrating that deflowering or defruiting may prolong the life of many monocarpic plants.[147,235] There are, however, some exceptions: in cocklebur[183] defloration has been reported to have no retarding effect on senescence of the whole plant and defruiting accelerates senescence of corn[6] and pepper.[118,119] In pea,[202] red kidney bean,[393] and soybean,[199] the developing seed (not the rest of the fruit) causes monocarpic senescence. It should be noted that anything that interferes with fruit development, for example, male sterility in soybeans,[409] disease, plant hormones,[202] etc. may delay or even prevent monocarpic senescence, but these are not direct effects on monocarpic senescence. Dioecious species such as hemp and spinach, which produce male plants with no seeds, can be monocarpic.[195,235,240] Sinçe defloration of male spinach plants delays whole plant senescence, developing seeds are not the only possible source of the senescence signal. Rather than viewing these differences as discrepancies, they should be considered as evidence that not only the sources of the senescence signal, but the causes of monocarpic senescence may differ. Inasmuch as monocarpy must have evolved numerous times, it should not be surprising that different mechanisms might be involved. In fact, monocarpy may have evolved through selection in a particular crop on more than one occasion,[325] thereby opening the possibility that some differences exist in the mechanisms even within a species.

In soybean and possibly also in some cucumbers and common beans, the senescence signal seems to be generated late in fruit development, after most of their growth has occurred.[261] Moreover, the senescence signal of soybean shows limited mobility, which is primarily downward, if at all.[199,261] This differs somewhat from the pea, where the influence of the fruit has been reported to move both upward and downward from the fruit, but apparently not from one major branch to another.[208,209]

Other types of correlative influences exist; one of particular importance in whole plants involves the roots. Treatments such as derooting,[42] as well as moisture stress,[332] and salinity,[26] which affect the health of the roots, also cause foliar senescence. On the other hand, induction of roots on leaf cuttings has long been known to prolong their life span.[53,235] Both lines of evidence point to the crucial role of the roots in maintaining the whole plant. Still another correlation involves the shoot apex; removal of the shoot apex or the whole shoot above a set of leaves may delay or prevent senescence of the remaining leaves.[184,235] Numerous other correlations control senescence of individual organs. To cite only a few, removing the head in barley may accelerates senescence of the flag leaf that subtends it,[211] and pollination of a flower may prevent senescence of the ovary, while it accelerates petal senescence.[99,235]

III. IS THE LOSS OF GENERATIVE CAPACITY A COMPONENT OF WHOLE PLANT SENESCENCE?

A. Vegetative Growth

Whole plant senescence is accompanied or preceded by a decrease or cessation of both shoot and root growth. Given the correlative influence of the roots on foliar senescence (Section II. D), it is of particular interest that root growth may slow or cease. The reduction in vegetative growth that occurs during the reproductive phase has been recognized for a long time in both monocarpic and polycarpic plants, and is generally attributed to a shift in distribution of growth-limiting nutrients within the plant.[191,205,293,390] Thus it is of importance to determine when the reduction in root or shoot growth occurs relative to flowering, fruit development, and senescence. Where the shoot apex is converted to an inflorescence (e.g., corn and cocklebur), vegetative node production ceases coincident with the appearance of flower buds, though stem elongation may continue. In other species (e.g., pea and cotton), where the apical meristem of the main shoot does not become an inflorescence, it may stop growing and senesce during their reproductive phase. In soybean,[33,198,257] cotton,[65] and pea,[202] stem elongation decreases but continues after flowering has started and stops while the fruit are growing. Using soybean, Bernard[27] has begun to dissect the genetic control of determinate stem growth and has two separate loci that alter determinateness. In the scarlet pimpernel, *Anagallis arvensis,* short days prevent flowering, and the plant produces lateral branches that run along the ground and establish their own root systems, but long days induce flowering, cessation of vegetative growth (including rooting), and senescence of the whole plant.[384] Flowering and vegetative growth are not always antagonistic, for under long days (16 hr), the shoot apex of the morning glory, *Ipomoea caerulea*, continues vegetative growth and flowers are produced from lateral buds, but short days cause the terminal bud to become a flower.[12] Similarly, Marx[214] has found a variety of pea (G2) that continues to grow and produce fruit under short days, whereas these plants terminate growth and die under long days. It is important to know if the rate of fruit development and nutrient demand (e.g., dry weight or nitrogen accumulation in seeds) is the same in senescing and non-senescing plants.

Just as the shoot apical meristems of flowering plants become less tive in supporting stem elongation, they produce fewer new leaves in pole bean,[116] cucumber,[67] gherkin,[122] and cowpea.[267] In the past, the change in morphology of the leaves produced during maturation in many species was taken as an index of senescence; however, in *I. caerulea,* senescence and death may be induced in the juvenile phase by short days (8 hr) without formation of the adult (lobed) leaf morphology.[12] Axillary buds may senesce and lose their capacity for growth if repressed through apical dominance for too long.[266,317]

In soybean, root growth has been reported to terminate at about the same time as shoot growth[231,369] or a bit sooner.[33,190] Cotton root growth also declines during the reproductive phase, apparently ahead of shoot growth.[35] The roots of radish cease growth at flowering.[255] The roots of common perilla, *Perilla frutescens,* continue to grow after node production ceases, but start to slow about the time when the first flowers open.[21] In rice[154] and wheat[11,37] the roots may stop growing later, during heading, and degenerate even later during grain development. Likewise, corn roots grow little after seed growth has started.[103] A plentiful supply of water and nutrients may, however, sustain wheat roots.[49] Pea nodules reach a maximum mass at flowering,[286] and substantial losses may occur at fruiting.[286,322] Soybean nodules also senesce during fruit development.[131,132]

Thus (at least in some cases) root growth may be affected early in reproductive development, apparently too early to be due to a nutrient deficiency resulting from massive diversion of the nutrient supplies to developing fruit. Another discrepancy in the nutrient diversion explanation is the stimulation of gherkin[122] and cucumber[75] stem elongation during the early stages of reproductive development (flowering). No doubt, fruit development influences the allocation of nutrients (see Section IV) and through this, alters growth; however, the possibility that vegetative growth may be controlled by more than nutrient supply has been under consideration for some time.[191] It remains to be determined how this control is exerted and what the control centers are.

The prime candidates for control centers for growth cessation are the reproductive structures. Indeed, the nutrient diversion explanation of the cessation of growth carries with it the prediction that removal of the reproductive sinks should restore vegetative growth, but this is only partially true. In the most clear-cut cases, Vienna wallflower[309] and mignonette,[235] simple defloration prevents growth cessation, and the plants become very large. At the other extreme, it has been reported that defloration does not restore vegetative growth or even prevent senescence in cocklebur.[183] In cotton,[66,83,84,215] okra,[138] tomato,[250] corn,[204] tobacco,[296] pea,[85,86,202,208-210,287] common perilla,[21] pepper,[118] and others (including some grasses that are used in lawns),[235] defloration or defruiting may prolong or promote vegetative growth. Defruiting[198] or even deflowering[257] only slightly promotes or prolongs stem elongation in soybean, although the treated plants have thicker, greener leaves and a more robust appearance (even when compared with untreated plants at their peak). On the other hand, flower removal may temporarily depress elongation of cucumber vine.[75]

The cessation of stem elongation[85,202] or decrease in size and loss of green color of the stem apex[70] have been used as measures of senescence and some interesting probes of their control have been made. It has long been known that grafting of shoots from older to younger woody plants can prolong the life of the shoot, and interspecific grafts can prolong (or shorten) the life of the shoot depending on the combination.[235] Stem apices from cotton plants could be induced to resume growth by grafting on to young plants[215] however, Lockhart and Gottschall[202] reported that the functional life of pea apices could not be extended by grafting to younger plants, and concluded that the factor responsible for apex senescence was not readily transmissible. Unfortunately, such experiments may be complicated by the diminished ability of older tissues to take in grafts.[390] In contrast, Malik and Berrie[209] report that this factor is transmissible. They also suggest that it may also affect the ability of grafts to join. Recent experiments with a special genetic line (G2) of pea with delayed apex senescence in short days indicate that G2 plants grown in short days produce a factor that can be transmitted across grafts to delay apex senescence of a photoperiod-insensitive line (I) under short days.[70,302] In any case, the apex senescence of the G2 plants may result from a gibberellin deficiency and the graft-transmissible substance from G2 plants may be a gibberellin.[303] Whether or not gibberellins are involved in other pea varieties is uncertain.[85,202] Beever and Woolhouse[21] speculate that a diminished supply of auxin from the shoot may decrease root growth and cytokinin production, which in turn may reduce shoot growth.

Cessation of growth, including stem growth, does seem to be related to monocarpic senescence, and in some plants, this may be a convenient measure of senescence. Cessation of apex growth, however, does not cause the death of most monocarpic plants, but may merely accompany monocarpic senescence. In fact, decapitation of a plant does not cause its senescence and may even have a preserving effect on the lower leaves (Section II. C). Even in peas detipping does not promote whole plant senescence.[209] Furthermore, deflowering[257] and depodding[198] produce only a slight increase (restoration) in soybean stem elongation even though these treatments prevent monocarpic senescence, which indicates they are under different controls, at least in soybeans. Thus making a distinction between apex senescence and monocarpic senescence may prevent a lot of confusion. Apex senescence may be a more important factor in polycarpic plants (where the plant may die because it is not able to renew and maintain its life support systems, leaves, and roots). It is also conceivable that in some determinate species or varieties (e.g., determinate varieties of tomato, cotton, pepper, etc.), cessation of shoot apex growth prevents replacement of old, inefficient leaves and roots thereby eventually causing the demise of the plant. Neither time nor size seem to be primary determinants of longevity, for in soybeans of the variety Biloxi (which have an obligate short day requirement for flower induction), prevention of flowering with illumination at night causes the soybeans to live much longer (more than a year) and grow larger (more than 23 feet) than normal.[257]

B. Reproductive Growth

Many plants produce far more flowers than fruit and initiate somewhat more fruit than they carry to maturation. Considerable evidence from a wide range of monocarpic and polycarpic plants indicates that the reproductive structures themselves exert an inhibitory effect on the development of other reproductive structures and may even induce their abortion, for deflowering or depodding can offset these effects.[146,205,221,268,287,364,405] Normally, this inhibition of excess flower and fruit development does not cause whole plant senescence. Nonetheless, it seems important to determine whether or not the flowers or fruit produce these effects via the same mechanism by which they induce senescence or inhibit vegetative growth. In the case of soybeans, at least, there seem to be some differences between these inhibitory effects. For example, continuously deflowered plants produce several times more flowers than control plants (260 and 100 flowers per plant respectively), and they are produced over a much longer period in deflowered plants, whereas pod removal has much less promotive effect on flower initiation.[257] Thus soybean flower initiation is controlled mainly by the flowers, or very young pods, in contrast to senescence which is induced by the pods during their later phase of development.[261] Since pod initiation and abortion do not seem to be altered by the auxin-cytokinin treatment which prevents monocarpic senescence in soybeans,[258] it appears that the senescence signal in soybeans may not repress pod initiation or promote abortion. In the soybean, the developing reproductive structures do not seem to control the termination of stem elongation,[198,257] but the pods do induce monocarpic senescence.[199] Thus a variety of different correlative signals seem to be involved, at least in soybeans.

IV. PHYSIOLOGICAL DECLINE

A. Introduction

As noted in Section I, senescence consists of a decline in physiological functions leading to death. In many respects, this decline is the most extensively investigated aspect of senescence, yet much of this information appears as a by-product of studies with other objectives and may not be correlated with other important changes. Furthermore, most of the biochemical work deals with detached organs rather than whole plants or attached organs which may represent a more natural situation (cf. Chapters 5, 6, and 7). Nonetheless, the studies on detached organs can, when set against a zroad background of whole plant studies, add quite a lot to our understanding of whole plant senescence. Since it seems more appropriate to review the biochemistry of senescence in depth in other chapters, this coverage will not attempt to be comprehensive but will only give an overview of some of the more important areas in terms of whole plant senescence.

The critical questions concerning whole plant senescence are (1) where does the degeneration start? and (2) what are the primary or central changes? Obviously, it is extremely important to have good kinetic studies (showing not only when the decline of a process starts, but also its magnitude at various times) and to relate these to other senescence parameters. By definition, all physiological processes are going to decline during senescence, but the kinetic studies ian help to unravel the sequence and the interrelations of this multitude of changes. As the exact nature of the correlative influences (at least partially hormonal) that control senescence is worked out, this will provide tools to use in identifying key physiological changes.

Because of their conspicuousness, the changes in leaves have received the most attention. Although justification has come in bits and pieces through the years, it now appears that the decline in photosynthesis in the leaves may be a critical change. Certainly, most plants cannot sustain themselves without photosynthesis, and likewise, plant parts deprived of photosynthate could be expected to degenerate eventually. This is not to say that the supply of photosynthate is the primary factor that limits the longevity of plant parts, but it does appear to play an important role.

One interesting, yet complicating aspect of foliar senescence is the wide variety of patterns that occur. Simultaneous (or almost simultaneous) senescence of the leaves occurs not only in monocarpic senescence, but in autumnal senescence of polycarpic plants. In addition, all of the leaves may senesce together when the shoot of certain perennials dies back to its roots, bulb, rhizome etc. (top senescence[193]). In many plants, there is also a progressive senescence where the basal (older) leaves senesce as new leaves are formed at the apex. Finally, the most widely used material in studies on foliar senescence is detached leaves. Here, detachment accelerates senescence, particularly if combined with special treatments such as darkness or heat.[259,376] Since these patterns may not be exactly identical to those in whole plant (e.g., monocarpic) senescence, some care has to be exercised in using these findings with detached leaves, etc. to explain whole plant senescence. Nonetheless, the studies on detached organs can be used to supplement the information we have from whole plant studies and can point the way to areas needing study in whole plants or attached organs.

At the risk of creating confusion, it should also be kept in mind that there may be some physiological differences in the controls of senescence in different species (Section II.), and these differences should not necessarily be viewed as conflicting results. Still, there seems to be a great deal of similarity in the metabolic and structural changes in different systems, and this similarity will be emphasized here.

In leaves, and presumably other organs, many components (protein, nucleic acid, chlorophyll, etc.) or processes (photosy thesis) may reach a maximum about the time maximum size (growth) is attained; this peak may be followed by an early, slow decline and then a later, rapid decline. The exact significance of the early, slow phase is still uncertain. The early, slow phase could be a metabolic step down following cessation of growth, but there is at present no reason to believe that it is not part of the continuum of degeneration leading to death.

B. Energy Metabolism

1. Respiration

Respiration measurements, like those of photosynthesis, have been performed throughout the senescence of both intact and detached plant organs, but little work exists to correlate these changes with other aspects of whole plant senescence. In general, the respiratory apparatus of senescing tissues remains active until late in senescence, and then declines rapidly.[128,161] The mitochondria often swell and decrease in number, but these changes generally occur late in senescence.[18,44] Changes in the respiratory quotient indicate a change in respiratory substrates, especially in detached leaves where amino acids accumulate.[161,376] A respiratory rise, climacteric, may appear in the senescence of both intact[128] and detached[114,207,376] leaves but may not occur in the senescence of all plants. The magnitude of the climacteric depends largely upon whether the leaf is attached or excised from the plant; the respiratory climax of attached, senescing oat leaves is less than half of that of excised leaves, possibly due to the transport of respirable materials out of the intact leaf[377] (see Chapter 5).

The fact that a) respiration decreases only late in senescence and that b) anoxia inhibits chlorophyll and protein breakdown[212,414] indicates that cellular energy is required during senescence, presumably for the synthesis of degradative enzymes. The respiratory climacteric of senescing intact *Perilla leaves*[128] correlates with a rise in RNA levels[129] and an increase in protease activity.[165] The respiratory climacteric of senescing, detached oat leaves peaks slightly before maximum proteolytic activity.[212,376] Neither the factors contributing to the final decline of respiratory activity, nor those that maintain respiration until late in senescence have been studied much. Increased RNase activity[126] or fatty acids[16] have been presented as possible direct causes of decline, but ABA[112] and other factors may also be important. Respiration could be maintained by hormones, e.g., cytokinin, or by demand and consumption of its products, as in photosynthesis (see Section V. B. 2). Although there is evidence that senescence requires oxidative metabolism, the respiratory decline apparently does not play an important role in the induction of senescence. Loss of respiratory capacity could be a factor in the final stages of senescence.

2. Photosynthesis

Numerous studies on the rate of photosynthesis as a function of individual leaf age[29,105,236,279,328,346,415,416] have led to the generalization that the net photosynthetic rate increases as leaves grow and then begins a gradual decline at the time of maximum leaf expansion. It is important to note that successive leaves of a plant complete their growth at different times, and thus may begin their gradual decline independently.[104] This decline is accelerated during the rapid phase of senescence of attached leaves.[128,415] During this rapid phase of whole plant senescence, the rate of photosynthesis in young leaves drops as rapidly as that of older leaves; young and old leaves degenerate more or less together.[20,339]

Thus with whole plant monocarpic senescence and autumnal senescence, the rapid senescence period is largely simultaneous in all the leaves of the plant. This suggests that the rapid phase of senescence results from a higher level of control (possible hormonal) which is superimposed on all organs independently of their age.

Although decreasing chlorophyll content is commonly used as a parameter of senescence, it is not necessarily an accurate measure of photosynthetic activity or senescence. For example, chlorophyll content usually decreases well after photosynthetic activity has begun to decline,[128,327,339] and in some cases, chlorophyll levels may show a steady increase while photosynthetic activity is decreasing[142] Furthermore, senescence, as measured by protein degradation, occurs without chlorophyll loss in a mutant of fescue.[378] In the situations where chlorophyll loss occurs during senescence, the visual method[198] of scoring senescence has the advantage of being buffered from minor fluctuations in chlorophyll content, thereby measuring only the large senescence-related chlorophyll loss. As with chlorophyll content, one cannot assume that a decline in photosynthetic activity is an infallible measure of senescence. For example, the depodding of soybean plants prevents their senescence and keeps them alive long beyond their normal time,[195,198] yet neither partial nor complete depodding prevents the decline in photosynthetic activity,[236,416] which normally occurs during the reproductive phase,[236] although depodding may slow it.[416] Even though depodding does not prevent the drop in photosynthetic rate, it does seem to stop the decrease in photosynthetic capacity (e.g., chlorophyll and reproductive ribulose-diphosphate carboxylase, RUDPCase capacity[236]). The problem is that photosynthesis, both the rate and capacity, is determined to some extent by the demand for photosynthate, and defruiting and/or cessation of vegetative growth removes the major sinks, thereby decreasing photosynthesis in these plants.[107,118,252]

As discussed under protein and nucleic acid changes (Section IV. D), the synthesis and activities of chloroplast enzymes appear to decline in some plants after the cessation of leaf growth. These changes parallel the decline in leaf photosynthesis rates[128] and may be an important factor in the loss of photosynthetic activity. Those CO_2 reduction cycle enzymes that are synthesized primarily on cytoplasmic ribosomes decline later in senescence than those synthesized on chloroplast ribosomes.[19] The enzymatic activity of RUDPCase is lost at a greater rate than total Fraction I protein,[46] indicating that conformational changes and/or loss of the active site may precede degradation of the protein. This breakdown of carboxylase contributes to increased mesophyll "resistance" to CO_2, which has been shown to be a controlling factor of photosynthesis during leaf ontogeny.[13,104,418] Careful treatment of the CO_2 uptake data from aging wheat leaves and CO_2 exchange data from aging cassava plants indicate that the rate of the carboxylation reactions is the limiting factor in photosynthetic decline, with little influence from stomatal resistance or the photochemical reactions.[13,279]

Increasing photorespiration could account for some of the decrease in net photosynthesis; however, photorespiration rates do not show a universal pattern of change with age. Measurements, which vary even within species, show increases with age,[320,321] decreases,[178,243] or no significant trends.[104,280] Photorespiration may increase greatly in senescing C_4 plants.[173,406] This change correlates with changes in the relative amounts of the two major carboxylases in the C_4 plant *Portulaca oleracea,* phosphoenolpyruvate carboxylase (PEPCase) decreases to a greater extent during senescence than RUDPCase.[173] Increased photorespiration in C_4 plants may be related to a shift to C_3 type CO_2 fixation in some plants; C_3 products increase in senescing *Portulaca*[174] but not in corn.[406]

Increasing stomatal resistance with leaf age[72,351,415,416] could be another possible factor in photosynthetic decline, but this apparently is not the usual case. In aging soybean leaves, both stomatal and mesophyll resistance increase as net photosynthesis decreases, but mesophyll resistance is shown to be the most limiting to CO_2 exchange.[416] Similarly, studies of aging bean,[104] wheat,[279] and soybean leaves[73] indicate that stomatal resistance changes cannot account for the overall decline of photosynthesis.

Some limited data suggest that the photochemical reactions may show their major decline in parallel with (but following) the CO_2-fixation enzymes.[57,133,142,346] The photochemical reactions may not contribute to the overall decrease in photosynthetic rate,[27]9 so the limiting factor in the decline of photosynthesis in aging and senescing leaves appears to be the CO_2-reduction reactions.

C. Nitrogen Fixation and Mineral Uptake

Just as root growth usually slows early in the reproductive phase (Section III. A), so do many root functions. Of particular importance is the decrease in mineral uptake (Section V. C. 1) and nitrogen fixation.

Symbiotic nitrogen fixation decreases with the aging and senescence of leguminous plants.[29,130,150,189,288,304] The initial decline in nitrogen fixation may occur during early pod fill in soybeans;[150,189] peas, and beans[29,188] however, some studies with soybeans place the decline later in pod filling.[132,135,368] The reason for these differences is unclear, but they may be caused by the environmental conditions. The decline in nitrogen fixation is associated with the senescence of individual nodules and bacteroids; [130,175,368] however, the structural changes may appear after the initial decline in nitrogen fixation.[180]

Numerous studies have demonstrated that symbiotic nitrogen fixation in legumes is directly dependent upon the amount of photosynthate available to the root nodules;[29,143,304] therefore the many factors that influence whole plant photosynthesis (including environmental conditions and the developmental status of the plant) will influence N_2 fixation. Of particular interest is the fact that depodding, which increases availability of photosynthate, can cause dramatic increases in nitrogen fixation and/or delay nodule senescence.[132,189,286,304] Likewise, the decline in nitrogen fixation with senescence parallels the decline in whole plant photosynthesis[29] (see Section IV. B. 2.) A mutant of soybean that maintains RUDPCase activity and chlorophyll throughout seed development also maintains its nitrogen-fixing ability in the root nodules.[3] Thus to a large extent, the decrease in nitrogen fixation may be viewed as a result of decreased photosynthesis in the whole plant. However, the specific trigger for the senescence of individual nodules and bacteroids is unknown and could involve hormones too.

In general, mineral uptake via the roots declines during the reproductive phase, but the exact timing may differ for different minerals, e.g., Ca^{+2} uptake may continue relatively late.[98,134,191,203,265,363,366,372] The loss of nitrogen-fixing and mineral-accumulating ability of roots appears to be an important (but not primary) component of whole plant, particularly monocarpic, senescence.

D. Protein and Nucleic Acids

The decline in protein and nucleic acid levels during plant senescence[129,153,226,227,241,273,284,347,417] is perhaps the most basic of all senescence-related events in that these changes could underlie all other aspects of the plant's physiological decline. To be fully relevant to this chapter, it is necessary to relate the changes in protein and nucleic acids to other changes in the whole plant, but this is often difficult.

Numerous studies demonstrate that while protein, RNA, and DNA of senescing tissues all decline, these three fractions, along with subfractions of protein and RNA, do not decline at the same rate, DNA being last to decrease.[46,47,129,165,226,227,241,273,347,417] More recently, it has been shown that an early gradual decline in protein and RNA, which begins about the time vegetative growth ceases, is due primarily to changes in the chloroplasts of cells, while the final rapid decline involves changes in both the chloroplast and the cytoplasm (including other organelles) of the cell.[46,47,281,375,417,418] However, the proteolytic enzymes themselves are cytoplasmic in origin (see Chapters 5 and 6). The initial decline of protein and nucleic acid components in the chloroplast involves a decrease in the major chloroplast enzyme RUDPCase[46,165,168] along with other enzymes known to be synthesized at least in part on chloroplast ribosomes.[19,353] The loss of these enzymes correlates with a reduction in the number of chloroplast polysomes,[47] a decline in chloroplast RNA levels,[46,47,375] and a major reduction in the ability of the chloroplast to synthesize protein and RNA.[132,281,352,375]

The gradual decline is followed by a period of more rapid plant senescence — usually marked by the initiation of leaf yellowing (chlorosis), and this involves many new changes in protein and nucleic acid metabolism. During the rapid decline, there is accelerated protein and nucleic acid degradation in both the cytoplasm and the chloroplast, along with the decrease in synthetic capacity in the whole cell.[19,47,129,165,413,419] In several plants, large quantities of hydrolytic enzymes appear to be synthesized on cytoplasmic ribosomes during the beginning stages of this accelerated decline.[17,74,80,94-96,126,217,411] (cf. Chapter 6). Release or activation of hydrolases compartmentalized in the vacuole and/or lysosomes (Section IV. E), a possible decrease in protease inhibitors,[313] and/or a decrease in protein and nucleic acid synthesis may contribute to the decrease in protein and nucleic acids. At the whole plant level, this increased degradation corresponds with the increased export of materials (amino acids, sugar, etc.) from the senescing leaves prior to their abscission (Section V). Thus it is of importance that some enzymes needed for synthesis of mobile compounds, e.g., arginine, also increase during senescence of leaves.[353]

Even though the reductions in chloroplast components and activities have been emphasized here, because of their relative earliness, the amount of information available on them, and their potential importance in whole plant senescence, it must be noted that many tissues which senesce do not have chloroplasts, and leaves which are etiolated also senesce, apparently as rapidly as green leaves.[376] Therefore, it cannot be concluded that cellular senescence necessarily starts in the chloroplast.

E. Membranes and Organelles

Most of the work on the sequence of ultrastructural changes has been done on leaves, but in general these follow a similar pattern in all higher plant cells.[44,45,306,333] Still there are some reports of significant differences between leaves senescing under different conditions, e.g., attached vs. detached leaves.[62,377]

As indicated earlier, the chloroplast shows some of the earliest signs of physiological decline. Of these, the decrease in RUDPCase activity is one of the first and most important, starting in the early, slow phase of decline. Correspondingly, there is an early loss of chloroplast ribosomes that appears to precede the decrease in cytoplasmic ribosomes. Massive ultrastructural changes, (e.g., in the thylakoids) seem to start with the rapid, later phase of senescence.[18,44,45,77,142,362]

Changes occurring at intermediate times include swelling, vesiculation, and disappearance of the endoplasmic reticulum (ER) and Golgi, along with the loss of ER-associated ribosomes.[18,44,334] The mitochondria seem to change relatively late.[18,44] The suggestion that the biochemical decline may occur relatively early in the chloroplast is supported by the observation that chloroplast lipids, (e.g., galacto- and sulfolipids) decrease before the phospholipids that predominate in nonchloroplast membranes,[79,253] but the relative stability of the isolated chloroplasts suggests that this degeneration may be imposed from outside the chloroplast.[57]. Although the chloroplasts start changing relatively early, they and other organelles may still be intact relatively late when the tonoplast ruptures.[18,44,269] Tonoplast breakdown may release hydrolytic enzymes, acids, and toxic compounds,[216,256,311,334] and this initiates the completion of cellular degradation. The final stage of cellular degradation involves disintegration of the plasmalemma and massive alteration of the nucleus,[18,44,334] but some organelles, (e.g., nuclei, chloroplasts, mitochondria) may persist in a recognizable, though highly modified, form.[18,44,45,77] Although membrane changes do occur, and no doubt play an important role in senescence, many key questions remain unanswered and will have to be understood before we will be able to understand fully the factors that cause a whole plant to die. The fact that calcium ions (which may preserve membrane integrity) can retard senescence of leaf disks[300] indicates the importance of the membranes. A related problem has to do with the possible isolation of hydrolytic enzymes in lysosomes or lysosome-like structures. It is generally acknowledged that most plant cells contain enough hydrolytic enzyme activity to break down most of their protein and RNA quickly, but obviously, this normally does not happen. This is probably due, at least partly, to isolation of the degradative enzymes in compartments such as lysosomes, but there has been some difficulty in demonstrating lysosomes in the classic sense in plants. Some investigators now consider the vacuole to be the lytic compartment in plant cells,[216,259] and isolated cotyledonary vacuoles are rich in hydrolases;[256] (Chapter 6) however, some recent data raise questions about this idea.[43] Undoubtedly, understanding the compartmentation of hydrolytic enzymes and the control of their access to their substrates is crucial to understanding whole plant senescence, and this will no doubt reveal a key aspect of control.

In summary, it is uncertain where whole plant senescence begins and which are the key changes. The changes occurring in the leaves do appear to be primary. Alterations in the photosynthetic apparatus, protein metabolism and membranes seem especially important, but more analyses of the kinetics of these changes in relation to each other in whole plants are needed. Although the degenerative changes described above occur in leaves during both monocarpic senescence and autumnal senescence of polycarpic plants, it is not yet clear why the polycarpic plant is able to survive. Evidently, the roots and the stem just become dormant or quiescent in the polycarpic plants, rather than dying.

V. NUTRIENT RESOURCE ALLOCATION

A. Introduction

The possibility that nutrient deficiencies may be important in the senescence of particular organs (or even the whole plant) has figured very prominently both in the research directed toward the causes of senescence and in the speculation on these causes. Even before biochemistry reached the stage where these critical nutrients could be identified and measured, it was observed that the senescence of monocarpic plants followed a period of fruit growth — which seemed to impose a massive demand on the vegetative parts — and that early removal of the reproductive structures could spare the life of the plant, or at least delay death, presumably by preventing the nutrient deficiency.[235] Later as our knowledge of plant biochemistry increased, it became possible to identify two general patterns of deficiency: (1) nutrient withdrawal (nutrient drain) from the senescing parts, and (2) diversion of the supply of nutrients needed for maintenance away from the senescing parts (nutrient diversion). In addition, we can now recognize several general categories of "nutrients" (photosynthetic assimilates, minerals and hormones, e.g., cytokinin); the remainder of this section will deal with the first two categories, while hormones (which may not be nutrients in the strict sense) will be considered in Section VI.

B. Allocation of Photosynthetic Assimilate

1. Pattern of Assimilate Movement

Extensive studies now show that the ^{14}C compounds derived from ^{14}C-CO_2 through photosynthesis are distributed differently as a plant ages or goes through different developmental stages particularly the vegetative compared with the reproductive phases.[264,289,373,381,382,394] As individual leaves develop and age, they switch over from net import of photosynthate to net export. The exported assimilate may be partitioned among the various sinks in different ways, with some quantitative differences in different species or environments. In general, the reproductive structures seem to be the strongest sinks taking priority over vegetative growing regions, and this may be an important factor (but probably not the only one) in the slowdown or cessation of vegetative growth cited in Section III. A. There is evidence that the photosynthate monopolization by the reproductive sinks creates a deficiency that interferes with the functioning of the roots in cotton,[65] soybean,[132,189] and cowpea.[143] In soybean and the common bean, very little photosynthate travels to the roots once podfill starts.[152,404] Assimilate movement also shows some restrictions which to some extent reflect vascular patterns. For example, in soybean, the photosynthate from a leaf moves to the pods at the same node or at the second node down;[10,25,30] apparently, the photosynthate follows the main vascular connections.[48,370] In wheat, the flag leaf is the main source of photosynthate for the developing grain, again favored by the vascular pattern.[92] At least in soybean, the pod and supplier leaf relationship can be changed by surgery, partial defoliation, and depodding.[199,261] The vascular architecture seems to impose a preferential pattern of translocation rather than an absolute restriction on movement.[230] The movement of photosynthate is not *always* directly from the producer leaf to the reproductive sink. Many species store some photosynthate during their vegetative phase for later use to support reproductive growth; for example, the roots of radish[255] or the leaf sheath and culm base in rice.[249,422]

Thus the distribution of photosynthate may change during development, and it may move along restricted channels. Reproductive sinks take priority over vegetative sinks, possibly at the expense of curtailing vegetative growth and root function. The resulting deficiencies may in fact play a role in the physiological decline (Section IV) leading to death in monocarpic plants.

2. Control of Photosynthetic Rate by Sink Demand

The rate of utilization of photosynthate does seem to govern the rate of photosynthesis; in numerous (but not all) cases, sink removal, (e.g., defruiting) decreases photosynthesis in the supplier leaves, while partial defoliation increases photosynthesis in the remaining leaves.[107,252,289,380] The modulation of photosynthetic rate by demand is usually not rapid and may be mediated by hormones and developmental changes rather than a direct metabolic feedback.[107] To some extent, leaves may even grow to meet increased demand, e.g., removal of the mature leaves from pea plants increases the growth of the remaining leaves.[206] Instead of depleting the leaves and causing senescence, high levels of demand actually sustain the leaves in some cases. Shading the ears (important contributors of photosynthate for the growing seeds) of barley increases their demands on the leaves and retards the senescence of the source leaves.[263,379,395] However, defruiting may accelerate foliar senescence in corn,[6] pepper,[118-120] and the flag leaf of barley.[211] Furthermore, net assimilation measured as total dry weight, is reduced as the size of the reproductive sink is decreased in soybean suggesting that decreased demand may lower net production.[88,182] Thus high levels of photosynthate demand by the developing fruit are not necessarily detrimental to the leaves (or the rest of the plant), and may even stimulate productive capacities. The monocarpic senescence response of soybean does not seem to be a simple deficiency effect, for it is saturated at a relatively low pod dose or sink size, (about 40%) in soybean.[261]

C. Mineral Uptake and Distribution

1. Patterns in Mineral Uptake

Since minerals may be critical, limiting resources, their uptake by and translocation from the roots will be considered here, though only those aspects that may have some, direct relevance to whole plant senescence will be covered. Usually, though not always, mineral uptake by roots decreases during the reproductive phase, and total uptake may parallel dry matter accumulation in a loose way.[98,191,203,265,363,366,372] The exact timing relative to cessation of growth may differ for different minerals, for different species, and even for different organs. This implies that mineral accumulation is to some extent controlled by demand, but it is clear in many cases[203] including cotton,[65] barley,[114] and soybean[265] that the decrease in uptake occurs just when reproductive growth and demand are greatest. N uptake declines in N_2-fixing species as well as in nonfixers, for N_2 fixation like other root activities diminishes during reproductive growth (see Section IV.C). Furthermore, certain minerals may accumulate faster than dry weight during the growth phase, e.g., Ca in soybean[134] or N, P, and Zn in three other legumes,[149] suggesting that sink demand is not the only regulator of mineral uptake. The manipulation of the reproductive sink size in soybeans by varying degrees of depodding shows that accumulation of the more mobile elements N, P, and K can be controlled by sink demand, whereas the less mobile elements Ca and Mg may not be.[182] As would be expected, essential minerals taken up during the reproductive phase move preferentially to the reproductive sinks, but many plants (especially monocarpic plants) may not take up enough to satisfy the requirements of reproductive growth. Thus the reproductive parts may need to withdraw at least some of the minerals they require from the vegetative parts, which could create a deficiency in the vegetative parts. Quantitative analysis of these demands, especially relative to photosynthesis, has been done for N in a number of crop plants;[338] however, some plants with relatively low N demands still die following fruiting.

2. Mineral Redistribution

Many studies have inventoried the contents of particular minerals in various parts as plants mature and produce fruit, but unfortunately these are generally not aimed at senescence. They usually do not show the relationship to senescence, omit roots and fallen leaves, and are expressed in units, (e.g., mg/g dry weight), which are not the most useful for analyzing senescence. Nonetheless, these data can add to our understanding of the senescence process, particularly in monocarpic plants.

Given the limiting supply of many minerals in the soil, it may be advantageous for a species to reutilize minerals already invested in vegetative structures for reproductive growth even if the parent is destroyed in the process. Redistribution of minerals probably occurs to some extent in all multicellular plants and during many stages of development, but it is particularly prominent during fruit growth.[203,238,265,408] In general, these studies are concerned with mineral inventories in relation to fruit development, but usually it is possible to guess at their relation to senescence. Mineral distribution in soybean has been studied most extensively.[33,76,124,127,238,265] N, P, K, and Mg, but not Fe, Ca, S, and Mn, seem to be redistributed from the leaves to the developing fruit of soybean. Mg, however, is not readily translocated. The relative ability of developing seeds to mobilize different minerals has been compared in a single study employing three species of legumes including pea.[149] The leaves, carpels, and seed coats lose 60 to 90% of their N, P, and K to the seeds, but they release only 20 to 60% of their Zn, Mn, Fe, and Cu and less than 20% of the Ca and Na. In soybean, reproductive sink removal (depodding) prevents mineral redistribution just as it prevents senescence.[76] It may be supposed that this would be true for monocarpic plants in general. Prevention of flower development in tobacco by cutting off the plant's top, however, does not completely prevent the loss of N from the lower leaves, for the N is redistributed to support additional growth of the roots, stem, and uppermost leaf group.[396] Thus in species where flower or fruit removal reinstates vegetative growth (see Section III.A), these new sinks can induce redistribution. The extent to which developing seeds depend on remobilized minerals (as opposed to minerals supplied directly from the soil via the roots) varies with the species and conditions. Under some conditions, the roots supply all the minerals needed for fruit development in corn.[58] Similarly, redistributed N may make only a small contribution to the developing seeds in broad bean.[63] At least for the mobile elements, almost all of the mineral in the seeds may be obtained by withdrawal from the vegetative parts. Although one might expect a plentiful supply of minerals in the soil to reduce internal redistribution, fertilization may produce exactly the opposite effect. P-deficient oats, for example, obtain only 30% of the P in their inflorescence from internal redistribution, whereas the inflorescences of plants given a surplus supply obtained 93% of their P by redistribution.[407] Likewise, redistribution of N may occur in wheat plants given an abundant supply of N.[222] These data suggest that redistribution may not be under a simple source-sink (or demand) control. Our studies on soybeans[164,258] suggest that given certain treatments (e.g., foliar applications of auxin and cytokinin), the N required by the growing seeds can be supplied by uptake from the external environment rather than redistribution.

There is no doubt that minerals may be taken out (or released) from senescing organs, particularly leaves, for use by developing fruit. On the surface, deficiency created in the donor organs provides a sufficient cause of their senescence. Yet, discrepancies exist; for example, in soybean, foliar applications of complete mineral nutrient solutions do not retard senescence, even though enough minerals penetrate to increase seed yield.[76] In addition, soybeans can be completely defoliated as late as full bloom (at which time regrowth of the foliage still occurs), but seed yield is reduced by only 20%.[397] This does not support the idea that fruit growth is totally dependent on a reservoir of nutrients that is accumulated in the leaves prior to the start of reproductive growth. Perhaps more important, manipulation of reproductive sink size (pod dose) in soybeans shows that the foliar senescence response is saturated at a relatively low sink size (< 50% of the maximum), rather than showing a good parallel[199,261].

D. Control of Resource Allocation

Implicit in the preceding discussion is the idea that metabolic sinks control nutrient distribution and translocation. The simplest and most popular explanation supposes that constituents released during the normal turnover in the source are circulated and then removed from circulation at the sink. The sink then reincorporates the constituents more actively than the source, thereby producing a gradual transfer from source to sink. No doubt this mechanism does exist, but it is probably not the entire explanation of control of nutrient distribution. For example, leaves may continue to move solutes toward the leaf base even after they are detached from the rest of the plant (the primary sink).[54,192,377] The arrangement of the vascular tissue could by channeling, determine the distribution of solutes, but generally the structure of the transport system itself does not control even though it may favor the movement from source to sink.[230] Another possibility is directed transport, which may, in effect, be the functional arrangement or geometry of the transport system.

A somewhat related question — can any solutes move away from a strong sink? — arises in connection with attempts to explain how developing fruit induce senescence of vegetative parts as in monocarpy (see Section II). Given the likelihood that the phloem is a major supplier of solutes to the developing fruit and the bidirectionality of the phloem,[64] there is no reason to suppose that some transport out of the fruit cannot occur. Several legume fruits, including soybeans,[290] exude solutes from their stylar tips or pedicels after cutting, but these solutes could be in transit to the developing fruit rather than in the process of export. Walker and Ho[388] have shown that tomato fruits (when cooled relative to the rest of the plant) actually become net exporters. Recently, ^{14}C from CO_2 fixed by pea fruits has been reported to move to the shoot apex.[108] More importantly, some hormones such as auxin are able to move out of the fruit or the seeds.[36,115] The possible movement of solutes out of the fruit has thus far generated little interest, but it will no doubt be studied extensively in the near future.

Plant hormones, especially cytokinins, can influence solute movement, possibly through alteration of sink (or source) activity, through directed transport, or through some yet unknown mechanism.[365] It is quite conceivable that hormones can alter phloem loading or unloading of a particular solute or perhaps some other critical transport phase. Numerous reports now show hormone effects on solute distribution, but in the most interesting cases, especially the relatively long term studies, it is not yet possible to be sure whether the hormone affects transport directly or indirectly via altered development, e.g. sink activity. Most intriguing are the situations where hormones are able to replace at least partially the effect of the reproductive structures[162,331,393] or vegetative stem apices.[31,245,291,301,326,329] In spite of the difficulties, there are a couple of cases where hormone-directed transport seems separable from hormone influences on sink activity. First, kinetin promotes the movement of γ-aminobutyric acid toward the site of application on a detached leaf, even though this amino acid is not incorporated into protein.[242] Second, auxin appears to stimulate transport to the site where it is applied to bean stems more than it promotes sink activity.[272] Third, auxin can induce the development of transport channels.[319] A hierarchy exists among hormone influences on solute movement, and it is of interest that in detached geranium leaves gibberellic acid is the most effective, with cytokinin and auxin less active.[294]

Thus while metabolic sink activity could be involved in controlling resource distribution, it seems reasonable to consider hormone-directed transport very important. The high cytokinin activity in fruit could, for example, be involved in producing their sink activity.[39] The changes in endogenous hormone levels to be described in Section

VI could, therefore, directly regulate transport as well as other senescence-related processes.

E. Is Nutrient Deficiency a Primary Cause of Senescence?

It is already implied in the preceding parts of this section that the nutrient deficiency explanations (nutrient drain or diversion, Section V.A) of monocarpic senescence may not be adequate explanations for the primary cause of monocarpic senescence. Deficiencies created by drain (mineral withdrawal) and diversion (minerals and cytokinin from the roots and photosynthate produced by the leaves) probably do play a role in monocarpic senescence, but it seems to be secondary.

Since our surgical manipulations of whole soybean plants test the nutrient deficiency theories of monocarpic senescence, we shall summarize these results here. The transmission of the senescence signal from the seeds to the leaves is quite limited, moving mainly downward if at all.[199,260,261] The movement of nutrients should not be this restricted. The localized nature of the senescence signal also argues against the idea that monocarpic senescence of soybean occurs through a primary effect on the roots (e.g., reduced cytokinin production or mineral accumulation). If the senescence signal first caused some change in the roots, which was then transmitted back to the leaves where it induced senescence, then one would expect a more widespread effect. In addition, we have been able to separate fruit development from monocarpic senescence in a variety of ways.[164,258,261] For example, single axis plants may be defoliated and depodded leaving only a single leaf and a single pod cluster separated by three internodes. When the pod cluster is below the leaf, the leaf (and the plant) does not senesce. In the reverse situation, the leaf senesces about 50% of the time, which is consistent with the limited downward movement of the senescence signal. The seed yield per pod cluster is the same in both arrangements. Since removal of the remaining leaf essentially stops fruit growth, this leaf, and not some storage depot elsewhere in the plant, must be supplying the growing fruit. Furthermore, the seed yield in these single pod clusters is actually a bit higher than in a comparable pod cluster on an unmodified plant. Thus *seed development* (and the required nutrient drain and diversion) *can occur without necessarily inducing monocarpic senescence.* Evidently, the needs of the developing fruit can be suppplied from current assimilation if these capacities are maintained.[257] Another case in point is occurrence of genetically altered soybeans which produce seed, but lack part or all of the monocarpic senescence process.[3,28,164] Finally, depodding surgically-modified plants at various stages of seed growth suggests that the senescence signal is generated quite late in seed development when most of the dry weight accumulation (nutrient accumulation) is finished.[261] Thus our data do not support the nutrient deficiency theories of monocarpic senescence at least for soybean. In addition, the nutrient drain and diversion hypotheses do not seem to hold for spinach and cocklebur.[183,195,259]

VI. HORMONAL CONTROL OF WHOLE PLANT SENESCENCE

A. Introduction

In working out the hormonal controls of plant senescence, several general approaches may be employed. The first and most widely used is external applications of plant hormones to promote or retard senescence. This approach can help to reveal the patterns of hormonal controls, but only if combined with other data. The second method is to correlate changes in endogenous hormone levels with senescence. Unfortunately, the majority of the relevant studies on endogenous hormones were not aimed at senescence per se and therefore were not closely correlated, if at all, with the kinetics of senescence in the plant. Sampling times for measurements are often expressed in terms that are difficult to relate to senescence, (e.g., days after germination, days after anthesis, etc.). Moreover, few studies have analyzed changes in both the fruit and leaves. This becomes especially important in trying to analyze the role of hormones in the correlative influences of the fruit on foliar senescence in monocarpic senescence. As in Section IV (Physiological Decline), many of the hormone studies deal with hormone effects — or changes in — detached leaves from both monocarpic and polycarpic plants. Nonetheless, these data seem to be useful in explaining whole plant senescence. Although the possible differences in the physiology of senescence suggested in Section II.D have been apparent in hormone physiology mainly in the senescence-retarding hormones, (Section VI.D), similarities in hormonal controls will be emphasized. The mechanism of hormone action is not well studied in relation to senescence, and this topic seems more appropriately considered in other chapters. Because (1) senescence, like many other plant processes such as dormant bud development[262] appears to involve a shift in the balance of senescence promoters and senescence retardants, and (2) the hormones studied so far seem to fit primarily into one of these two categories, this section is subdivided accordingly into senescence promoters and retardants. Each subdivision is also split into effects of exogenous hormones and changes in endogenous hormones.

B. Effects of Exogenous Senescence Promoters

1. Abscisic Acid

The senescence-accelerating ability of abscisic acid (ABA) is well documented. ABA promotes chlorophyll loss in leaf discs collected outdoors from seventeen tree species plus bracken fern.[89] ABA also accelerates yellowing of detached leaves in radish, barley, nasturtium, and *Rumex pulcher*.[14,15,23,167,355] Different sensitivities to ABA have been found within a species; leaf discs of some cultivars of soybean respond to ABA more rapidly than others.[345] The physiological status of the leaf material is also important; leaves aged by floating on water show a greater response to ABA than freshly excised material.[274]

The effect of ABA on attached leaves is generally less pronounced. In many cases, ABA has no effect on attached leaves of the same species for which it promotes the senescence of excised leaves or leaf discs.[89,345,349] ABA sprays have no effect on wheat seedling leaf senescence.[410] Translocation, inactivation, poor uptake, and competition from endogenous hormones are all possible reasons for this difference between attached and detached leaves. In fact, attached leaves sprayed once (or even several times) with ABA solutions may still receive less of the hormone than leaves floating on the solutions. Nevertheless, ABA sprays can accelerate senescence of attached leaves including tree of heaven, Brussels sprouts, and soybean.[89,197,200,345] ABA hastens fruit maturation in soybeans, but not enough to account for the acceleration of senescence.[197,200] ABA does not induce the senescence of depodded soybeans, indicating that ABA alone cannot completely replace the influence of the pod upon senescence.[197,200]

The function of ABA as a promoter of flower tissue senescence, including such parameters as wilting, color fading or "blueing", and protein loss, has been established.[32,218,219] Here, ABA may act through hastening of autocatalytic ethylene production[219] (see Chapter 7).

In some cases, ABA actually exhibits a senescence-retarding effect. ABA sprays cause persistence of grape inflorescences, which mimics cytokinin sprays and have been termed the "cytokinin sparing" effects.[245] At 10^{-4} *M*, ABA retards senescence in radish leaf discs, but this retardation may be due to the disruption of the normal senescence process.[60] In addition, ABA may cause short term inhibition of senescence of cut flowers by closing the stomata, thereby reducing water loss.[117]

Acceleration of RNA and protein loss in senescing tissues by ABA has been reported several times.[23,60,181,281] ABA elevates protease and nuclease activity through increased synthesis (or the activation of stored enzyme), but not all hydrolases are increased. ABA has been found to block the rise of β-1,3-glucan hydrolase in tobacco leaf disks, while this enzyme increases during the senescence of the water controls.[237] Overall, ABA decreases protein synthesis and increases protein degradation, but the relative contributions of these processes and their responses to exogenous ABA are not clear. Not all of the studies on protein synthesis effects are corrected for changes in the specific activity of the precursor pool or decreased precursor uptake over time.[259,315]

ABA (10^{-5} *M*) increases the permeability of *Rhoeo* leaf sections far above that of water controls.[74] The acceleration of ultrastructural changes (which accompanies ABA-promoted senescence) may be similar to the normal process[233] or somewhat different.[62]

In summary, exogenous ABA can promote a large number of biochemical changes commonly associated with senescence, but these changes may sometimes differ a bit from those that would have occurred in senescing tissues without the exogenous ABA.

2. Ethylene

Ethylene plays a very prominent role in the senescence of certain plant parts, particularly fruit and petals, and in the abscission process. Ethylene is generally implicated as an inducer in the senescence of flower tissue, which responds to lower concentrations than leaves do. Ethylene at 1 ppm induces chlorosis in rose leaves,[424] but up to 25 ppm it has no effect on the detached leaves of *Nymphoides*.[111] Ethrel (which breaks down to ethylene[2]) sprays have no effect on foliar senescence in wheat seedlings.[410] Exogenous ethylene (applied as Ethrel or the gas) increases endogenous ethylene and ABA production and overcomes the senescence-retarding effects of exogenous cytokinins in flowers.[40,171,172,219,350]

Ethylene has well-characterized effects on the abscission process, stimulating the synthesis and secretion of cellulases and polygalacturonases.[2,259] Ethylene (7 ppm) has no effect on leucine incorporation in *Rhoeo* leaf disks, but there may be a slight inhibition of uridine incorporation into RNA.[316] In morning glory corollas, ethylene accelerates both fading and the rise in RNase, though the fading comes first.[171] Treatment of cut carnation and chrysanthemum flowers with ethylene, 3 ppm, causes the acceleration of senescence-associated changes in protein electrophoretic patterns; the proteins become a more diverse mixture with lower molecular weights than those of fresh tissue (reflecting partial degradation during senescence?).[285]

Ethylene causes no detectable changes in permeability of *Rhoeo* leaf sections,[316] but does accelerate the loss of $^{36}Cl^-$ and $^{86}Rb^+$, as well as ^{14}C-labeled metabolites from CO_2 and glucose in Japanese morning glory flower tissue.[125] Exogenous ethylene can therefore promote a variety of processes that could be important in whole plant senescence.

3. Other Senescence Promoters

(cf. also Chapter 5) L-serine promotes the senescence of detached oat leaves — possibly through stimulation of the formation of two proteases, which normally increase throughout the senescence of the leaf.[212,335] Since cycloheximide prevents this senescence, and L-serine has only a very slight effect on the senescence of isolated chloroplasts, the protease synthesis is assumed to occur on cytoplasmic polysomes.[56,212] As usual, the chloroplast and its protein are attacked first in the intact cell. L-arginine (0.3 M) by itself does not inhibit senescence, but does counteract serine.[335] Kinetin, of course, retards the senescence of detached oat leaves; in this case by prevention of proteolysis perhaps by inhibition of protease synthesis. Enigmatically kinetin plus serine causes greater promotion of senescence than serine alone.[212] L-cysteine, L-glycine, and L-alanine also promote senescence and protease activity, but they act only at higher concentrations than serine.[213] The role of endogenous serine in control of senescence remains to be clarified.

C. Changes in Endogenous Senescence Promoters

1. Changes in Abscisic Acid

Detached, senescing nasturtium leaves show increases in ABA-like activity measured by bioassay, and these changes start two days before visible senescence (chlorophyll loss).[55] ABA (measured by gas-liquid chromatography) rises sharply in detached tobacco leaves, and then rapidly declines,[93] a pattern similar to that in attached soybean leaves. Diffusible ABA increases in aging Coleus leaves.[34] Increases in endogenous ABA controlled by the photoperiod may play an important role in the natural induction of leaf senescence, especially in tree species undergoing autumn leaf senescence.[262] The rise in ABA levels during senescence might make this hormone an attractive candidate for the role of general senescence inducer, but not all studies support this. ABA does not increase significantly in attached rice leaves during monocarpic senescence,[270] and young cocklebur leaves contain relatively high ABA concentrations that *decrease* throughout the life of the leaf.[308] The primary leaves of the common bean show widely varying levels of ABA during senescence but no consistent change.[61] In attached bean cotyledons, Dumbroff et al.[82] have attributed the changes in ABA levels to changes in water potential of the senescing tissue, rather than to differences in senescence activity (see the discussion of stress effects below). Senescent bean leaves contain less free ABA than young leaves; however, the senescent leaves contain more bound ABA, thereby equalizing the total ABA in the young and senescent leaves.[278]

In analyzing the senescence-related changes in ABA levels, it is essential to consider and control environmental factors, particularly including water stress, which is now known to increase ABA levels. Leaf levels of ABA increase rapidly in response to water stress, seemingly through *de novo* synthesis.[232,420,423] An 8-hr period of water stress in perennial rye *(Lolium temulentum)* produces a 30-fold rise of ABA in the leaves.[177] Water stress has produced increases in ABA that have been correlated with accelerated senescence in tobacco.[93] The work of Dumbroff et al.[82] on senescing bean cotyledons shows that the changing ABA levels are better correlated with internal changes in water potential than with senescence. The acceleration of senescence in tobacco, bean, and sunflower leaves by a brief exposure to high temperature (47°C, 2 min) is accompanied by increased ABA.[156] Nutrient deprivation, particularly nitrogen stress (which is known to accelerate senescence)[228] also increases ABA in the leaves of sunflower, especially the older leaves.[109] Water stress-induced ABA appears in a different subcellular compartment (cytoplasm?) from the nonstress ABA (chloroplast).[206a] The stress-induced increases in ABA may be large enough to mask the senescence-related changes, or if variable, they could obscure them. Clearly, stress-induced senescence and ABA need to be controlled if other patterns, especially monocarpic senescence, are to be studied.

ABA levels can be altered by other hormones. For example, endogenous ABA increases with age in cut rose flowers, and this can be triggered by both endogenous and exogenous ethylene.[219] Cytokinin can also promote the transformation of free ABA to a bound form.[93]

The endogenous concentrations of ABA in developing fruits of barley, wheat, soybean, and cherry tomato change in a similar pattern; ABA rises during the rapid growth stage of the fruit or seed, peaks at the point of maximum fresh and dry weight of the fruit, and then declines rapidly with senescence of the plant and with ripening or drying of the fruit.[1,110,176,224,305,307] Developing cotton and pea fruit show a slightly different pattern, with low levels of ABA during rapid fruit growth and a rise during fruit maturation.[71,87] During the final maturation of cotton fruit, there is a rise in ABA content corresponding to a period of fruit wall senescence.[71] The magnitude of the increase in fruit may be quite large. Wheat grains have 40 times as much ABA (per gram fresh weight) at maximum fresh weight than in early development, and soybeans have 15 times as much.[176,305] Soybeans, however, contain 30 times more ABA at maximum fresh weight than wheat.[224,305] Thus the later stages of fruit development are accompanied or preceded by a rise in ABA levels.

So far, relatively little consideration has been given to the sites of synthesis and translocation of ABA, which may be involved in senescence. Leaves, roots, fruits, and probably flowers, are able to synthesize their own ABA.[228,423] Excised wheat and tomato fruits develop ABA levels as high as those of attached fruits.[223,229] Both young and mature leaves of castor bean have the capacity to synthesize ABA.[423] Roots appear to be able to synthesize ABA,[383] but this ABA does not seem to move into the aerial parts.[137] Studies by Mizrahi et al.[234] indicate little or no translocation of ABA induced by water stress. Some ABA, however, can move in the phloem and xylem,[7,81,148] and, in castor bean, ABA is translocated from mature leaves to shoot tips.[424] The loss of ABA from fruits need not be via translocation; excised wheat grains lose ABA during dehydration at the same rate as those attached to the plant.[224]

Thus increased ABA levels in pre-senescent and in senescing leaves can be due to local synthesis, although the translocation of ABA from the fruit or seed to the leaves is still possible. On the whole, while many pieces of evidence suggest that ABA may play an important role as an inducer of senescence, including monocarpic senescence, the picture is far from complete.

2. Changes in Ethylene Production

The senescence of flower petals (fading, rolling, etc.) is discussed in detail in Chapter 7. In general, it is accompanied by the increased production of ethylene.[5,171,172,254] Increased ethylene evolution increases the levels of ABA, which may then decrease further production of ethylene.[219] In this case, the ABA could serve to slow the rapid rise in endogenous ethylene, which occurs due to its autocatalytic mode of synthesis.[171] Work on morning glory flowers suggests that ethylene may not be the initial trigger of petal senescence, but only an accelerator.[172]

Increased ethylene production in leaves generally seems to be related more to abscission processes than to other aspects of senescence. Leaves have been reported to produce more ethylene[185,275] or the same amount[123] during senescence. Unfortunately, little has been done to examine the involvement of ethylene in whole plant senescence; perhaps that is because it appears not to play a very important part. At least in soybeans 2% (v/v) CO_2, which is known to antagonize ethylene action, can delay monocarpic senescence by only a couple of days, and even this small delay seems to be a result of delayed fruit development.[323] The limited mobility of the monocarpic senescence signal produced by the soybean seeds also argues against the participation of ethylene, which should be able to move readily in the intercellular gas spaces.[199]

3. Other Senescence Promoters

Senescence Factor (SF), produced by senescing or damaged leaves of Coleus or the common bean may function as an abscission promoter.[159,271,278] SF is physiologically and chemically different from ABA or ethylene.[278] Cotton fruits also contain a senescence factor which promotes abscission but differs from the above SF and ABA.[348] Another unidentified promoter of senescence is produced by the axis of pea seedlings and acts on the cotyledons.[386] The physiological roles of these senescence factors and serine (Section B.3) remain to be determined; however, it seems likely that more than just ABA and ethylene will be found to function as senescence promoters. It would be a mistake at this early stage to abandon the search for new classes of senescence promoters and to try to explain senescence only in terms of ABA and ethylene.

D. Effects of Exogenous Senescence Retardants

As with the senescence promoters, external applications (primarily to the foliage) have provided much of the most encouraging evidence for the physiological role of senescence retardants. The primary hormones involved here are auxin, gibberellin, and especially cytokinin, but a large number of other compounds, many of which are higher plant constituents, can retard senescence.[259]

Following the initial reports that certain synthetic cytokinins could retard senescence of detached leaves,[295,310] this effect of cytokinins has been substantiated many times and extended both to other organs and to other cytokinins[24, 259] (see Chapter 5). Like most generalizations, the generality of the cytokinin effect has its exceptions; some detached leaves do not seem to respond to cytokinin,[259,276,412] and the cytokinin may not retard senescence when applied only to the adaxial surface.[247] Soon after the initial reports on cytokinins, it was demonstrated that auxins too could retard senescence of detached leaves from some species,[259,276,277,314] but it was clear right from the start that auxin treatments do not work with all species.[259,310] About the same time, gibberellic acid was demonstrated to delay development of autumn coloration when applied to several species of deciduous trees.[38] Subsequently, gibberellins were shown to delay senescence of detached leaves and leaf segments, but not all species respond to gibberellins.[22,89,102,259,403] While species that respond to gibberellins often respond to cytokinins too, there seems to be some distinctive differences in the ability of different species to respond to exogenous hormones. Some species, e.g., detached *Nicotiana rustica* leaves, have a complex requirement, both a cytokinin and an auxin being needed for maximal retardation of chlorosis.[91]

Applications of these hormones (especially for cytokinins) to attached organs may be a somewhat different matter from using detached organs. Often, cytokinins have no effect on attached leaves that have a normal connection to a root system.[185,239,244,251] The exact reasons for this remain unclear, but the hormones are generally applied to the adaxial surfaces of the leaves, sometimes an ineffective procedure.[247] Nonetheless foliar applications of cytokinins have now been shown to retard senescence in attached leaves of several species including common bean,[4,100] oat,[377] soybean,[200, 323] and wheat.[410] Spraying with the auxin NAA delays senescence of attached soybean leaves,[160,164] and gibberellic acid sprays delay autumnal senescence in the attached leaves of some woody species.[38] Most dramatic, the combination of an auxin, NAA, and a cytokinin, BA, is actually able to stop monocarpic senescence in soybeans.[258]

E. Changes in Endogenous Senescence Retardants

1. Correlation of Senescence-Retarding Hormone Levels with Senescence

Cytokinin activity in the leaves decreases during senescence and may parallel chlorophyll loss.[197,200,270,356] Cultivars of rice and soybean with early senescence show correspondingly earlier decreases in foliar cytokinin.[270] While free zeatin declines in rice and soybean leaves, the zeatin derivatives dropped less rapidly, or even rose.[270] Similarly, senescing *Streptocarpus* leaves show a decline in butanol-extractable cytokinins (free zeatin and nucleosides), while the residual water-soluble cytokinins (zeatin nucleotides) increase.[356]

Auxin activity drops to low levels in the senescing first-trifoliate leaves of the common bean,[336] but an increase is reported in senescing primary leaves of the same species.[399] Measurements of auxin by modern physical methods show that the IAA content in leaves of cotton and common bean decreases with age,[374] thereby supporting the former report on bean leaves. Endogenous gibberellin-like components decline during the senescence of detached nasturtium and dandelion leaves, species in which pretreatment with GA_3 or kinetin delays senescence.[55,101]

Cytokinins, auxin-like activity, and gibberellin-like activity have all been found to decrease before autumn leaf senescence in several temperate tree species.[7,78,90,144,299] Senescing leaves of poplar and maple show a loss in free zeatin, with a corresponding increase of zeatin mononucleotide.[90] During autumnal senescence of willow leaves, zeatin and zeatin ribosides decrease, while the corresponding glucosides increase.[359]

2. Transport and Sites of Synthesis of Senescence-Retarding Hormones

The ability of root formation to halt senescence of detached leaves has been recognized by horticulturists for at least a century.[235] It has been suggested that the roots may synthesize and translocate a senescence-inhibiting factor.[53,54] After the discovery that exogenous cytokinins could also retard senescence of detached leaves, Kulaeva[185] showed that root exudate contained a kinetin-like factor that retarded leaf senescence. Subsequently, numerous studies have shown root apices to be major sites of endogenous cytokinin synthesis, primarily zeatin and its glycoside derivatives.[50,169,170,179,337,343,383,421] While free zeatin may be the "active" endogenous cytokinin, zeatin riboside may be the major translocation form.[113,140,144,196,342] Zeatin glucosides are probably storage or inactivation products.[145,186,282,361,389]

The movement of root-produced cytokinins in the xylem is well known.[179,342] Increases in cytokinin in the xylem and shoot have been related to root growth.[81,318] However, the large increase in cytokinins that occurs following floral induction in common Perilla is not due to greater root growth.[21] Cytokinin content increases of 5- and 7-fold in the root exudate occur with flowering induction in common Perilla and sunflower, respectively.[20,340] The decrease in cytokinin flux from sunflower roots correlates with the cessation of root growth,[340] while the large increase in Perilla occurs after root growth has ceased. Tomato is a bit different in that the cytokinin content of the root exudate declines during early flower development and then increases when the flowers open.[68] Removal of the apical and lateral buds of vegetative Perilla plants increases the cytokinin content of xylem sap.[20] Removal of all other shoot organs causes high levels of zeatin glucoside to accumulate in the primary leaves of common bean.[389] The removal of all leaves longer than 1 cm increases cytokinin levels in the buds of cockelbur.[140] Shoot cytokinin levels may increase as a result of fewer sinks in these cases. Such cytokinin diversions can be used to explain some of the senescence-retarding effects of organ removal, such as the removal of some leaves retarding the senescence of others.[419] Removal of the roots from soybean seedlings accelerates senescence of the primary leaves, and the roots can be replaced with cytokinin solution (benzyladenine), thereby maintaining the leaves.[151] Although root production of cytokinins may be important in controlling foliar senescence, this does not appear to be a primary causal factor in monocarpic senescence of soybean.[260,261]

The appearance of cytokinins in the phloem suggests possible *de novo* synthesis of cytokinins in the leaves or redistribution of previously imported cytokinin.[121,297,358,387] Recent work by Henson and Wareing[141] suggests that detached leaves (and buds) of cocklebur are not capable of independent cytokinin production. It is, however, possible that they glucosylate zeatin to make zeatin glucoside. The appearance of various cytokinins at different stages of development (especially the senescence-related decrease in free zeatin, with parallel increases in zeatin derivatives, see above) probably results from conversion of zeatin in the leaf, but may reflect transport of different cytokinin derivatives from the roots.[68,360] In any case, analyses of endogenous cytokinin levels should include not only zeatin, but also its major derivatives, since the qualitative changes may be more important in senescence regulation than changes in total cytokinins.

Buds have also been suggested as cytokinin synthesis sites;[144,166] but the work on cocklebur mentioned above[141] places this in doubt.

The synthesis of auxin in shoot meristems and young leaves coupled with polar transport out of these structures is well known. Gibberellins are synthesized in shoot meristems or young leaves,[163,201] and also in root apices with translocation in the transpiration stream.[51,163,298,383]

Fruit development is a complex process, which could involve all the known plant hormones,[87]* thus it is difficult to distinguish those hormonal changes that are important to whole plant senescence from those related only to fruit maturation and senescence. In general, fruit cytokinin levels are high during cell division and rapid fruit growth, but subsequently drop to low levels.[1,39,42,52,69,197,200,400] Not all cytokinins change in parallel, for in cherry tomato, cytokinin ribotide increases as total cytokinin activity decreases.[1] The cytokinin decrease in soybean fruit begins before visible leaf senescence and parallels the decrease in leaves and shoots.[197,200] Auxins and gibberellins also peak at various stages of fruit growth, but fall off before final fruit maturation and ripening.[1,52,87,248,307,341,400] Since fruit, or at least seeds, may be active sites of auxin, cytokinin, and gibberellin synthesis, it is not clear why they would need to import large quantities of these hormones, especially cytokinin. In fact, fruits, particularly the seeds, are able to export some materials including auxin.[9,36,115]

At least two of the physiological roles of endogenous cytokinins can be deduced from the exogenous application experiments; first, the maintenance of key metabolic processes, e.g., the protein synthesis and photosynthetic machinery, and second, direction of metabolite transport. Root formation on excised leaves maintains protein and chlorophyll levels in the same way as kinetin or reapplied root exudate, thus it appears that roots maintain leaves through supplying cytokinin (but nutritional factors may also be involved).[54,185,331] In general, as zeatin declines, derivatives (mainly glycosides) are built up at the expense of free zeatin. Since senescence of attached plant organs involves decreased protein synthesis and the redistribution of metabolites (primarily amino nitrogen),[76] it would appear that these cytokinin derivatives are inactive in maintenance of protein synthesis and as metabolite attractants. Finally, there is little direct evidence to indicate that nutrient transport to developing seeds is a function of their high cytokinin content, although many studies with exogenous hormone demonstrate this possibility. The physiological role of endogenous auxins and gibberellins in senescence regulation seems to be similar to that of the cytokinins for plant systems in which they are active (maintenance of protein levels, directed transport, etc.). The argument made against nutrient deficiency as a primary causative factor in monocarpic senescence of soybean (Section V.E)[257,260,261] can be applied to cytokinin deficiency, yet cytokinins may play an important secondary role.

VII. GENERAL CONCLUSIONS

At this time, it appears that senescence of monocarpic plants is more amenable to

* Their roles, and the uncertainties involved in some of the data, especially those dealing with exogenous application of hormones, are set out in detail in Section 4 of Chapter 8.

analysis than senescence of polycarpic plants — and therefore we know more about monocarpic senescence. Still, many fundamental questions remain unanswered. Usually (though not always) the fruits, particularly the seeds, induce monocarpic senescence. A wide range of degenerative changes occur during monocarpic senescence and changes in one part of the plant alter other parts of the plant. In most monocarpic plants, the decline in leaf function and shedding of the leaves seem to be particularly crucial. These correlative effects, e.g., seed effects on the leaves, may be mediated by hormones, as well as other factors.

The four major theories to explain monocarpic senescence are (1) limited longevity or growth potential, i.e., the plant dies at a particular age or because it can no longer grow and replace its leaves, etc.; (2) nutrient withdrawal, e.g., the growing fruit remove critical nutrients such as nitrogen from the vegetative parts, thereby killing them; (3) nutrient diversion, e.g., nutrients such as photosynthate from the leaves are diverted to the growing fruit causing a deficiency in other vegetative parts, especially in the roots; and (4) the developing fruit produce senescence hormone(s), which causes the rest of the plant to degenerate. Our studies on soybean argue against (1) (see Section III.A), as well as (2) and (3) (see Section V.E). Thus we favor (4), the senescence hormone theory, mainly by default. It also seems unlikely that developing fruit would exert such dramatic and rapid effects simply through passive actions as in (2) and (3). Although rather little study has been made of the ability of the fruit to export at the same time they are accumulating massive amounts of assimilates, there are several reports of exported materials, including some hormones.[9,36,108,115]

Although we favor (4), the senescence hormone theory, as a primary control or trigger, there is no doubt that other factors are also important in monocarpic senescence. Indeed, nutrient withdrawal and diversion do occur and must play some role in the senescence process. Furthermore, a deficiency in senescence retarding hormones, especially cytokinins, may be important in senescence, though our data on soybeans favor secondary roles for nutrient and cytokinin deficiencies in soybeans. As mentioned in Section II.D, the mechanisms for monocarpic senescence may differ from one species to the next, so that the soybean data may not apply in detail to all monocarpic species.

Finally, regarding the significance of the senescence phenomenon, senescence is now recognized as an important developmental process, which increases its theoretical interest. Perhaps more important, senescence contributes to a decline in productivity of many economically important plants; monocarpic crop plants actually lose their productive capacities at the time when they are needed most. Thus understanding and controlling senescence could be of great economic value. Both hormonal treatment and genetic alterations offer promise in the control of monocarpic senescence.

ACKNOWLEDGMENTS

The extensive editorial assistance of Sarah Noodén and the participation of Jon Hosler in the early draft of parts of Sections IV and VI is gratefully acknowledged. The author's research mentioned here and some of the literature search in preparing this review was supported in part by research grant 416-15-79 from the USDA Cooperative State Research Service under P.L. 89-106.

REFERENCES

1. **Abdel-Rahman, M., Thomas, T. H., Doss, G. J., and Howell, L.**, Changes in endogenous plant hormones in cherry tomato fruits during development and maturation, *Physiol. Plant.*, 34, 39—43, 1975.
2. **Abeles, F. B.**, *Ethylene in Plant Biology*, Academic Press, New York, 1973.
3. **Abu-Shakra, S. S., Phillips, D. A., and Huffaker, R. C.**, Nitrogen fixation and delayed leaf senescence in soybeans, *Science*, 199, 973—975, 1978.
4. **Adedipe, N. O., Hunt, L. A., and Fletcher, R. A.**, Effects of benzyladenine on photosynthesis, growth, and senescence of the bean plant, *Physiol. Plant.*, 25, 151—153, 1971.
5. **Akamine, E. K.**, Ethylene production in fading *Vanda* orchid blossoms, *Science*, 140, 1217—1218, 1963.
6. **Allison, J. C. S. and Weinmann, H.**, Effect of the absence of the developing grain on carbohydrate content and senescence of maize leaves, *Plant Physiol.*, 46, 435—436, 1970.
7. **Alvim, R., Hewett, E. W., and Saunders, P. F.**, Seasonal variation in the hormonal content of willow. I. Changes in abscisic acid content and cytokinin activity in the xylem sap, *Plant Physiol.*, 57, 474—476, 1976.
8. **Anderson, J. W. and Rowan, K. S.**, Activity of peptidase in tobacco leaf tissue in relation to senescence, *Biochem. J.*, 97, 741—746, 1965.
9. **Antoszewksi, R. and Lis, E.**, Translocation of some radioactive compounds from the strawberry receptacle to the mother plant, *Bull. Acad. Pol. Sci., Ser. Sci. Biol.*, 16, 443—446, 1968.
10. **Aronoff, S.**, Translocation from soybean leaves, *Plant Physiol.* , 30, 184—185, 1955.
11. **Asana, R. D. and Singh, D. N.**, On the relation between flowering time, root, growth, and soil moisture extraction in wheat under non-irrigated cultivation, *Indian J. Plant Physiol.*, 10, 154—169, 1967.
12. **Ashby, E.**, Studies in the morphogenesis of leaves. VI. Some effects of length of day upon leaf shape in *Ipomoea caerulea*, *New Phytol.*, 49, 375—387, 1950.
13. **Aslam, M., Lowe, S. B., and Hunt, L. A.**, Effect of leaf age on photosynthesis and transpiration of cassava *(Manihot esculenta)*, *Can. J. Bot.*, 55, 2288—2295, 1977.
14. **Aspinall, D., Paleg, L., and Addicott, F.**, Abscisin and some hormone-regulated plant responses, *Aust. J. Biol. Sci.*, 20, 869—882, 1967.
15. **Back, A., Bittner, S., and Richmond, A. E.**, The effect of abscisic acid on the metabolism of kinetin in detached leaves of *Rumex pulcher*, *J. Exp. Bot.*, 23, 744—750, 1972.
16. **Baddeley, M. S. and Simon, E. W.**, An inhibitor of mitochondrial oxidations in extracts prepared from senescing leaves, *J. Exp. Bot.*, 20, 94—99, 1969.
17. **Balz, H. P.**, Intracellulare Lokalization und Funktion von hydrolytischen Enzymen bei Tabak, *Planta*, 70, 207—236, 1966.
18. **Barton, R.**, Fine structure of mesophyll cells in senescing leaves of *Phaseolus*, *Planta*, 71, 314—325, 1966.
19. **Batt, T. and Woolhouse, H. W.**, Changing activities during senescence and sites of synthesis of photosynthetic enzymes in leaves of Labiate, *Perilla frutescens (L.) Britt. J. Exp. Bot.*, 26, 569—579, 1975.
20. **Beever, J. E. and Woolhouse, H. W.**, Increased cytokinin export from the roots of *Perilla frutescens* following disbudding or floral induction, in *Mechanisms of Regulation of Plant Growth*, Bielski, R. L., Ferguson, A. R., and Cresswell, M. M., Eds., The Royal Society of New Zealand, Wellington, 1974, 681—686.
21. **Beever, J. E. and Woolhouse, H. W.**, Changes in the growth of roots and shoots when *Perilla frutescens* is induced to flower, *J. Exp. Bot.*, 26, 451—464, 1975.
22. **Beevers, L.**, Effect of gibberellic acid on the senescence of leaf discs of nasturtium *(Tropaeolum majus)*, *Plant Physiol.*, 41, 1074—1076, 1966.
23. **Beevers, L.**, Growth regulator control of senescence in leaf discs of nasturtium *(Tropaeolum majus)*, in *Physiology of Plant Growth Substances*, Wightman, F. and Setterfield, G., Eds., Runge, Ottawa, 1968, 1417—1435.
24. **Beevers, L.**, Senescence, in *Plant Biochemistry*, 3rd ed., Bonner, J. and Varner, J. E., Eds., Academic Press, New York, 1976, 771—794.
25. **Belikov, I. F.**, The local utilization of photosynthetic products in soybeans, *Dokl. Akad. Nauk SSSR*, 102, 379—381, 1955.
26. **Ben-Zioni, A., Itai, C., and Vaadia, Y.**, Water and salt stresses, kinetin and protein synthesis in tobacco leaves, *Plant Physiol.*, 42, 361—365, 1967.
27. **Bernard, R. L.**, Two genes affecting stem termination in soybeans, *Crop Sci.*, 12, 235—239, 1972.
28. **Bernard, R. L., and Weiss, M. G.**, Qualitative genetics, in *Soybeans: Improvement, Production and Uses*, American Society of Agronomists, Madison, 1973, 117—154.

29. **Bethlenfalvay, G. J. and Phillips, D. A.,** Ontogenetic interactions between photosynthesis and symbiotic nitrogen fixation in legumes, *Plant Physiol.*, 60, 419—421, 1977.
30. **Blomquist, R. V. and Kust, C. A.,** Translocation pattern of soybeans as affected by growth substances and maturity, *Crop Sci.*, 11, 390—393, 1971.
31. **Booth, A. Moorby, J., Davies, C. R., Jones, H., and Wareing, P. F.,** Effects of indolyl-3-acetic acid on the movements of nutrients within plants, *Nature (London)*, 194, 204—205, 1962.
32. **Borohov, A., Tirosh, T., and Halevy, A. H.,** Abscisic acid content of senescing petals on cut rose flowers as affected by sucrose and water stress, *Plant Physiol.*, 58, 175—178, 1976.
33. **Borst, M. L. and Thatcher, L. E.,** L Life history and composition of the soybean plant, *Ohio Agric. Exp. Stn. Res. Bull.*, 494, 1—96, 1931.
34. **Böttger, M.,** Die hormonale Regulation des Blattfals bei *Coleus rehneltianus* Berger. II. Die naturliche Rolle von Abscisinsäure im Blattfallproze ss, *Planta*, 93, 205—213, 1970.
35. **Boughey, A. S.,** Physiological cotton wilt in the Sudan Gezira, *Ann. Appl. Biol.*, 31, 12—18, 1944.
36. **Bourbouloux, A. and Bonnemain, J. L.,** Transport de l'auxine-^{14}C en provenance de jeunes gousses de *Vicia faba* L., *Planta*, 115, 161—172, 1972.
37. **Brenchley, W. E. and Jackson, V. G.,** Root development in barley and wheat under different conditions of growth, *Ann. Bot. (London)*, 35, 533—556, 1921.
38. **Brian, P. W., Petty, J. H. P., and Richmond, P. T.,** Effects of gibberellic acid on development of autumn colour and leaf-fall of deciduous woody plants, *Nature (London)*, 183, 58—59, 1959.
39. **Bruinsma, J.,** Rolle der Cytokinine bei Bluten- und Früchteentwicklung, *Z. Pflanzenernaehr. Bodenkd.*, 140, 15—23, 1977.
40. **Burg, S. P.,** The physiology of ethylene formation, *Annu. Rev. Plant Physiol.*, 13, 265—302, 1962.
41. **Burrows, W. J. and Carr, D. J.,** Effects of flooding the root system of sunflower plants on the cytokinin content in the xylem sap, *Physiol. Plant.*, 22, 1105—1112, 1969.
42. **Burrows, W. J. and Carr, D. J.,** Cytokinin content of pea seeds during their growth and development, *Physiol. Plant.*, 23, 1064—1070, 1970.
43. **Butcher, H. C., Wagner, G. J., and Siegelman, H. W.,** Localization of acid hydrolases in protoplasts: examination of the proposed lysosomal function of the mature vacuole, *Plant Physiol.*, 59, 1098—1103, 1977.
44. **Butler, R. D.,** The fine structure of senescing cotyledons of cucumber, *J. Exp. Bot.*, 18, 535—543, 1967.
45. **Butler, R. D. and Simon, E. W.,**Ultrastructural aspects of senescence in plants, *Adv. Gerontol. Res.*, 3, 73—129, 1971.
46. **Callow, J. A.,** Ribosomal RNA, fraction I protein synthesis and ribulose diphosphate carboxylase activity in developing and senescing leaves of cucumber, *New Phytol.*, 73, 13—20, 1974.
47. **Callow, J. A., Callow, M. E., and Woolhouse, H. W.,** *In vitro* protein synthesis, RNA, and polyribosomes in senescing leaves of *Perilla, Cell Differ.*, 1, 79—90, 1972.
48. **Carlson, J. B.,** Morphology, in *Soybeans: Improvement, Production and Uses*, Caldwell, B. E., Ed., American Society of Agronomists, Madison, 1973, 17—95.
49. **Carpenter, R. W., Haas, H. J., and Miles, E. F.,** Nitrogen uptake by wheat in relation to nitrogen content of soil, *Agron. J.*, 44, 420—423, 1952.
50. **Carr, D. J. and Burrows, W. J.,** Evidence of the presence in xylem sap of substances with kinetin-like activity, *Life Sci.*, 5, 2061—2077, 1966.
51. **Carr, D. J., Reid, D. M., and Skene, K. G. M.,** The supply of gibberellins from root to the shoot, *Planta*, 63, 382—392, 1964.
52. **Chatterjee, A. R., Mandal, R. K., and Sircar, S. M.,** Changes in the levels of growth substances during grain filling of rice, *Indian J. Plant Physiol.*, 19, 254—258, 1976.
53. **Chibnall, A. C.,** *Protein Metabolism in the Plant*, Yale University Press, New Haven, 1939.
54. **Chibnall, A. C.,** Protein metabolism in rooted runner bean leaves, *New Phytol.* 53, 31—37, 1954.
55. **Chin, T. and Beevers, L.,** Changes in endogenous growth regulators in Nasturtium leaves during senescence, *Planta*, 92, 178—188, 1970.
56. **Choe, H. T. and Thimann, K. V.,** The metabolism of oat leaves during senescence. III. The senescence of isolated chloroplasts, *Plant Physiol.*, 55, 828—834, 1975.
57. **Choe, H. T. and Thimann, K. V.,** The retention of photosynthetic activity by senescing chloroplasts of oat leaves, *Planta*, 135, 101—108, 1977.
58. **Clark, R. B.,** Mineral element concentrations in corn leaves by position on the plant and age, *Commun. Soil Sci. Plant Anal.*, 6, 439—450, 1975.
59. **Cole, L. C.,** The population consequences of life history phenomena, *Q. Rev. Biol.*, 29, 103—137, 1954.
60. **Colquhoun, A. J. and Hillman, J. R.,** The effects of abscisic acid on senescence in leaf discs of radish, *Raphanus sativus* L., *Planta*, 105, 213—224, 1972.

61. **Colquhoun, A. J. and Hillman, J. R.,** Endogenous abscisic acid and the senescence of leaves of *Phaseolus vulgaris* L., *Z. Pflanzenphysiol.*, 76, 326—332, 1975.
62. **Colquhoun, A. J., Hillman, J. R., Crewe, C., and Bowes, B. G.,** An ultrastructural study of the effects of abscisic acid on senescence of leaves of radish *(Raphanus sativus* L.) *Protoplasma*, 84, 205—222, 1975.
63. **Cooper, D. R., Hill-Cottingham, D. G., and Lloyd-Jones, C. P.,** Absorption and re-distribution of nitrogen during growth and development of field bean, *Vicia faba, Physiol. Plant.*, 38, 313—318, 1976.
64. **Crafts, A. S. and Crisp, C. E.,** *Phloem Transport in Plants*, W. H. Freeman, San Francisco, 1971.
65. **Crowther, F.,** Studies in growth analysis of the cotton plant under irrigation in the Sudan. I. The effects of different combinations of nitrogen applications and water supply, *Ann. Bot., (London)*, 48, 877—913, 1934.
66. **Dale, J.,** Some effects of the continuous removal of floral buds on the growth of the cotton plant, *Ann. Bot. (London)*, 92, 636—649, 1959.
67. **Danielson, L. L.,** Effect of daylength on growth and reproduction of the cucumber, *Plant Physiol.*, 19, 638—648, 1944.
68. **Davey, J. E. and van Staden, J.,** Cytokinin translocation: changes in zeatin and zeatin-riboside levels in the root exudate of tomato plants during their development, *Planta*, 130, 69—72, 1976.
69. **Davey, J. E. and van Staden, J.,** A cytokinin complex in the developing fruits of *Lupinus albus, Physiol. Plant.*, 39, 221—224, 1977.
70. **Davies, P. J., Probesting, W. M., and Gianfagna, T. J.,** Hormonal relationships in whole plant senescence, in *Plant Growth Regulation, 9th International Conference on Plant Growth Substances*, Pilet, P. E., Ed., Springer-Verlag , Berlin, 1977, 273—280.
71. **Davis, L. A. and Addicott, F. T.,** Abscisic acid: correlations with abscission and with development in the cotton fruit, *Plant Physiol.*, 49, 644—648, 1972.
72. **Davis, S. D., van Bavel, C. H. M., and McCree, K. J.,** Effect of leaf aging upon stomatal resistance in bean plants, *Crop Sci.*, 17, 640—645, 1977.
73. **Davis, S. D. and McCree, K. J.,** Photosynthetic rate and diffusion conductance as a function of age in leaves of bean plants, *Crop Sci.*, 18, 280—282, 1978.
74. **De Leo, P., and Sacher, J. A.,** Control of ribonuclease and acid phosphatase by auxin and abscisic acid during senescence of *Rheo* leaf sections, *Plant Physiol.*, 46, 806—811, 1970.
75. **Dearborn, R. B.,** Nitrogen nutrition and chemical composition in relation to growth and fruiting of the cucumber plant, *Mem. Cornell Agr. Expt. Station*, 192, 1936.
76. **Derman, B. D., Rupp, D. C., and Noodén, L. D.,** Mineral distribution in relation to fruit development and monocarpic senescence in Anoka soybeans, *Am. J. Bot.*, 65, 205—213, 1978.
77. **Dodge, J. D.,** Changes in chloroplast fine structure during the autumnal senescence of *Betula* leaves, *Ann. Bot., (London)*, 34, 817—824, 1970.
78. **Dörffling, K.,** Uber das Wuchsstoff-Hemmstoff-system von *Acer pseudoplutanus* L. I. Der Jahresgang der Wuchs und Hemmstoffe in Knospen, Blattern und im Kambium, *Planta*, 60, 390—412, 1963.
79. **Draper, S. R.,** Lipid changes in senescing cucumber cotyledons, *Phytochemistry*, 8, 1641—1647, 1969.
80. **Drivdahl, R. H. and Thimann, K. V.,** Proteases of senescing oat leaves. II. Reaction to substrates and inhibitors, *Plant Physiol.*, 61, 501—505, 1978.
81. **Dumbroff, E. B. and Brown, D. C. W.** Cytokinin and inhibitor actipity in roots and stems of sugar maple seedlings through the dormant season, *Can. J. Bot.*, 54, 191—197, 1976.
82. **Dumbroff, E. B., Brown, O. C. W., and Thompson, J. E.,** Effect of senescence on levels of free abscisic acid and water potentials in cotyledons of bean, *Bot. Gaz. (Chicago)*, 138, 261—265, 1977.
83. **Eaton, F. M.,** Early defloration as a method of increasing cotton yields and the relation of fruitfulness to fiber and boll characters, *J. Agric. Res. (Washington, D.C.,)* 42, 447—462, 1931.
84. **Eaton, F. M. and Ergle, D. R.,** Fiber properties and carbohydrate and nitrogen levels of cotton plants as influenced by moisture supply and fruitfulness, *Plant Physiol.*, 27, 541—562, 1952.
85. **Ecklund, P. R. and Moore, T. C.,** Quantitative changes in gibberellin and RNA correlated with senescence of the shoot apex in the "Alaska" pea, *Am. J. Bot.*, 55, 494—503, 1968.
86. **Ecklund, P. R. and Moore, T. C.,** RNA and protein metabolism in senescent shoot apices of "Alaska" peas, *Am. J. Bot.*, 56, 327—334, 1969.
87. **Eeuwens, C. J. and Schwabe, W. W.,** Seed and pod wall depelopment in *Pisum sativum*, L. in relation to extracted and applied hormones, *J. Exp. Bot.*, 26, 1—14, 1975.
88. **Egli, D. B. and Leggett, J. E.,** Rate of dry matter accumulation in soybean seeds with varying source-sink ratios, *Agron. J.*, 68, 371—374, 1976.
89. **El-Antably, H. M. M., Wareing, P. F., and Hillman, J.,** Some physiological responses to DL-abscisin (Dormin), *Planta*, 73, 74—90, 1967.

90. **Engelbrecht, L.**, Cytokinins in buds and leaves during growth, maturity, and ageing (with a comparison of two bioassays), *Biochem. Physiol. Pflanz.*, 162, 547—558, 1971.
91. **Engelbrecht, L. and Conrad, K.**, Vergleichende Untersuchungen zur Wirkung von Kinetin und Auxin, *Ber. Dtsch. Bot. Ges.*, 74, 42—46, 1961.
92. **Evans, L. T. Wardlaw, I. F., and Fischer, R. A.**, Wheat, in *Crop Physiology: Some Case Histories*, Evans, L. T., Ed., Cambridge University Press, London, 1975, 101—149.
93. **Even-Chen, Z. and Itai, C.**, The role of abscisic acid in senescence of detached tobacco leaves, *Physiol. Plant.*, 34, 97—100, 1975.
94. **Farkas, G. L.**, Hormonal control of nucleases in plants, *Acta Bot. Acad. Sci. Hung.*, 20, 31—36, 1974.
95. **Feller, U. K.**, Proteolytic enzymes in leaves of field grown beans and wheat, *Plant Physiol.*, 61, (Suppl.), 27, 1978.
96. **Feller, U. K., Soong, T. S. T., and Hageman, R. H.**, Leaf proteolytic activities and senescence during grain development of field grown corn, *(Zea mays* L.) *Plant Physiol.*, 59, 290—294, 1977.
97. **Ferguson, C. and Simon, E. W.**, Membrane lipids in senescing green tissues, *J. Exp. Bot.*, 24, 307—316, 1973.
98. **Fischer, H.**, Der Mineralstofftransport, *Encycl. Pl. Physiol.*, XIII, 1967, 200—268.
99. **Fitting, H.**, Untersuchungen uber die vorzeitige Entblätterung von Blüten, *Jahrb. Wiss. Bot.*, 49, 187—263, 1911.
100. **Fletcher, R. A.**, Retardation of leaf senescence by benzyladenine in intact bean plants, *Planta*, 89, 1—8, 1969.
101. **Fletcher, R. A., Oegema, T., and Horten, R. F.**, Endogenous gibberellin levels and senescence in *Taraxacum officinale*, *Planta*, 86, 98—102, 1969.
102. **Fletcher, R. A. and Osborne, D. J.**, Regulation of protein and nucleic acid synthesis by gibberellin during leaf senescence, *Nature (London)*, 207, 1176—1177, 1965.
103. **Foth, H. D.**, Root and top growth of corn, *Agron. J.*, 54, 49—52, 1962.
104. **Fraser, D. E. and Bidwell, R. G. S.**, Photosynthesis and photorespiration during the ontogeny of the bean plant, *Can. J. Bot.*, 52, 2561—2570, 1974.
105. **Freeland, R. O.**, Effect of age of leaves upon the rate of photosynthesis in some conifers, *Plant Physiol.*, 27, 685—690, 1952.
106. **Gautheret, R. J.**, Sur la possibilité de réaliser la culture indéfinie des tissus de tubercules de carrote, *C. R. Acad. Sci.*, 208, 118—120, 1939.
107. **Geiger, D. R.**, Effects of translocation and assimilate demand on photosynthesis, *Can. J. Bot.*, 54, 2337—2345, 1976.
108. **Gianfagna, T. J. and Davies, P. J.**, Fruit-induced apical senescence in a genetic line of peas, *Plant Physiol.*, 61 (Suppl.), 26, 1978.
109. **Goldbach, E., Goldbach, H., Wagner, H., and Michael, G.**, Influence of N-deficiency on the abscisic acid content of sunflower plants, *Physiol. Plant.*, 34, 138—140, 1975.
110. **Goldbach, H. and Michael, G.**, Abscisic acid content of barley grains during ripening as affected by temperature and variety, *Crop Sci.*, 16, 797—799, 1976.
111. **Goldney, D. C. and van Steveninck, R. F. M.**, Ethylene production and biochemical changes in detached leaves of *Nymphoides indica*, in *Plant Growth Substances, 1970*, Carr, D. J., Ed., Springer-Verlag, Berlin, 1972, 604—610.
112. **Goldthwaite, J. J.**, Energy metabolism of *Rumex* leaf tissue in the presence of senescence-regulating hormones and sucrose, *Plant Physiol.*, 54, 399—403, 1974.
113. **Gordon, M. E., Letham, D. S., and Parker, C. W.**, Regulators of cell division in plant tissues. XVIII. The metabolism and translocation of zeatin in intact radish seedlings, *Ann. Bot. (London)*, 38, 809—825, 1974.
114. **Gregory, F. G.**, Mineral nutrition of plants, *Annu. Rev. Biochem.*, 6, 557—578, 1937.
115. **Grochowska, M. J.**, Translocation of indole-3-acetic acid-2-^{14}C injected into seeds of five-week-old apple fruits, *Bull. Acad. Pol. Sci., Sec. Sci. Biol.*, 16, 577—580, 1968.
116. **Gustafson, F. G.**, Growth studies on fruits. An explanation of the shape of the growth curve, *Plant Physiol.*, 2, 153—161, 1927.
117. **Halevy, A. H., Mayak, S., Tirosh, T., Spiegelstein, H., and Kofranek, A. M.**, Opposing effects of abscisic acid on senescence of rose flowers, *Plant Cell Physiol.*, 15, 813—821, 1974.
118. **Hall, A. J.**, Assimilate source-sink relationships in *Capsicum annuum* L. I. The dynamics of growth in fruiting and deflorated plants, *Aust. J. Plant Physiol.*, 4, 623—626, 1977.
119. **Hall, A. J. and Brady, C. J.**, Assimilate source-sink relationships in *Capsicum annuum* L. II. Effects of fruiting and defloration on the photosynthetic capacity and senescence of the leaves, *Aust. J. Plant Physiol.*, 4, 771—783, 1977.
120. **Hall, A. J. and Milthorpe, F. L.**, Assimilate source-sink relationships in *Capsicum annuum* L. III. The effects of fruit excision on photosynthesis and leaf and stem carbohydrates, *Aust. J. Plant Physiol.*, 5, 1—13, 1978.

121. **Hall, S. M. and Baker, D. A.,** The chemical composition of *Ricinus* phloem exudate, *Planta (Berlin),* 106, 131—140, 1972.
122. **Hall, W. C.,** Effects of photoperiod and nitrogen supply on growth and reproduction in the gherkin, *Plant Physiol.,* 24, 753—769, 1949.
123. **Hall, W. C., Truchelut, G. B., Leinweber, C. L., and Herrero, F. A.,** Ethylene production by the cotton plant and its effect under experimental and field conditions, *Physiol. Plant.,* 10, 306—317, 1957.
124. **Hammond, L. C., Black, C. A., and Norman, A. G.,** Nutrient uptake by soybeans on two Iowa soils, *Iowa Agr. Home Econ. Expt. Station Res. Bull.,* 384, 1951.
125. **Hanson, A. D. and Kende, H.,** Ethylene-enhanced ion and sucrose efflux in the morning glory flower tissue, *Plant Physiol.,* 55, 663—669, 1975.
126. **Hanson, J. B., Wilson, C. M., Chrispeels, M. J., Krueger, W. A., and Swanson, H. R.,** Ribonuclease and other factors involved in the respiratory senescence of maize scutellum, *J. Exp. Bot.,* 16, 282—293, 1965.
127. **Hanway, J. J. and Weber, C. R.,** N, P, and K percentages in soybean [*Glycine max* (L.)] plant parts, *Agron. J.,* 63, 286—290, 1971.
128. **Hardwick, K., Wood, M., and Woolhouse, H. W.,** Photosynthesis and respiration in relation to leaf age in *Perilla frutescens* (L.) Britt, *New Phytol.,* 67, 79—86, 1968.
129. **Hardwick, K. and Woolhouse, H. W.,** Foliar senescence in *Perilla frutescens* (L.) Britt., *New Phytol.,* 66, 545—552, 1967.
130. **Hardy, R. W. F., Burns, R. C., Herbert, R. R., Holsten, R. D., and Jackson, E. K.,** Biological nitrogen fixation: a key to world protein, in *Biological Nitrogen Fixation in Natural and Agricultural Habitats,* Lie, T. A. and Mulder, E. G., Eds., Martinus Nijhoff, The Hague, 1971, 561—590.
131. **Hardy, R. W. F., Burns, R. C., and Holsten, R. D.,** Applications of the acetylene-ethylene assay for measurement of nitrogen fixation, *Soil Biol. Biochem.,* 5, 47, 1973.
132. **Hardy, R. W. F., Holsten, R. D., Jackson, E. K., and Burns, R. C.,** The acetylene-ethylene assay for N_2 fixation: laboratory and field evaluation, *Plant Physiol.,* 43, 1185—1207, 1968.
133. **Harnischfeger, G.,** Studies on chloroplast degeneration *in vivo.* III. The effect of magnesium on electron transport and phosphorylation in plastids isolated from young and ageing cotyledons, *J. Exp. Bot.,* 25, 269—276, 1974.
134. **Harper, J. E.,** Seasonal nutrient uptake and accumulation patterns in soybeans, *Crop Sci.,* 11, 347—350, 1971.
135. **Harper, J. E. and Hageman, R. H.,** Canopy and seasonal profiles of nitrate reductase in soybeans [*Glycine max.* (L.) Merr.], *Plant Physiol.,* 49, 146—154, 1972.
136. **Harper, J. L.,** *Population Biology of Plants,* Academic Press, London, 1977.
137. **Hartung, W.,** Der Transport von [2-^{14}C] Abscisinsäure aus dem Wurzelsystem intakter Bohnenkeimlinge in die oberirdischen Teile der Pflanze, *Z. Pflanzenphysiol.,* 83, 81—84, 1977.
138. **Harvey, E. M.,** A preliminary report on the vegetative growth of the okra *(Hibiscus esculentus,* Linn.) in relation to the production of varying amounts of reproductive tissues, *Ore. Agric. Exp. Stn. Bull.,* 284, 1931.
139. **Henson, I. E. and Wareing, P. F.,** Cytokinins in *Xanthium strumarium* L.: distribution in the plant and production in the root system, *J. Exp. Bot.,* 27, 1268—1278, 1976.
140. **Henson, I. E. and Wareing, P. F.** An effect of defoliation on the cytokinin content of buds of *Xanthium strumarium,Plant Sci. Lett.,* 9, 27—31, 1977.
141. **Henson, I. E. and Wareing, P. F.,** Cytokinins in *Xanthium strumarium* L.: the metabolism of cytokinins in detached leaves and buds in relation to photoperiod, *New Phytol.,* 78, 27—33, 1977.
142. **Hernändez-Gil, R. and Schaedle, M.,** Functional and structural changes in senescing *Populus deltoides* (Bartr.) chloroplasts, *Plant Physiol.,* 51, 245—249, 1973.
143. **Herridge, D. F. and Pate, J. S.,** Utilization of net photosynthate for nitrogen fixation and protein production in an annual legume, *Plant Physiol.,* 60, 759—764, 1977.
144. **Hewett, E. W. and Wareing, P. F.,** Cytokinins in *Populus × robusta:* changes during chilling and bud burst, *Physiol. Plant,* 28, 393—399, 1973.
145. **Hewett, E. W. and Wareing, P. F.,** Cytokinins in *Populus × robusta:* qualitative changes during development, *Physiol. Plant,* 29, 386—389, 1973.
146. **Hicks, D. R. and Pendleton, J. W.,** Effect of floral bud removal on performance of soybeans, *Crop Sci.,* 9, 435—437, 1969.
147. **Hildebrand, F.,** Die Lebensdauer und Vegetationsweise der Pflanzen, ihre Ursachen und ihre Entwicklung, *Bot. Jahrb.,* 2, 51—135, 1882.
148. **Hoad, G. V.,** (±) abscisin II, [(±)dormin] in phloem exudate of willow, *Life Sci.,* 6, 1113—1118, 1967.
149. **Hocking, P. J. and Pate, J. S.,** Mobilization of minerals to developing seeds of legumes, *Ann. Bot. (London),* 41, 1259—1278, 1977.

150. **Howell, R. W.**, Physiology of the soybean, *Adv. Agron.*, 12, 265—310, 1960.
151. **Hsia, C. P. and Kao, C. H.**, The importance of roots in regulating the senescence of soybean primary leaves, *Physiol. Plant.*, 43, 385—389, 1978.
152. **Hume, D. J. and Criswell, J. G.**, Distribution and utilization of ^{14}C-labelled assimilates in soybeans, *Crop Sci.*, 13, 519—524, 1973.
153. **Hurst, P. R. and Gahan, P. B.**, Turnover of DNA in ageing tissues of *Lycopersicon esculentum, Ann. Bot. (London)*, 39, 71—76, 1975.
154. **Inada, K.**, Physiological characteristics of rice roots, especially with the view-point of plant growth stage and root age, *Bull. Natl. Inst. Agr. Sci. Jpn. Ser. D.*, 16, 19—156, 1967.
155. **Incoll, L. D. and Neales, T. F.**, The stem as a temporary sink before tuberization in *Helianthus tuberosus* L., *J. Exp. Bot.*, 21, 469—476, 1970.
156. **Itai, C. and Benzioni, A.**, Regulation of senescence and related metabolic processes, in *Plant Growth Substances, 1973*, Hirokawa Publishing, Tokyo, 1974, 1022—1026.
157. **Jackson, B. D.**, *A Glossary of Botanic Terms*, 4th ed., G. Duckworth and Co., London, 1928.
158. **Jackson, M. B. and Osborne, D. J.**, Ethylene, the natural regulator of leaf abscission, *Nature (London)*, 225, 1019—1022, 1970.
159. **Jacobs, W. P., Shield, J. A., and Osborne, D. J.** Senescence factor and abscission of Coleus leaves, *Plant Physiol.*, 37, 104—106, 1962.
160. **James, A. L., Anderson, I. C., and Greer, H. A. L.**, Effects of naphthaleneacetic acid on field-grown soybeans, *Crop Sci.*, 5, 472—474, 1965.
161. **James, W. O.**, *Plant Respiration*, Clarendon Press, Oxford, 1953.
162. **Jeffcoat, B. and Harris, G. P.**, Hormonal regulation of ^{14}C-labelled assimilates in the flowering shoot of carnation, *Ann. Bot. (London)*, 36, 353—361, 1972.
163. **Jones, R. L. and Phillips, I. D. J.**, Organs of gibberellin synthesis in light-grown sunflower plants, *Plant Physiol.*, 41, 1381—1386, 1966.
164. **Kahanak, G. M., Okatan, Y., Rupp, D. C., and Noodén, L. D.**, Hormonal and genetic alteration of monocarpic senescence in soybeans, *Plant Physiol.*, 61 (Suppl.), 26, 1978.
165. **Kannangara, C. G. and Woolhouse, H. W.**, Changes in the enzyme activity of soluble protein fractions in the course of foliar senescence in *Perilla frutescens* (L.) Britt., *New Phytol.*, 67, 533—542, 1967.
166. **Kannangara, T. and Booth, A.**, Diffusible cytokinins in shoot apices of *Dahlia variabilis, J. Exp. Bot.*, 25, 459—467, 1974.
167. **Karanov, E. N. and Pogoncheva, E. M.**, Effect of abscisic acid on chlorophyll destruction in detached leaves and its interaction with other growth regulators, *Fiziol. Rast. (Sofia)*, 2, 3—12, 1976.
168. **Kawashima, N., Imai, A., and Tamaki, E.**, Studies on protein metabolism in higher plants. IV. Changes in protein components in detached leaves of tobacco, *Plant Cell Physiol.*, 8, 595—604, 1967.
169. **Kende, H.** , Kinetinlike factors in root exudate of sunflower, *Proc. Natl. Acad. Sci. U.S.A.*, 53, 1302—1307, 1965.
170. **Kende, H.**, The cytokinins, *Int. Rev. Cytol.*, 31, 301—338, 1971.
171. **Kende, H. and Baumgartner, B.**, Regulation of aging in flowers of *Ipomoea tricolor* by ethylene, *Planta*, 116, 279—289, 1974.
172. **Kende, H. and Hanson, A. D.**, Relationship between ethylene evolution and senescence in morning glory flower tissue, *Plant Physiol.*, 57, 523—527, 1976.
173. **Kennedy, R. A.**, Relationship between leaf development, carboxylase enzyme activities and photorespiration in the C_4-plant *Portulaca oleracea* L., *Planta*, 128, 149—154, 1976.
174. **Kennedy, R. A. and Laetsch, W. M.**, Relationship between leaf development and primary photosynthetic products in the C_4-plant *Portulaca oleracea* L., *Planta*, 115, 113—124, 1973.
175. **Kijne, J. W.**, The fine structure of pea root nodules. 2. Senescence and disintegration of the bacteroid tissue, *Physiol. Plant Pathol.*, 7, 17—21, 1975.
176. **King, R. W.**, Abscisic acid in developing wheat grains and its relationship to grain growth and maturation, *Planta*, 132, 43—51, 1976.
177. **King, R. W. and Evans, L. T.**, Inhibition of flowering in *Lolium temulentum* L. by water stress: a role for abscisic acid, *Aust. J. Plant Physiol.*, 4, 225—233, 1977.
178. **Kisaki, T. S., Hirabayashi, S., and Yano, N.**,Effect of the age of tobacco leaves on photosynthesis and photorespiration , *Plant Cell Physiol.*, 14, 505—514, 1973.
179. **Klämbt, D.**, Cytokinine aus *Helianthus annuus, Planta*, 82, 170—178, 1968.
180. **Klucas, R. V. and Arp, D.**, Physiological and biochemical studies on senescing tap root nodules of soybeans, *Can. J. Microbiol.*, 23, 1426—1432, 1977.
181. **Knypl, J. S. and Mazureyk, W.**, Arrest of chlorophyll and protein breakdown in senescing leaf discs of kale by cycloheximide and vanillin, *Curr. Sci.*, 40, 294—295, 1971.

182. **Kollman, G. E., Streerer, J. G., Jeffers, D. L., and Curry, R. B.,** Accumulation and distribution of mineral nutrients, carbohydrate, and dry matter in soybean plants as influenced by reproductive sink size, *Agron. J.*, 66, 549—554, 1974.
183. **Krizek, D. T., McIlrath, W. J., and Vergara, B. S.,** Photoperiodic induction of senescence in *Xanthium* plants, *Science*, 151, 95—96, 1966.
184. **Krul, W. R.,** Nucleic acid and protein metabolism in senescing and regenerating soybean cotyledons, *Plant Physiol.*, 54, 36—40, 1974.
185. **Kulaeva, O. N.,** The effect of roots on leaf metabolism in relation to the action of kinetin on leaves, *Fiziol. Rast.*, 9, 182—189, 1962.
186. **Laloue, M.,** Cytokinins: 7-glucosylation is not a prerequisite of the expression of their biological activity, *Planta*, 134, 273—275, 1977.
187. **LaMarche, V. C.,** Environment in relation to age of bristlecone pines, *Ecology*, 50, 53—59, 1969.
188. **LaRue, T. A. G. and Kurz, W. G. W.,** Estimation of nitrogenase in intact legumes, *Can. J. Microbiol.*, 19, 304—305, 1973.
189. **Lawn, R. J. and Brun, W. A.,** Symbiotic nitrogen fixation in soybeans, I. Effect of photosynthetic source-sink manipulations, *Crop Sci.* 14, 11—16, 1974.
190. **Leggett, J. E. and Frere, M. H.,** Growth and nutrient uptake by soybean plants in nutrient solutions of graded concentrations, *Plant Physiol.*, 48, 457—460, 1971.
191. **Leonard, E. R.,** Interrelations of vegetative and reproductive growth with special reference to indeterminate plants, *Bot. Rev.*, 28, 353—410, 1962.
192. **Leonard, O. A.,** Translocation of carbohydrates in the sugar beet, *Plant Physiol.*, 14, 55—74, 1939.
193. **Leopold, A. C.,** Senescence in plant development, *Science*, 134, 1727—1732, 1961.
194. **Leopold, A. C.,** Aging, senescence, and turnover in plants, *BioScience*, 25, 659—662, 1975.
195. **Leopold, A. C.,** Niedergang-Kamien, E., and Janick, J., Experimental modification of plant senescence, *Plant Physiol.*, 34, 570—573, 1959.
196. **Letham, D. S.,** Regulators of cell division in plant tissues. XX. The cytokinins of coconut milk, *Physiol. Plant*, 32, 66—70, 1974.
197. **Lindoo, S. J. and Noodén, L. D.,** Studies on the role of cytokinin- and ABA-like substances in monocarpic senescence of soybeans, *Plant Physiol.*, 57, (Suppl.), 27, 1976.
198. **Lindoo, S. J. and Noodén, L. D.,** The interrelation of fruit development and leaf senescence in "Anoka" soybeans, *Bot. Gaz. (Chicago)*, 137, 218—223, 1976.
199. **Lindoo, S. J. and Noodén, L. D.,** Behavior of the soybean senescence signal, *Plant Physiol.*, 59, 1136—1140, 1977.
200. **Lindoo, S. J. and Noodén, L. D.,** Correlations of cytokinins and abscisic acid with monocarpic senescence in soybean, *Plant Cell Physiol.*, 19, 997—1006, 1978.
201. **Lockhart, J. A.,** Studies of the organ of production of the natural gibberellin factor in higher plants, *Plant Physiol.*, 32, 204—207, 1957.
202. **Lockhart, J. A. and Gottschall, V.,** Fruit-induced and apical senescence in *Pisum sativum*, *Plant Physiol.*, 36, 389—398, 1961.
203. **Loehwing, F. W.,** Mineral nutrition in relation to the ontogeny of plants, in *Mineral Nutrition of Plants*, Troug, E., Ed., University of Wisconsin Press, Madison, 1951, 343—358.
204. **Loomis, W. E.,** The translocation of carbohydrates in maize, *Iowa State Coll. J. Sci.*, 9, 295—306, 1935.
205. **Loomis, W. E.,** Growth correlation, in *Growth and Differentiation of Plants*, Loomis, W. E., Ed., Iowa State College Press, Ames, 1953, 197—217.
206. **Lovell, P. H., Oo, H. T., and Sagar, G. R.,** An investigation into the rate and control of assimilate movement from leaves in *Pisum sativum*, *J. Exp. Bot.*, 23, 255—266, 1972.
206a. **Loveys, B. R.,** The intracellular location of abscissic acid in stressed and nonstressed leaf tissue, *Physiol. Plant*, 40, 6—10, 1977.
207. **MacNicol, P. K., Young, R. E., and Biale, J. B.,** Metabolic regulation in the senescing tobacco leaf. I. Changes in pattern of ^{32}P incorporation into leaf disc metabolites, *Plant Physiol.*, 51, 793—797, 1973.
208. **Malik, N. S. A.,** Endogenous Control of Senescence in Garden Pea (*Pisum sativum* L.) Ph.D. thesis, Glasgow University, Scotland, 1973.
209. **Malik, N. S. A. and Berrie, A. M. M.,** Correlative effects of fruits and leaves in senescence of pea plants, *Planta*, 124, 169—175, 1975.
210. **Malik, N. S. A. and Davies, P. J.,** The effect of fruit development on the growth capacity of apical meristems, *Planta*, 129, 191—192, 1976.
211. **Mandahar, C. L. and Garg, I. D.,** Effect of ear removal on sugars and chlorophylls of barley leaves, *Photosynthetica*, 9, 407—409, 1975.
212. **Martin, C. and Thimann, K. V.,** The role of protein synthesis in the senescence of leaves. I. The formation of protease, *Plant Physiol.*, 49, 64—71, 1972.

213. **Martin, C. and Thimann, K. V.,** Role of protein synthesis in the senescence of leaves. II. The influence of amino acids on senescence, *Plant Physiol.*, 50, 432—437, 1972.
214. **Marx, G. A.,** Influence of genotype and environment on senescence in peas, *BioScience,* 18, 505—506, 1968.
215. **Mason, T. G.,** Growth and abscission in sea island cotton, *Ann. Bot. (London),* 36, 454—484, 1922.
216. **Matile, P.,** Vacuoles, in *Plant Biochemistry,* 3rd ed., Bonner, J. and Varner, J., Eds., Academic Press, New York, 1976.
217. **Matile, P. and Winkenbach, F.,** Function of lysosomes and lysosomal enzymes in senescing corolla of the morning glory *(Ipomoea purpurea), J. Exp. Bot.*, 122, 759—771, 1971.
218. **Mayak, S. and Dilley, D. R.,** Regulation of senescence in carnation (*Dianthus caryophyllus*). Effect of abscisic acid and carbon dioxide on ethylene production, *Plant Physiol.*, 58, 663—665, 1976.
219. **Mayak, S. and Halevy, A. H.,** Interrelationships of ethylene and abscisic acid in the control of rose petal senescence, *Plant Physiol.*, 50, 341—346, 1972.
220. **Mayak, S., Halevy, A. H., and Katz, M.,** Correlative changes in phytohormones in relation to senescence processes in rose petals, *Physiol. Plant.*, 27, 1—4, 1972.
221. **McAlister, D. F. and Krober, O. A.,** Response of soybeans to leaf and pod removal, *Agron. J.*, 50, 674—677, 1958.
222. **McCalla, A. G.** The effect of nitrogen nutrition on the protein and nonprotein nitrogen of wheat, *Can. J. Res.*, 9, 542—570, 1933.
223. **McGlasson, W. D. and Adato, I.,** Changes in the concentrations of abscisic acid in fruits of normal and Nr, rin, and nor mutant tomatoes during growth, maturation, and senescence, *Aust. J. Plant Physiol.*, 3, 809—817, 1976.
224. **McWha, J. A.,** Changes in abscisic acid levels in developing grains of wheat *(Triticum aestivum* L.), *J. Exp. Bot.*, 26, 823—827, 1975.
225. **Medawar, P. B.,** *The Uniqueness of the Individual,* Methuen, London, 1957.
226. **Meyer, A.,** Eiweisstoffwechsel und Vergilben der Laubbätter von *Tropaeolum majus, Flora (Jena),* 111, 85—127, 1918.
227. **Michael, G.,** Über die Beziehungen zwischen Chlorophyll und Eiweissabbau in vergilbenden Laubblatt von *Tropaeolum,Z. Bot.*, 29, 385—425, 1936.
228. **Milborrow, B. V.,** The chemistry and physiology of abscisic acid, *Annu. Rev. Plant Physiol.*, 25, 259—307, 1974.
229. **Milborrow, B. V. and Robinson, D. R.,** Factors affecting the biosynthesis of abscisic acid, *J. Exp. Bot.*, 24, 537—548,1973.
230. **Milthorpe, F. L. and Moorby, J.,** Vascular transport and its significance in plant growth, *Annu. Rev. Plant Physiol.*, 20, 117—138, 1969.
231. **Mitchell, R. L. and Russell, W. J.,** Root development and rooting patterns of soybeans [*Glycine max* (L.) Merrill] evaluated under field conditions, *Agron. J.*, 63, 313—316, 1971.
232. **Mittelheuser, C. J. and van Steveninck, R. F. M.,** Stomatal closure and inhibition of transpiration induced by (RS)-abscisic acid, *Nature (London),* 221, 281—282, 1969.
233. **Mittelheuser, C. J. and van Steveninck, R. F. M.,** Effect of ABA and kinetin on ultrastructure of senescing wheat leaves, in *Plant Growth Substances, 1970,* Carr, D. J., Ed., Springer-Verlag, Berlin, 1972, 618—623.
234. **Mizrahi, Y., Blumenfeld, A., Bittner, S., and Richmond, A. E.,** Abscisic acid and cytokinin content of leaves in relation to salinity and relative humidity, *Plant Physiol.*, 48, 752—755, 1971.
235. **Molisch, H.,** *The Longevity of Plants,* Fulling, H., (Transl.), Science Press, Lancaster, Pa., 1938.
236. **Mondal, M. H., Brun, W. A., and Brenner, M. L.,** Effects of sink removal on photosynthesis and senescence in leaves of soybean *(Glycine max* L.) plants, *Plant Physiol.*, 61, 394—397, 1978.
237. **Moore, A. E. and Stone, B. A.,** Effect of senescence and hormone treatment on the β-1,3-glucan hydrolase in *Nicotiana glutinosa* leaves, in *Plant Growth Substances, 1970,* Carr, D. J., Ed., Springer-Verlag, Berlin, 1972, 598—603.
238. **de Mooy, C. J., Pesek, J., and Spaldon, E.,** Mineral nutrition, in *Soybeans: Improvement, Production and Uses,* Caldwell, B. E., Ed., American Society Agronomy, Madison, 1973, 267—352.
239. **Mothes, K.,** Über das Altern der Blatter und die Möglichkeit ihrer Wiederverjungung, *Naturwissenschaften,* 47, 337— 35, 1960.
240. **Mothes, K. and Engelbrecht, L.,** Über geschlechtsverschieden Stoffwechsel zweihäusiger einjähriger Pflanzen I. Untersuchungen uber den Stickstoffumsatz beim Hauf (*Cannabis sativa* L.), *Flora (Jena),* 139, 1—27, 1952.
241. **Mothes, K., Engelbrecht, L., and Kulaeva, O.,** Über die Wirkung des Kinetins auf Stickstoffverteilung und Eiweissynthese in isolierte Blattern, *Flora (Jena),* 147, 445—464, 1959.
242. **Mothes, K., Engelbrecht, L., and Schutte, H. R.,** Über die Akkumulation von α-aminoisobuttersaure im Blattgewebe unter dem Einfluss von Kinetin, *Physiol. Plant,* 14, 72—75, 1961.

243. **Mulchi, C. L., Volk, R. S. and Jackson, W. A.,** Oxygen exchange of illuminated leaves at carbon dioxide compensation, in *Photosynthesis and Photorespiration,* Hatch, M. D., Osmund, C. B., and Slayter, R. O., Eds., Interscience, Toronto, 1971, 35—50.
244. **Müller, K. and Leopold, A. C.,** Correlative aging and transport of ^{32}P in corn leaves under the influence of kinetin, *Planta,* 68, 167—185, 1966.
245. **Mullins, M. E. and Osborne, D. J.,** Effect of abscisic acid on growth correlation in *Vitis vinifera* (L.), *Aust. J. Biol. Sci.,* 23, 479—485, 1970.
246. **Mullins, M. G.,** Hormone-directed transport of assimilates in decapitated internodes of *Phaseolus vulgaris* L., *Ann. Bot. London,* 34, 897—909, 1970.
247. **Mullins, M. G., Harsono, S., and Batten, D. J.,** Sources of variation in cytokinin-induced bud formation in leaf disks of *Nictoiana tabacum, Phytomorphology,* 26, 176—182, 1976.
248. **Murakami, Y.,** Changes in endogenous gibberellins in ripening fruits of cucumber, in *Plant Growth Substances, 1973,* Horikawa Publishing Co., Tokyo, 1974, 275.
249. **Murata, Y. and Matsushima, S.,** Rice, in *Crop Physiology: Some Case Histories,* Evans, L. T., Ed., Cambridge University Press, London, 1975, 73—99.
250. **Murneek, A. E.,** Effects of correlation between vegetative and reproductive functions in the tomato (*Lycopersicon esculentum* Mill.), *Plant Physiol.,* 1, 3—56, 1926.
251. **Naito, K., Tsuji, H., and Hatakeyama, I.,** Effect of benzyladenine on DNA, RNA, protein, and chlorophyll contents in intact bean leaves: differential responses to benzyladenine according to leaf age, *Physiol. Plant.,* 43, 367—371, 1978.
252. **Neales, T. F. and Incoll, L. D.,** The control of leaf photosynthesis rate by the level of assimilate concentration in the leaf. A review of the hypothesis, *Bot. Rev.,* 34, 107—125, 1968.
253. **Newman, D. W., Rowell, B. W., and Byrd, K.,** Lipid transformations in greening and senescing leaf tissue, *Plant Physiol.,* 51, 229—233, 1973.
254. **Nichols, R.,** Ethylene production during senescence of flowers, *J. Hortic. Sci.,* 41, 279—290, 1966.
255. **Nightingale, G. T.,** Light in relation to the growth and chemical composition of some horticultural plants, *Proc. Am. Soc. Hortic. Sci.,* 19, 18—29, 1922.
256. **Nishimura, M. and Beevers, H.,** Hydrolases in vacuoles from castor bean endosperm, *Plant Physiol.,* 62, 44—48, 1978.
257. **Noodén, L. D.,** Regulation of senescence, *World Soybean Research Conference,* Vol. II, 1977.
258. **Noodén, L. D., Kahanak, G. M., and Okatan, Y.,** Prevention of monocarpic senescence in soybeans with auxin and cytokinin: an antidote for self-destruction, *Science,* 206, 851—843, 1979.
259. **Noodén, L. D. and Leopold, A. C.,** Phytohormones and the endogenous regulation of senescence and abscission, in *Phytohormones and Related Compounds — A Comprehensive Treatise,* Vol. 2, Letham, D. S., Goodwin, P. B., and Higgins, T. J., Eds., Elsevier, Amsterdam, 1978, 329—368.
260. **Noodén, L. D. and Lindoo, S. J.,** Monocarpic senescence, *What's New in Plant Physiology,* 9, 25—28, 1978.
261. **Noodén, L. D., Rupp, D. C., and Derman, B. D.,** Separation of seed development from monocarpic senescence in soybeans, *Nature (London),* 271, 354—357, 1978.
262. **Noodén, L. D. and Weber, J. A.,** Environmental and hormonal control of dormancy in terminal buds of plants, in *Dormancy and Developmental Arrest,* Clutter, M. E., Ed., Academic Press, New York, 1978, 221—268.
263. **Nösberger, J. and Thorne, G. N.,** Effects of removing florets or shading the ear on barley production and distribution of dry matter, *Ann. Bot. (London),* 29, 635—644, 1965.
264. **Ogren, W. L. and Rinne, R. W.,** Photosynthesis and seed metabolism, in *Soybeans: Improvement, Production and Uses,* American Society Agronomy, Madison, 1973, 391—416.
265. **Ohlrogge, A. J.,** Mineral nutrition of soybeans, *Adv. Agron.,* 12, 229—263, 1960.
266. **Oizumi, H.,** Studies on the mechanism of branching and its agronomic considerations in soybean plants, *Tohoku Nat. Agr. Exp. Sta. Bull.* 25, 1—95, 1962.
267. **Ojehomon, O. O.,** A comparison of the vegetative growth, development, and seed yield of three varieties of cowpea, *Vigna unguiculata* (L.) Walp., *J. Agric. Sci.,* 74, 363—374, 1970.
268. **Ojehomon, O. O.,** Effect of continuous removal of open flowers on the seed yield of two varieties of cowpea, *Vigna unguiculata* (L.), Walp., *J. Agric. Sci.,* 74, 375—381, 1970.
269. **Oliveira, L. and Bisalputra, T.,** Ultrastructural studies in the brown alga *Ectocarpus* in culture: ageing, *New Phytol.,* 78, 131—138, 1977.
270. **Oritani, T. and Yoshida, R.,** Studies on nitrogen metabolism in crop plants. XII. Cytokinins and abscisic acid-like substances levels in rice and soybean leaves during their growth and senescence, *Proc. Crop Sci. Soc. Jpn.,* 42, 280—287, 1973.
271. **Osborne, D. J.,** Acceleration of abscission by a factor produced in senescent leaves, *Nature (London),* 176, 1161—1163, 1955.
272. **Osborne, D. J.,** Identity of the abscission-accelerating substance in senescent leaves, *Nature (London),* 183, 1593, 1959.

273. **Osborne, D. J.**, Effect of kinetin on protein and nucleic acid metabolism in *Xanthium* leaves during senescence, *Plant Physiol.*, 37, 595—602, 1962.
274. **Osborne, D. J.**, Hormonal mechanisms regulating senescence and abscission, in *Biochemistry and Physiology of Plant Growth Substances*, Wightman, F. and Setterfield, G., Eds., Runge Press, Ottawa, 1968, 815—840.
275. **Osborne, D. J.**, Internal factors regulating abscission, in *Shedding of Plant Parts*, Kozlowski, T. T., Ed., Academic Press, New York, 1973, 125—147.
276. **Osborne, D. J. and Hallaway, H. M.**, Auxin control of protein levels in detached autumn leaves, *Nature (London)*, 188, 240—241, 1960.
277. **Osborne, D. J. and Hallaway, H. M.**, The auxin, 2,4-dichlorophenoxyacetic acid, as a regulator of protein synthesis and senescence in detached leaves of *Prunus*, *New Phytol.*, 63, 334—347, 1964.
278. **Osborne, D. J., Jackson, M., and Milborrow, B. V.**, Physiological properties of abscission accelerator from senescent leaves, *Nature (London), New Biol.*, 240, 98—101, 1972.
279. **Osman, A. M. and Milthorpe, F. L.**, Photosynthesis of wheat leaves in relation to age, illuminance, and nutrient supply. II. Results, *Photosynthetica*, 5, 61—70, 1971.
280. **Ozbun, J. L., Volk, R. J., and Jackson, W. A.**, Effect of light and darkness on gaseous exchange of bean leaves, *Plant Physiol.*, 39, 523—527, 1964.
281. **Paranjothy, K. and Wareing, P. F.**, The effects of abscisic acid, kinetin, and 5-fluorouracil on ribonucleic acid and protein synthesis in senescing radish leaf disks, *Planta*, 99, 112—119, 1971.
282. **Parker, C. W. and Letham, D. S.**, Regulators of cell division in plant tissues. XVI. Metabolism of zeatin by radish cotyledons and hypocotyls, *Planta*, 114, 199—218, 1973.
283. **Parthier, B.**, Untersuchungen über den Aminosäure-Einbau in die Blatteiweiss des Tabaks, *Flora (Jena)*, 151, 368—397, 1961.
284. **Parthier, B.**, Proteinsynthese in grünen Blättern, *Flora (Jena)*, 154, 230—244, 1964.
285. **Parups, E. V.**, Disc electrophoresis of proteins of senescing and fresh leaves and petals of certain ornamental plants, *J. Am. Soc. Hortic. Sci.*, 96, 168—171, 1971.
286. **Pate, J. S.**, Nodulation studies in legumes. I. The synchronization of host and symbiotic development in the field pea, *Pisum arvense* L., *Aust. J. Biol. Sci.*, 11, 366—381, 1958.
287. **Pate, J. S.**, Nodulation studies in legumes. II. The influence of various environmental factors on symbiotic expression in the vetch (*Vicia sativa* L.) and other legumes, *Aust. J. Biol. Sci.*, 11, 496—515, 1958.
288. **Pate, J. S.**, Uptake, assimilation, and transport of nitrogen compounds by plants, *Soil Biol. Biochem.*, 5, 109—119, 1973.
289. **Pate, J. S. and Flinn, A. M.**, Fruit and seed development, in *The Physiology of the Garden Pea*, Sutcliffe, J. F. and Pate, J. S., Eds., Academic Press, London, 1977, 431—453.
290. **Pate, J. S., Sharkey, P. J., and Lewis, O. A. M.**, Phloem bleeding from legume fruits — a technique for study of fruit nutrition, *Planta*, 120, 229—243, 1974.
291. **Patrick, J. W. and Wareing, P. F.**, Experiments on the mechanism of hormone-directed transport, in *Plant Growth Substances, 1970*, Carr, D. J., Ed., Springer-Verlag, Berlin, 1972, 695—700.
292. **Patrick, J. W. and Wareing, P. F.**, Auxin-promoted transport of metabolites in stems of *Phaseolus vulgaris* L., *J. Exp. Bot.*, 27, 969—982, 1976.
293. **Pearsall, W. H.**, Studies in growth. IV. Correlations in development, *Ann. Bot. (London)*, 37, 261—276, 1923.
294. **Penot, M. and Beraud, J.**, Migrations orientées et phytohormones — Valeur de la feuille détachée comme matérial expérimental, *Physiol. Plant.*, 42, 14—20, 1978.
295. **Person, C., Samborski, D. J., and Forsyth, F. R.**, Effect of benzimidazole on detached wheat leaves, *Nature (London)*, 180, 1294—1295, 1957.
296. **Petrie, A. H. K., Watson, R., and Ward, E. D.**, Physiological ontogeny in the tobacco plant, I., *Aust. J. Exp. Biol. Med. Sci.*, 17, 93—122, 1939.
297. **Phillips, D. A. and Cleland, C. F.** Cytokinin activity from the phloem sap of *Xanthium strumarium* L., *Planta*, 102, 173—178, 1972.
298. **Phillips, I. D. J. and Jones, R. L.**, Gibberellin-like activity in bleeding sap of root systems of *Helianthus annus* detected by a new dwarf pea epicotyl assay and other methods, *Planta*, 63, 269—278, 1964.
299. **Phillips, I. D. J. and Wareing, P. F.**, Studies in dormancy of sycamore. I. Seasonal changes in the growth-substance content of the shoot, *J. Exp. Bot.*, 9, 350—364, 1958.
300. **Poovaiah, B. W. and Leopold, A. C.**, Deferral of leaf senescence with calcium, *Plant Physiol.*, 52, 236—239, 1973.
301. **Prochazka, S., Schraudolf, H., and Sonka, J.**, Transport of ^{14}C-kinetin in intact and decapitated pea seedlings (*Pisum sativum* L.), *Z. Pflanzenphysiol.*, 81, 308—313, 1977.
302. **Proebsting, W. M., Davies, P. J., and Marx, G. A.**, Evidence for a graft-transmissible substance which delays apical senescence in *Pisum sativum* L., *Planta*, 135, 93—94, 1977.

303. **Proebsting, W. M., Davies, P. J., and Marx, G. A.,** Photoperiod-induced changes in gibberellin metabolism in relation to apical growth and senescence in genetic lines of peas (*Pisum sativum* L.), *Planta,* in press, 1978.

304. **Quebedeaux, B., Havelka, U. D., Livak, K. L., and Hardy, R. W. F.,** Effect of altered pO_2 in the aerial part of soybean on symbiotic N_2 fixation, *Plant Physiol.,* 56, 761—764, 1975.

305. **Quebedeaux, B., Sweetser, P. B., and Rowell, J. C.,** Abscisic acid levels in soybean reproductive structures during development, *Plant Physiol.,* 58, 363—366, 1976.

306. **Quick, W. A., and Shaw, M.,** The physiology of host-parasitic relations. XIV. The effect of rust infection on the nucleic acid content of wheat leaves, *Can. J. Bot.,* 42, 1531—1540, 1964.

307. **Radley, M.,** The development of wheat grain in relation to endogenous growth substances, *J. Exp. Bot.,* 27, 1009—1021, 1976.

308. **Raschke, K. and Zeevaart, J. A. D.,** Abscisic acid content, transpiration, and stomatal conductance as related to leaf age in plants of *Xanthium strumarium* L., *Plant Physiol.,* 58, 169—174, 1976.

309. **Reichert, C.,** Lund und Gartenschatz, *Praktisches Handbuch für den Blumen und Zierpflanzen Gartenbau,* 6 Aufl., 5 Teil., 1821, cited in **Molisch., H.,** *The Longevity of Plants,* Science Press, 1938.

310. **Richmond, A. E. and Lang, A.,** Effect of kinetin on protein content and survival of detached *Xanthium* leaves, *Science,* 125, 650—651, 1957.

311. **Robards, A. W.,** *Electron Microscopy and Plant Ultrastructure,* McGraw-Hill, London, 1970.

312. **Rosenfeld, A.,** *Prolongevity,* Alfred A. Knopf, New York, 1976.

313. **Ryan, C. A.,** Synthesis of chymotrypsin inhibitor I protein in potato leaflets induced by detachment, *Plant Physiol.,* 43, 1859—1865, 1968.

314. **Sacher, J. A.,** Studies on auxin-membrane permeability relations in fruit and leaf tissues, *Plant Physiol.,* 34, 365—372, 1959.

315. **Sacher, J. A.** Studies on permeability, RNA, and protein turnover during ageing of fruit and leaf tissues, *Symp. Soc. Exp. Biol.,* 21, 269—303, 1967.

316. **Sacher, J. A. and Salminen, S. O.,** Comparative studies of effect of auxin and ethylene on permeability and synthesis of RNA and protein, *Plant Physiol.,* 44, 1371—1377, 1969.

317. **Sachs, T.,** Senescence of inhibited shoots of peas and apical dominance, *Ann. Bot. (London),* 30. 447—456, 1966.

318. **Sachs, T.,** A possible basis for apical organization in plants, *J. Theor. Biol.,* 37, 353—361, 1972.

319. **Sachs, T.,** The induction of transport channels by auxin, *Planta,* 127, 201—206, 1975.

320. **Salin, M. L. and Homann, P. H.,** Changes in photorespiratory activity with leaf age, *Plant Physiol.,* 48, 193—196, 1971.

321. **Salin, M. L. and Homann, P. H.,** Glycolate metabolism in young and old tobacco leaves, and effects of α-hydroxy-2-pyridinemethanesulfonic acid, *Can. J. Bot.,* 51, 1857—1865, 1973.

322. **Salter, P. J. and Drew, D. H.,**Root growth as a factor in the response of *Pisum sativum* L. to irrigation , *Nature (London),* 206, 1063—1064, 1965.

323. **Schreyer, S. L. and Noodén, L. D.,** Screening nutrients and growth promoters or inhibitors for ability to mimic or antagonize the death signal in soybeans, *Plant Physiol.,* 56 (Suppl.), 83, 1975.

324. **Schulman, E.,** Longevity under adversity in conifers, *Science,* 119, 396—399, 1954.

325. **Schwanitz, F.,** *The Origin of Cultivated Plants,* Harvard University Press, Cambridge, Mass., 1966.

326. **Sêbânek, J.,** Einfluss von Pasten mit verschiedenen Indolylessigsaure-Konzentrationen auf den Transport von ^{32}P in dekapitierten Erbsenepikotylen, *Planta,* 75, 283—285, 1967.

327. **Sestâk, Z.,** Changes in the chlorophyll content as related to photosynthetic activity and age of leaves, *Photochem. Photobiol.,* 2, 101—110, 1963.

328. **Sestâk, Z. and Catsky, J.,** Sur les relations entre le contenu en chlorophylle et l'activite photosynthetique pendant la croissance et le vieillissement des feuilles, in *Le Chloroplaste, Croissance et Vieillissement,* Sironval, C., Ed., Masson, Paris, 1967, 213—262.

329. **Seth, A. K., Davies, C. R., and Wareing, P. F.,** Auxin effects on the mobility of kinetin in the plant, *Science,* 151, 587—588, 1966.

330. **Seth, A. and Wareing, P. F.,** Isolation of a kinin-like root-factor in *Phaseolus vulgaris, Life Sci.,* 4, 2275—2280, 1965.

331. **Seth, A. K. and Wareing, P. F.,** Hormone-directed transport of metabolites and its possible role in plant senescence, *J. Exp. Bot.,* 18, 65—77, 1967.

332. **Shah, C. B. and Loomis, R. S.,** Ribonucleic acid and protein metabolism in sugar beet during drought, *Physiol. Plant.,* 18, 240—254, 1965.

333. **Shaw, M., Bhattacharya, P. K., and Quick, W. A.,** Chlorophyll, protein, and nucleic acid levels in detached, senescing wheat leaves, *Can. J. Bot.,* 43, 739—746, 1965.

334. **Shaw, M. and Manocha, M. S.,** Fine structure in detached, senescing wheat leaves, *Can. J. Bot.,* 43, 747—755, 1965.

335. **Shibaoka, H. and Thimann, K. V.,** Antagonisms between kinetin and amino acids. Experiments on the mode of action of cytokinins, *Plant Physiol.,* 46, 212—220, 1970.

336. **Shoji, K., Addicott, F. T., and Swets, W. A.,** Auxin in relation to leaf blade abscission, *Plant Physiol.*, 26, 189—191, 1951.
337. **Short, K. C. and Torrey, J. G.,** Cytokinins in seedling roots of pea, *Plant Physiol.*, 49, 155—160, 1972.
338. **Sinclair, T. R. and de Wit, C. T.,** Photosynthate and nitrogen requirements for seed production by various crops, *Science*, 189, 565—567, 1975.
339. **Singh, B. N. and Lal, K. N.,** Investigations of the effect of age on assimilation of leaves, *Ann. Bot. (London)*, 49, 291—307, 1935.
340. **Sitton, D., Itai, C., and Kende, H.,** Decreased cytokinin production in the roots as a factor in shoot senescence, *Planta*, 73, 296—300, 1967.
341. **Skene, K. G. M.,** The gibberellins of developing bean seeds, *J. Exp. Bot.*, 21, 236—246, 1970.
342. **Skene, K. G. M.,** Cytokinins in the xylem sap of grape vine canes: changes in the activity during cold storage, *Planta*, 104, 89—92, 1972.
343. **S Skene, K. G. M.,** Cytokinin production by roots as a factor in the control of plant growth, in *The Development and Function of Roots*, Torrey, J. G. and Clarkson, D. T., Eds., Academic Press, London, 1975, 365—396.
344. **Slife, F. W. and Early, E. B.,** Effect of Ethrel on growth and yield of soybeans, *Agron. J.*, 62, 434—435, 1970.
345. **Sloger, C. and Caldwell, B. E.,** Response of cultivars of soybean to synthetic abscisic acid, *Plant Physiol.*, 45, 634—635, 1970.
346. **Smillie, R. M. and Krotkov, G.,** Enzymatic activities of subcellular particles from leaves. IV. Photosynthetic phosphorylation and photosynthesis by isolated chloroplasts from pea leaves, *Can. J. Bot.*, 37, 1217—1225, 1959.
347. **Smillie, R. M. and Krotkov, G.,** Changes in dry weight, protein, nucleic acid, and chlorophyll content of growing pea leaves, *Can. J. Bot.*, 39, 891—900, 1961.
348. **Smith, O. E.,** Changes in abscission-accelerating substances with development of cotton fruit, *New Phytol.*, 68, 313—332, 1968.
349. **Smith, O. E., Lyon, J. L., Addicott, F. T., and Johnson, R. E.,** Abscission physiology of abscisic acid, in *Physiology of Plant Growth Substances*, Wightman, F. and Setterfield, G., Eds., Runge Press, Ottawa, 1968, 1547—1560.
350. **Smith, W. H., Parker, J. C., and Freeman, J. C.,** Exposure of cut flowers to ethylene in the presence and absence of carbon dioxide, *Nature (London)*, 211, 99—100, 1966.
351. **Solarova, J.,** Changes in minimal diffusive resistances of leaf epidermis during aging of primary leaves of *Phaseolus vulgaris* L., *Biol. Plant.*, 15, 237—240, 1973.
352. **Spencer, P. W.,** Incorporation of ^{14}C leucine into apple leaf protein and its inhibition by protein synthesis inhibitors during growth and senescence, *Plant Physiol.*, 52, 151—155, 1973.
353. **Spencer, P. W. and Titus, J. S.,** Biochemical and enzymatic changes in apple leaf tissue during autumnal senescence, *Plant Physiol.*, 49, 746—750, 1972.
354. **Spencer, P. W. and Titus, J. S.,** The occurrence and nature of ornithine carbamoyltransferase in senescing apple leaf tissues, *Plant Physiol.*, 54, 382—385, 1974.
355. **Srivastava, B. I. S.,** Acceleration of senescence and of the increase of chromatin-associated nucleases in excised barley leaves by abscisin II and its reversal by kinetin, *Biochim. Biophys. Acta*, 169, 534—536, 1968.
356. **van Staden, J.,** Changes in endogenous cytokinin levels during abscission and senescence of *Streptocarpus* leaves, *J. Exp. Bot.*, 24, 667—673, 1973.
357. **van Staden, J.,** Occurrence of a cytokinin glucoside in the leaves and in honeydew of *Salix babylonica*, *Physiol. Plant.*, 36, 225—228, 1976.
358. **van Staden, J.,** Seasonal changes in the cytokinin content of *Ginkgo biloba* leaves, *Physiol. Plant.*, 38, 1—5, 1976.
359. **van Staden, J.,** Seasonal changes in the cytokinin content of leaves of *Salix babylonica*, *Physiol. Plant.*, 40, 296—299, 1977.
360. **van Staden, J. and Davey, J. E.,** Cytokinin translocation in xylem sap of herbaceous plants, *Z. Pflanzenphysiol.*, 77, 377—382, 1976.
361. **van Staden, J. and Davey, J. E.,** The metabolism of zeatin and zeatin riboside by soybean callus, *Ann. Bot. (London)*, 41, 1041—1048, 1977.
362. **Stearns, M. E. and Wagenaar, E. B.,** Ultrastructural changes in chloroplasts of autumn leaves, *Can. J. Genet. Cytol.*, 13, 550—560, 1971.
363. **Stenlid, G.,** Salt losses and redistribution of salts in higher plants, *Encycl. Plant Physiol.*, IV, 615—637, 1958.
364. **van Steveninck, R. F. M.,** Factors affecting the abscission of reproductive organs in yellow lupin (*Lupinus luteus* L.) I. The effect of different patterns of flower removal, *J. Exp. Bot.*, 8, 373—381, 1957.

365. **van Steveninck, R. F. M.**, Effect of hormones and related substances on ion transport, *Encycl. Plant Physiol.*, IIB, 307—342, 1976.
366. **Steward, F. C. and Millar, F. K.**, Salt accumulation in plants: a reconsideration of the role of growth and metabolism, *Symp. Soc. Exp. Biol.*, 8, 367—406, 1954.
367. **Street, H. E.**, The ageing of root meristems, *Symp. Soc. Exp. Biol.*, 21, 517—542, 1967.
368. **Streeter, J. G.**, Nitrogen nutrition of field-grown soybean plants. I. Seasonal variations in soil nitrogen and nitrogen composition of stem exudate, *Agron. J.*, 64, 311—314, 1972.
369. **Suetsugu, I., Anaguchi, I., Saito, K., and Kumano, S.**, Developmental process of the root- and top-organs in the soybean varieties, *Hokuriku Agr. Exp. Sta. Bull.*, 3, 89—96, 1962.
370. **Sun, C. N.**, Histogenesis of the leaf and structure of the shoot apex in *Glycine max* (L.) Merrill, *Bull. Torrey Bot. Club*, 84, 163—174, 1957.
371. **Sutcliffe, J. F.**, *Mineral Salts Absorption in Plants*, Pergamon Press, Oxford, 1962.
372. **Sutcliffe, J. F.**, Regulation in the whole plant, *Encycl. Plant Physiol.*, IIB, 394—417, 1976.
373. **Swanson, C. A.**, Translocation of organic solutes, in *Plant Physiology, a Treatise*, Vol. 2, Steward, F. C., Ed., Academic Press, New York, 1959, 481—551.
374. **Sweetser, P. B. and Swartzfager, D. G.**, Indole-3-acetic acid levels of plant tissue as determined by a new high performance liquid chromatographic method, *Plant Physiol.*, 61, 254—258, 1978.
375. **Takegami, T.**, A study on senescence in tobacco leaf disks. II. Chloroplast and cytoplasmic rRNAs, *Plant Cell Physiol.*, 16, 417—425, 1975.
376. **Tetley, R. M. and Thimann, K. V.**, The metabolism of oat leaves during senescence. I. Respiration, carbohydrate metabolism, and the action of cytokinins, *Plant Physiol.*, 54, 294—303, 1974.
377. **Thimann, K. V., Tetley, R. M., and Thanh, T. V.**, The metabolism of oat leaves during senescence. II. Senescence in leaves attached to the plant, *Plant Physiol.*, 54, 859—862, 1974.
378. **Thomas, H. and Stoddart, J. L.**, Separation of chlorophyll degradation from other senescence processes in leaves of a mutant genotype of meadow fescue *(Festuca pratensis* L.), *Plant Physiol.*, 56, 438—441, 1975.
379. **Thorne, G. N.**, Survival of tillers and distribution of dry matter between ear and shoot of barley varieties, *Ann. Bot. (London)*, 26, 37—40, 1962.
380. **Thorne, J. H. and Koller, H. R.**, Influence of assimilate demand on photosynthesis, diffusive resistances, translocation, and carbohydrate levels of soybean leaves, *Plant Physiol.*, 54, 201—207, 1974.
381. **Thornley, J. H. M.**, *Mathematical Models in Plant Physiology*, Academic Press, London, 1976.
382. **Thrower, S. L.**, The pattern of translocation during leaf ageing, *Symp. Soc. Exp. Biol.*, 21, 483—506, 1967.
383. **Torrey, J. G.**, Root hormones and plant growth, *Annu. Rev. Plant Physiol.*, 27, 435—459, 1976.
384. **Trippi, V. S. and Brulfert, J.**, Organization of the morphophysiologic unit in *Anagallis arvensis* and its relation with the perpetuation mechanism and senescence, *Am. J. Bot.*, 60, 641—647, 1973.
385. **Trippi, V. and Montaldi, E.**, The aging of sugar cane clones, *Phyton (Buenos Aires)*, 14, 79—91, 1960.
386. **Varner, J. E., Balce, L. V., and Huang, R. C.**, Senescence of cotyledons of germinating peas. Influence of axis tissue, *Plant Physiol.*, 38, 88—92, 1963.
387. **Vonk, C. R.**, Studies on phloem exudation from *Yucca flacida*, Haw. XIII. Evidence for the occurrence of a cytokinin nucleotide in the exudate, *Acta Bot. Neerl.*, 23, 541—548, 1974.
388. **Walker, A. J. and Ho, L. C.**, Carbon translocation in the tomato: effects of fruit temperature on carbon metabolism and the rate of translocation, *Ann. Bot. (London)*, 41, 825—832, 1977.
389. **Wang, T. L., Thompson, A. G., and Horgan, R.**, Cytokinin glucoside from the leaves of *Phaseolus vulgaris* L., *Planta, (Berlin)*, 135, 285—288, 1977.
390. **Wangermann, E.**, Longevity and ageing in plants and plant organs, in *Handbuch der Pflanzenphysiologie*, Vol. 15/2, Ruhland, W., Ed., Springer-Verlag, Berlin, 1965, 1037—1057.
391. **Wangermann, E. and Ashby, E.**, Studies in the morphogenesis of leaves. VII. Part I. Effects of light intensity and temperature on the cycle of ageing and rejuvenation in the vegetative life history of *Lemna minor*, *New Phytol.*, 50, 186—199, 1951.
392. **Wangermann, E. and Lacey, H. J.**, Studies in the morphogenesis of leaves. X. Preliminary experiments on the relation between nitrogen nutrition, rate of respiration, and rate of ageing of fronds of *Lemna minor*, *New Phytol.*, 54, 182—198, 1955.
393. **Wareing, P. F. and Seth, A. K.**, Ageing and senescence in the whole plant, *Symp. Soc. Exp. Biol.*, 21, 543—558, 1967.
394. **Wardlaw, I. F.**, The control and pattern of movement of carbohydrates in plants, *Bot. Rev.*, 34, 79—105, 1968.
395. **Watson, D. J., Thorne, G. N., and French, S. A. W.**, Physiological cause of differences in grain yield between varieties of barley, *Ann. Bot. (London)*, 22, 321—352, 1958.
396. **Watson, R. and Petrie, A. H. K.**, Physiological ontogeny in the tobacco plant, *Aust. J. Exp. Biol. Med. Sci.*, 18, 313—339, 1940.

397. **Weber, C. R.**, Effects of defoliation and topping simulating hail injury to soybeans, *Agron. J.*, 47, 262—266, 1955.
398. **Westing, A. H.**, The longevity and aging of trees, *Gerontologist*, 4, 10—15, 1964.
399. **Wheeler, A. W.**, Changes in auxin in expanding and senescent primary leaves of dwarf french beans *(Phaseolus vulgaris)*, *J. Exp. Bot.*, 19, 102—107, 1968.
400. **Wheeler, A. W.**, Changes in growth substance contents during growth of wheat grains, *Ann. Appl. Biol.*, 72, 327—334, 1972.
401. **White, P. R.** Potentially unlimited growth of excised tomato root tips in a liquid medium, *Plant Physiol.*, 9, 585—600, 1934.
402. **White, P. R.**, Potentially unlimited growth of excised plant callus in an artificial nutrient, *Am. J. Bot.*, 26, 59—64, 1939.
403. **Whyte, P. and Luckwill, L. C.**, A sensitive bioassay for gibberellins based on retardation of leaf senescence in *Rumex obtusifolius* (L.), *Nature (London)*, 210, 1360, 1966.
404. **Wien, H. C., Altschuler, S. L., Ozbun, J.L., and Wallace, D. H.**, ⁴C-assimilate distribution in *Phaseolus vulgaris* L. during the reproductive period, *J. Am. Soc. Hortic. Sci.*, 101, 510—513, 1976.
405. **Wien, H. C., Sansted, R. F., and Wallace, D. H.**, The influence of flower removal on growth and seed yield of *Phaseolus vulgaris* L., *J. Am. Soc. Hortic. Sci.*, 98, 45—49, 1973.
406. **Williams, L. E. and Kennedy, R. A.**, Relationship between early photosynthetic products, photorespiration, and stage of leaf development in *Zea mays*, *Z. Pflanzenphysiol.*, 81, 314—322, 1977.
407. **Williams, R. F.**, The effects of phosphorus supply on the rates of intake of phosphorus and nitrogen and upon certain aspects of phosphorus metabolism in gramineous plants, *Austr. J. Sci. Res. Ser. B*, 1, 333—361, 1948.
408. **Williams, R. F.**, Redistribution of mineral elements during development, *Annu. Rev. Plant Physiol.*, 6, 25—42, 1955.
409. **Wilson, R. F., Burton, J. W., Buck, J. A., and Brim, C. A.**, Studies on genetic male-sterile soybeans I. Distribution of plant carbohydrate and nitrogen during development, *Plant Physiol.*, 61, 838—841, 1978.
410. **Wittenbach, V. A.** Induced senescence of intact wheat seedlings and its reversibility, *Plant Physiol.*, 59, 1039—1042, 1977.
411. **Wittenbach, V. A.**, Changes in proteolytic activity and the level of RUDPCase in the flag leaf of wheat during grain development and senescence, *Plant Physiol.*, 61, (Suppl.), 25, 1978.
412. **Wollgiehn, R.**, Untersuchungen über den Einfluss des Kinetins auf den Nucleinsäure — und Proteinstoffwechsel isolierter Blatter, *Flora (Jena)*, 151, 411—437, 1961.
413. **Wollgiehn, R.**, Nucleic acid and protein metabolism of excised leaves, *Symp. Soc. Exp. Biol.*, 21, 231—246, 1967.
414. **Wood, J. G. and Cruickshank, D. H.**, The metabolism of starving leaves. 5. Changes in amounts of some amino acids during starvation of grass leaves; and their bearing on the nature of the relationship between proteins and amino acids, *Aust. J. Exp. Biol. Med. Sci.*, 22, 111—123, 1944.
415. **Woodward, R. G.**, Photosynthesis and expansion of leaves of soybean grown in two environments, *Photosynthetica*, 10, 274—279, 1976.
416. **Woodward, R. G. and Rawson, H. M.**, Photosynthesis and transpiration in dicotyledonous plants. II. Expanding and senescing leaves of soybean, *Aust. J. Plant Physiol.*, 3, 257—267, 1976.
417. **Woolhouse, H. W.**, The nature of senescence in plants, in *Symp. Soc. Exp. Biol.*, 21, 179—213, 1967.
418. **Woolhouse, H. W.** Longevity and senescence in plants, *Sci. Prog. (Oxford)*, 61, 123—147, 1974.
419. **Woolhouse, H. W.**, Senescence processes in the life cycle of flowering plants, *BioScience*, 28, 25—31, 1978.
420. **Wright, S. T. C. and Hiron, R. W. P.**, (+)−Abscisic acid, the growth inhibitor induced in detached wheat leaves by a period of wilting, *Nature (London)*, 224, 719—720, 1969.
421. **Yoshida, S.**, Physiological aspects of grain yield, *Annu. Rev. Plant Physiol.*, 23, 437—464, 1972.
422. **Yoshida, R., Oritani, T., and Nishi, A.**, Kinetin-like factors in the root exudate of rice plants, *Plant Cell Physiol.*, 12, 89—94, 1971.
423. **Zeevaart, J. A. D.**, Sites of abscisic acid synthesis and metabolism in *Ricinus communis* L., *Plant Physiol.*, 59, 788—791, 1977.
424. **Zimmerman, P. W., Hitchcock, A. E., and Crocker, W.**, The effect of ethylene and illuminating gas on roses, *Contrib. Boyce Thompson Inst.*, 3, 459—481, 1931.

INDEX

A

B

C

D

E

G

H

I

J

K

L

M

O

P

Q

R

S

T

U

V

W

X

Y

Z